PRECALCULUS MATHEMATICS

SECOND EDITION

Merrill Mathematics Series

Erwin Kleinfeld, *Editor*

PRECALCULUS MATHEMATICS

SECOND EDITION

F. Lane Hardy

Chicago State University

Charles E. Merrill Publishing Company

A Bell & Howell Company

Columbus, Ohio

International Standard Book Number: 0-675-09251-5

Library of Congress Catalog Card Number: 75-148246

1 2 3 4 5 6 7 8 9 10—79 78 77 76 75 74 73 72 71

Printed in the United States of America

For Knolie, Pole, and Phodal Knut

PREFACE

This new edition of *Precalculus Mathematics* is best described by saying that it more nearly reflects the CUPM recommendations for "Mathematics 0" than did the original. Topics presented in this edition but not in the original include numerical trigonometry, computations by logarithms, theory of equations, polar coordinates and 3-dimensional analytic geometry. The rather formal development of the real numbers has been replaced by a brief descriptive treatment which omits entirely the field axioms. The central theme of the text, however, remains the same as before: functions.

Many new problems have been included in this edition as well as sample solutions. These sample solutions have been inserted within the problem sets and should increase the book's usefulness as a text for self-study.

This book was originally conceived as preparation for students who wish to study calculus and the experience of my colleagues and myself at Armstrong State College and Chicago State University indicates that students who successfully complete a course based on this material are usually successful in calculus.

There is enough material for a two-semester course if one wishes to include all topics. However, several arrangements for a one-semester course are possible, depending upon the needs of the students involved. For example, the following might easily be covered in one term:

Chapter 1: all	6 lessons
Chapter 2: Sections 2.1 — 2.2	2 lessons
Chapter 3: all	4 lessons
Chapter 4: Sections 4.1 — 4.6, 4.8	7 lessons
Chapter 5: all	8 lessons
Chapter 6: all	3 lessons
Chapter 8: 8.1	1 lesson

I appreciate very much the comments and criticisms of those mathematicians who have used the 1st edition of this text. My colleague, Professor Dale Underwood, has made many valuable suggestions after reading the entire manuscript and working all the exercises. To him I extend my sincere appreciation.

F. Lane Hardy
Chicago, Illinois

CONTENTS

Contents

SETS
AND
FUNCTIONS

Of fundamental importance in the study of mathematics is the theory of sets. For the purposes of this book we require only the language of sets and, in this chapter, we set forth the language of sets which will be useful throughout the remainder of this exposition.

Our concept of set is that of many things conceived as a single object. For example, we think of Mr. and Mrs. Jones, John, and Jane (their children) as the Jones family. (The term "family" is sometimes used as a synonym for "set.") Also, we think of certain individual advisors to the President collectively and call this group the cabinet. Such terms as "family," "team," "group," "duo," etc. suggest the common use of the set concept.

We shall use the words "set" and "collection" interchangeably, and the individuals which comprise a set or collection will be called the *members* of the set. In the examples above, Mrs. Jones is a member of the Jones family, and the Secretary of HEW is a member of the cabinet.

The customary practice of naming or denoting sets by capital letters and members of sets by lower case letters will be followed here. Usually Latin letters will be employed: $A, B, C, \ldots$ will be used for sets and $a, b, c, \ldots$ for members of sets. As a shorthand notation for the phrase "a is a member of the set A," we use the symbolism

$$a \in A$$

— the symbol "$\in$" standing for the phrase "is a member of." In case a is not a member of set A, we write

$$a \notin A.$$

A common method of specifying or describing a set is to tabulate its members or elements. The set that we call the Jones family is described by its members: Mr. and Mrs. Jones, John, Jane. Similarly, we specify the President's cabinet by listing its members: Secretary of State, Secretary of the Treasury, Secretary of Defense, Attorney General, Secretary of HEW, Secretary of the Interior, Secretary of Agriculture, etc. When this method is used to describe a set, it is common practice to enclose the list of members in braces. As an example of this, we say that the Jones family is

$$\{\text{Mr. Jones, Mrs. Jones, John, Jane}\}.$$

Generally, if a, b, c, ... are members and the only members, of a set A, we write

$$A = \{a, b, c, \ldots\}.$$

We read this: "the set A is the set whose members are a, b, c, ..."

Quite often it is not practical to list or tabulate all members of a set. This will always be the case when a large number of objects is involved. In such cases an alternative procedure for specifying the set is to use a property which the members of the set share. Rather than list all colleges in the U.S., for example, we describe the set by the property: "is a college in the U.S." In other words, we speak of the set whose members are colleges of the U.S. Symbolically, we express this using the brace notation as follows:

$$\{x \mid x \text{ is a college in the U.S.}\}.$$

We read this: "the set of all objects, say x, such that x is a college in the U.S." There is obviously nothing special about the use of the letter "x" here, and any other letter of the alphabet would have served just as well. If P stands for any property whatever, we express the fact that an object, say a, has property P by writing $P(a)$. For example, P may be the property: "is a whole number." Then $P(10)$ expresses the fact that 10 is a whole number. We can then, for each such property P, describe a set:

$$\{x \mid P(x)\}$$

i.e., the set of *all* objects, say x, having the property P. If P is the property "is a whole number," then $\{x \mid P(x)\}$ describes the collection of whole numbers.

Example 1–1

The set whose members are 1, 2, 3 is denoted by "$\{1, 2, 3\}$" and also by "$\{y \mid y \text{ is one of the first three whole numbers}\}$" and also by "$\{x \mid x \text{ is 1 or } x \text{ is 2 or } x \text{ is 3}\}$." The statements "$1 \in \{1, 2, 3\}$," "$2 \in \{1, 2, 3\}$" and "$3 \in \{1, 2, 3\}$" are true, while "$4 \in \{1, 2, 3\}$" is false.

Suppose that $P(x)$ expresses the fact that x is a number, and that $x^2 = 1$. Then "$\{x \mid P(x)\}$" describes the set $\{1, -1\}$.

Example
1–2

The sets A, B are the *same* or *equal* provided their members are the same.

"Set A is equal to set B" is expressed by "$A = B$." From our definition of equal sets, we know that $\{1, 2, 3\}$ is the same set as $\{y \mid y$ is one of the first three whole numbers$\}$ and also that

$$\{1, -1\} = \{x \mid x \text{ is a number and } x^2 = 1\}.$$

It should be emphasized that *a set and its members are distinct from each other*. This, no doubt, is obvious for sets of more than one member, but is equally true for sets of only one member: The set whose only member is 8, *i.e.* $\{8\}$, is not the same as 8. Generally, $\{a\} \neq a$ for any object a. Sets of only one element are sometimes called *singletons* or *unit sets*.

Consider the sets $A = \{t \mid t$ is a triangle$\}$ and $B = \{r \mid r$ is a right triangle$\}$. It is clear that every member of B is in particular a triangle, and hence, a member of A. Another way of saying this is: "if $x \in B$, then $x \in A$." This relation between the sets A, B may be expressed by saying that "B is a subset of A."

A set B is said to be a *subset* of set A provided every member of B is a member of A.

If B is a subset of A, this is symbolically expressed by writing "$B \subseteq A$." Thus, we have

$B \subseteq A$ if and only if every member of B is a member of A, or
$B \subseteq A$ if and only if $x \in B$ implies $x \in A$.

From our definition of equality for sets (they have the same members) and our definition of subset, we conclude that

$$A = B \text{ if and only if } A \subseteq B \text{ and } B \subseteq A. \qquad \textbf{(1-1)}$$

Let $P(x)$ express: "x is a triangle of Euclidean geometry and the sum of the interior angles of x is 180°," and let $Q(y)$ express: "y is a triangle of Euclidean geometry." Then if $A = \{x \mid P(x)\}$ and $B = \{y \mid Q(y)\}$, we have $A = B$. This is easily established by using **(1-1)** above, and observing

Example
1–3

that $A \subseteq B$ and $B \subseteq A$. It is clear that every member of A is a member of B so that $A \subseteq B$; and also, since in Euclidean geometry *every* triangle has an angle sum of 180°, $B \subseteq A$.

Example 1–4

If we insist that every property P defines a set, we are led to the conclusion that there are sets having no elements. Consider, for example, the property $P(x)$: "x is a triangle of Euclidean geometry having angle sum less than 180°." Then the set $\{x \mid P(x)\}$ has no elements since there are no objects having the property in question. One can think of many properties of this kind. The role of this set which has no elements is analogous to that of the number zero; this analogy will become clearer as we proceed. We call such a set a *null set*, and use the symbol "$\varnothing$" to denote any such set. As an immediate consequence of our definition of subset, we have

$$\varnothing \subseteq X \text{ for all sets } X.$$

This is true because there are *no* elements of $\varnothing$ which are *not* in X. This also shows that there can be only one null set since if $\varnothing_1$, $\varnothing_2$ are both null sets, then $\varnothing_1 \subseteq \varnothing_2$ and $\varnothing_2 \subseteq \varnothing_1$ from the above, and hence by **(1-1)** $\varnothing_1 = \varnothing_2$.

Exercises

1. Decide which of the following statements are true and which are false.
   (a) $4 \in \{1, 2, 4, 7\}$
   (b) $5 \in \{1, 2, 3, 7\}$
   (c) $100 \in \{x \mid x \text{ is an even number}\}$
   (d) $12 \notin \{w \mid w \text{ is an odd number}\}$
   (e) $7 \in \{t \mid t + 8 = 15\}$
   (f) $3 \in \{y \mid y^2 - 9 = 0\}$
   (g) $1 \in \{q \mid q^2 + 1 = 0\}$
   (h) $5 \in \{m \mid m^2 + m + 1 = 0\}$

2. List the members of each of the following sets.
   (a) $\{y \mid y + 1 = 10\}$
   (b) $\{a \mid 2a - 1 = 0\}$
   (c) $\{q \mid q \text{ is a positive factor of } 5\}$
   (d) $\{m \mid m \text{ is a positive factor of } 12\}$
   (e) $\{x \mid 2x = 0\}$
   (f) $\{b \mid (b - 1)b = 0\}$
   (g) $\{z \mid z^2 - 16 = 0\}$

3. Decide which of the following statements are true and which are false.
   (a) $2\frac{1}{2} \in \{\frac{1}{2}, \frac{3}{2}, \frac{5}{2}\}$
   (b) $\{2\frac{1}{2}, 3\frac{1}{2}\} \subseteq \{0.25, 1.3, 2.5\}$
   (c) $\{2.5, \frac{2}{3}\} \subseteq \{x \mid x$ is a positive number$\}$
   (d) $\{y \mid y$ is 2$\} \subseteq \{x \mid x$ is an even whole number$\}$
   (e) $\{x \mid x$ is the square of a whole number$\} \subseteq$
       $\{r \mid r$ is a whole number larger than 2$\}$
   (f) $\emptyset \subseteq \{5, 13\}$
   (g) $\{5, 29, 12\} \subseteq \{5, 12, 29\}$
   (h) $\{1, 2, 8\} \subseteq \{x \mid x$ is a whole number$\}$
   (i) $\{w \mid w$ is a whole number$\} \subseteq \{1, 2, 8\}$
   (j) $\{1, 2, 8\} = \{y \mid y$ is a whole number$\}$

4. Let Y be a particular point in a plane, and let

   $A = \{c \mid c$ is a circle of radius 4, 6, or 8, and center at $Y\}$
   $B = \{d \mid d$ is a circle of radius 4, 6 or 8$\}$.

   (a) How many members has set A?
   (b) How many members has set B?
   (c) Which of the statements $A \subseteq B$, $B \subseteq A$, $A = B$ are true?

5. Denote by W the set whose members are letters of the Latin alphabet:
   $a, b, c, \ldots$. Let V denote the vowels.
   (a) Which of the statements below are true?

   (i) $V \subseteq W$ (ii) $W \subseteq V$ (iii) $W = V$

   (b) If $Q = \{y \mid y$ is a letter of the word "algebra"$\}$, discuss the
   following statements:

   (i) $Q \subseteq V$ (ii) $Q \subseteq W$ (iii) $Q = \{r, g, l, b, e, a\}$

6. (a) If $X = \{\emptyset, \{\emptyset\}\}$, discuss the following statements:

   (i) $\emptyset \subseteq X$ (ii) $\emptyset \in X$ (iii) $\{\emptyset\} \in X$ (iv) $\{\emptyset\} \subseteq X$

   (b) If $X = \{\emptyset\}$, discuss the statements in (a).
   (c) With $X = \emptyset$ discuss the statements in (a).

7. Prove that if W is a set, $W \subseteq W$.

8. (a) If $A = \{1, 2\}$, how many subsets does A have?
   (b) If $A = \{1, 2, 3\}$, how many subsets does A have?
   (c) Can you guess the number of subsets of a set of 4 elements? Of 5?

9. From the definition of subset, show that $\{a\} \subseteq A$ if, and only if,
   $a \in A$.

10. Let A, B, C denote sets.
   (a) Prove that if $A \subseteq B$ and $B \subseteq C$, then $A \subseteq C$.
   (b) Prove that if $A \subseteq B$, $B \subseteq C$ and $C \subseteq A$, then $A = B = C$.

11. Prove that for sets X, Y if $X \subseteq Y$, then $a \notin Y$ implies that $a \notin X$.

1.2

Union, Intersection, Complement

If one starts with sets A, B, there are several ways of describing sets in terms of A and B. The sets so described will sometimes be different from both A and B and thus produce for us new sets. Two such descriptions are embodied in the following definitions.

Definition

If A and B are sets, the *union* of A, B is

$$\{y \mid y \in A \text{ or } y \in B\}.$$

The *intersection* of A, B is

$$\{q \mid q \in A \text{ and } q \in B\}.$$

The word "or" as used in the definition of union is used in its *inclusive* sense; *i.e.*, the statement "$y \in A$ or $y \in B$" is true if and only if either *one* or *both* of the statements "$y \in A$," "$y \in B$" is true. The word "or" will consistently be used in this way.

The notation commonly used for union and intersection is "$\cup$" and "$\cap$" respectively; *i.e.*,

$$A \cup B = \{t \mid t \in A \text{ or } t \in B\}$$

and

$$A \cap B = \{r \mid r \in A \text{ and } r \in B\}.$$

Example
1–5

For the sets $A = \{1, 2, 3\}$, $B = \{3, 4, 5\}$, we have $1 \in (A \cup B)$ since it is true that $1 \in A$ or $1 \in B$. Similarly, $2 \in (A \cup B)$, $3 \in (A \cup B)$, and $5 \in (A \cup B)$. Quite obviously, the statement "$x \in A$ or $x \in B$" is false if $x \notin A$ and $x \notin B$. We conclude from the definition of union that 1, 2, 3, 4, 5, are the members and the only members of $A \cup B$:

$$A \cup B = \{1, 2, 3\} \cup \{3, 4, 5\} = \{1, 2, 3, 4, 5\}.$$

Now to consider $A \cap B = \{1, 2, 3\} \cap \{3, 4, 5\}$, we note that the statement "$y \in A$ and $y \in B$" is true if and only if $y = 3$. Hence, from the definition of intersection,

$$A \cap B = \{1, 2, 3\} \cap \{3, 4, 5\} = \{3\}.$$

If $X = \{y \mid y$ is an even whole number$\}$
     $= \{2, 4, 6, 8, \ldots\},$

and

     $Y = \{w \mid w$ is an odd whole number$\} = \{1, 3, 5, 7, \ldots\},$

then

   $X \cup Y = \{1, 2, 3, 4, 5, \ldots\} = \{z \mid z$ is a whole number$\}.$

Also

   $X \cap Y = \varnothing.$

Example 1–6

If W, Z are sets for which the statement "$a \in W$ and $a \in Z$" is false for every object a, then $W \cap Z = \varnothing$ from the definition of intersection. See Example 1–6.

Example 1–7

Suppose that "$a \in A$" is a true statement for some set A and object a. Then the statement "$a \in A$ or $a \in B$" is true for every set B. This shows that if $a \in A$, then $a \in (A \cup B)$ by the definition of union. Therefore, from the definition of subset,

$$A \subseteq (A \cup B) \text{ for all sets } A, B.$$

In a similar manner it follows that

$$B \subseteq (A \cup B) \text{ for all sets } A, B.$$

Example 1–8

Under what conditions does the union of two sets A, B not result in a set different from both A and B? For instance, what can we conclude from "$A \cup B = A$"? We can conclude that $B \subseteq A$. This is true because of our definitions of union, equality, and subset. The definition of subset requires that each element of B be a member of A for $B \subseteq A$ to be true. But if $x \in B$, $x \in (A \cup B)$ by Example 1–8, and since we have $A \cup B = A$, the definition of equality of sets requires that $x \in A$. Thus, if $x \in B$, $x \in A$. Hence, if $A \cup B = A$, $B \subseteq A$. These considerations then show that we obtain a set different from both A and B by taking the union $A \cup B$, only if $B \not\subseteq A$ and $A \not\subseteq B$.

Example 1–9

Definition

If X, Y are sets, the *difference* of X, Y is

$$\{p \mid p \in X \quad \text{and} \quad p \notin Y\}.$$

For the difference of X, Y we adopt the notation $X - Y$; *i.e.*,

$$X - Y = \{p \mid p \in X \quad \text{and} \quad p \notin Y\}.$$

(We read $X - Y$: "X less Y.")

To illustrate this definition, let Z be the set of whole numbers and let X be the set of even whole numbers. Then the statement "$x \in Z$ and $x \notin X$" is true if and only if x is a whole number and x is not even; *i.e.*, if and only if x is a whole number and x is an odd whole number. We conclude that $Z - X$ is the set of odd whole numbers.

It is easy to prove that for all sets X, Y

$$X \cup Y = Y \cup X \quad \text{and} \quad X \cap Y = Y \cap X.$$

(See Exercise 16.) This usually is expressed by saying that the union and intersection of sets are *commutative* operations. This property of the union and intersection is not shared by the difference operation, however. By this we mean that it is *not* true that for all sets X, Y

$$X - Y = Y - X.$$

To see this, let $X = \{1, 2, 3\}$ and $Y = \{3, 4, 5\}$. Then by definition of "$-$"

$$X - Y = \{1, 2\} \quad \text{and} \quad Y - X = \{4, 5\}.$$

Therefore, $X - Y \neq Y - X$.

In particular discussions, the sets under consideration will all be subsets of some set, say U. The set U may be referred to as the *universal set* and, in this situation, the following abbreviation is usually adopted: instead of writing $U - A$ we write A'. This is read "A-complement." Since it is understood that $x \in U$, we have

$$A' = \{x \mid x \notin A\}.$$

It always will be understood in the following that each set A is a subset of some set U whenever we use the notation A'.

From the definition of A', $x \notin A$ if and only if $x \in A'$. Hence, we may express the difference $X - A$ as follows:

Example
1–10

$$X - A = \{t \mid t \in X \text{ and } t \notin A\}$$
$$= \{t \mid t \in X \text{ and } t \in A'\}$$
$$= X \cap A'.$$

12. Describe the following sets by listing their members, if

$X = \{7, 8, 11, 13, 4\}$, $Y = \{2, 4, 8, 17, 9\}$, $Z = \{1, 7, 5, 11, 19\}$.

(a) $X \cup Y$ (b) $X \cup Z$ (c) $Y \cup Z$ (d) $X \cap Y$
(e) $X \cap Z$ (f) $Y \cap Z$ (g) $X - Y$ (h) $Y - X$
(i) $X - Z$ (j) $Z - X$ (k) $Y - Z$ (l) $Z - Y$

13. Make the following computations; all complements are to be considered relative to the set $N = \{1, 2, 3, \ldots\}$.
 (a) $\{\frac{1}{2}, 3, 5\} \cup \{2, 7\}$
 (b) $\{\frac{1}{2}, 3, 5\} \cap \{2, 7\}$
 (c) $\{\frac{1}{2}, 3, 5\} - \{2, 7\}$
 (d) $\{2, 4, 6, 8\} \cap \{3, 6, 9\}$
 (e) $\{3, 6, 9\} \cup \{2, 4, 6, 8\}$
 (f) $\{3, 6, 9\} - \{2, 4, 6, 8\}$
 (g) $\{s \mid s = 2m \text{ for } m = 2, 3, 4\} \cap \{p \mid p = 3x \text{ for } x = 1, 2, 3, 4\}$
 (h) $\{a \mid a = 2b \text{ for } b = 1, 2, 3, 4, 5, 6\} \cap$
     $\{c \mid c = 3d \text{ for } d = 1, 2, 3, 4, 5, 6\}$
 (i) $\{1, 2, 3, 4, 5, 6\}'$
 (j) $\{1, 2, 5, 6\}'$
 (k) $\{m \mid m \text{ is a whole number greater than } 4\}'$

14. In each of the following you are to draw a conclusion from the given information. Several conclusions are often possible. For example, given: $\{1, a\} \subseteq \{1, 2, 3\}$, $a \neq 1$, the conclusion is $a = 2$ or $a = 3$.
 (a) Given: $X \subseteq \{1, 2\}$
 (b) Given: $X \subseteq \{1, 2\}$ and $\{1, 2\} \subseteq X$
 (c) Given: $X \cup Y \subseteq Y$
 (d) Given: $A \subseteq \emptyset$
 (e) Given: $X \cup \{5, 4\} = \{2, 3, 4, 5\}$
 (f) Given: $M \cap \{1, 2\} = \{1\}$
 (g) Given: $a \in \{1, 2, 3\} \cap \{2, 3, 4\}$

(h) Given: $\{1, 2, 7\} - X = \{7\}$
(i) Given: $Y - \{1, 2\} = \{3, 5, 9\}$
(j) Given: $A - M = \varnothing$
(k) Given: $z \in W'$

15. Prove that for sets S, T

$$(S \cap T) \subseteq S \quad \text{and} \quad (S \cap T) \subseteq T.$$

16. For all sets X, Y prove that

$$(a) \; X \cap Y = Y \cap X.$$
$$(b) \; X \cup Y = Y \cup X.$$

These are called the *commutative laws* for intersection and union.

17. Prove that for all sets L, M, N

$$(a) \; L \cup (M \cup N) = (L \cup M) \cup N.$$
$$(b) \; L \cap (M \cap N) = (L \cap M) \cap N.$$

These are called the *associative laws* for "$\cup$" and "$\cap$," and show that we may write $L \cup M \cup N$ and $L \cap M \cap N$ without ambiguity.

18. Prove that for all sets L, M, N

$$(a) \; L \cap (M \cup N) = (L \cap M) \cup (L \cap N).$$
$$(b) \; L \cup (M \cap N) = (L \cup M) \cap (L \cup N).$$

These are called the *distributive laws*. We now shall prove the first of these to illustrate a method of proof commonly used in showing the equality of two sets.

To prove (a) we use the definition of equality for sets and show that

$$L \cap (M \cup N) \subseteq (L \cap M) \cup (L \cap N)$$

and

$$(L \cap M) \cup (L \cap N) \subseteq L \cap (M \cup N).$$

It then will follow from the definition of equality for sets that statement (a) is true.

From the definition of "$\subseteq$," to show that $L \cap (M \cup N) \subseteq (L \cap M) \cup (L \cap N)$, we must show that if x is a member of $L \cap (M \cup N)$, then x is a member of $(L \cap M) \cup (L \cap N)$. If $x \in L \cap (M \cup N)$, then from the definition of "$\cap$," $x \in L$ and $x \in (M \cup N)$. But

$x \in (M \cup N)$ implies, from the definition of "$\cup$," that $x \in M$ or $x \in N$. Therefore, we know that if $x \in L \cap (M \cup N)$, then $x \in L$ and $x \in M$ or $x \in L$ and $x \in N$. The definition of "$\cap$" gives $x \in L \cap M$, if $x \in L$ and $x \in M$; $x \in L \cap N$, if $x \in L$ and $x \in N$. So, if $x \in L \cap (M \cup N)$, then $x \in (L \cap M)$ or $x \in (L \cap N)$. Finally, the definition of "$\cup$" implies that $x \in (L \cap M) \cup (L \cap N)$, if $x \in (L \cap M)$ or $x \in (L \cap N)$. Hence, if $x \in L \cap (M \cup N)$, then $x \in (L \cap M) \cup (L \cap N)$. This concludes the part of the proof that $L \cap (M \cup N) \subseteq (L \cap M) \cup (L \cap N)$.

Now, if $x \in (L \cap M) \cup (L \cap N)$, then $x \in (L \cap M)$ or $x \in (L \cap N)$ from the definition of "$\cup$." In each of these cases $x \in L$; in the first case, $x \in M$ (by the definition of "$\cap$"), and in the second case, $x \in N$ (by the definition of "$\cap$"). Therefore, by definition of "$\cup$," $x \in (M \cup N)$. Hence, $x \in L \cap (M \cup N)$, if $x \in (L \cap M) \cup (L \cap N)$.

19. If S, T, L, M are sets such that $S \subseteq L$ and $T \subseteq M$, prove that

$$(S \cup T) \subseteq (L \cup M) \qquad \text{and} \qquad (S \cap T) \subseteq (L \cap M).$$

20. Prove that if L, M are sets
    (a) $L = (L - M) \cup (L \cap M)$
    (b) $L = L \cap (L \cup M)$
    (c) $(L \cap M)' = L' \cup M'$ }
    (d) $(L \cup M)' = L' \cap M'$ } DeMorgan's Laws.
    (e) $(L - M)' = L' \cup M$
    (f) $L \cap M = (L' \cup M')'$
    (g) $(L - M) \cup (M - L) = (L \cup M) - (L \cap M)$

21. If A, B are sets such that $A \cap B = A$, what relation between A, B can be deduced? Prove your answer correct. Compare this with Example 1–9 and draw an analogous conclusion.

A very convenient device for picturing sets geometrically is illustrated by the figures below. In a plane we represent a universal set U by the interior of some quadrilateral (or other closed figures, for that matter). Then the subsets of U are represented by closed figures interior to U as in Fig. 1–1. Letting the representative figures overlap, we may represent both the union and the intersection of sets by shading appropriate areas. In Fig. 1–1, $S \cap T$ is represented by the shaded area and the shaded area of Fig. 1–2 represents $S \cup T$.

1.3
Geometric
Representation
of Sets

Figure
1–1
Figure
1–2

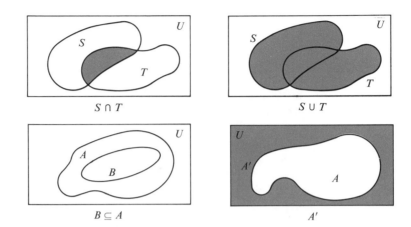

$S \cap T$

$S \cup T$

Figure
1–3
Figure
1–4

$B \subseteq A$

A'

We may represent the relation $B \subseteq A$ as in Fig. 1–3 simply by taking the representative figure for B entirely interior to the figure for A. Consider Example 1–9 of Section 1.2 with the aid of this representation. Other representations are suggested in Figs. 1–5, 1–6, and 1–7.

Figure
1–5
Figure
1–6

$S - T$

$T - S$

Figure
1–7

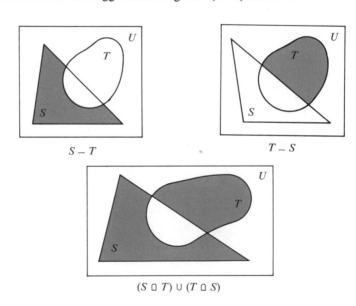

$(S \cup T) \cup (T \cup S)$

Example
1–11

To illustrate the validity of Exercise 20(c), we construct separate figures for $(L \cap M)'$ and $L' \cup M'$. In the figures of the left-hand column of Fig. 1–8, we build up $(L \cap M)'$ by stages; in the right-hand column we do a similar thing for $L' \cup M'$. In the two separate columns the final stage is the same shaded area. This, then, suggests the validity of the statement $(L \cap M)' = L' \cup M'$.

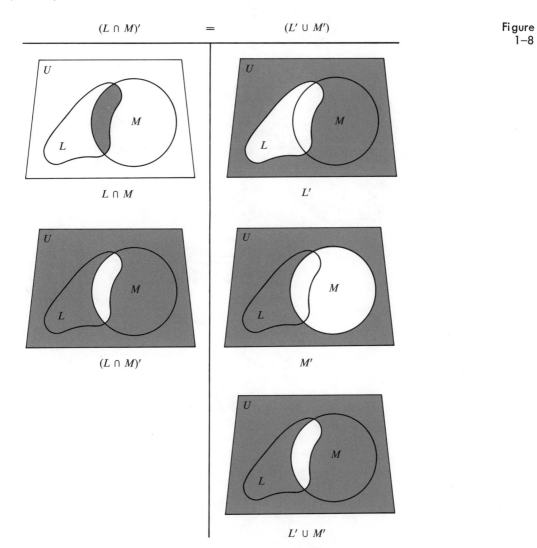

$(L \cap M)'$ $=$ $(L' \cup M')$

$L \cap M$

L'

$(L \cap M)'$

M'

$L' \cup M'$

Figure 1–8

22. In the following problems, use appropriate set notation to express the shaded areas.

Exercises

13

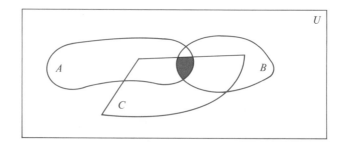

For example, this shaded area may be expressed by $A \cap B \cap C$.

(a)

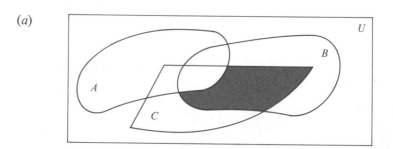

(b)

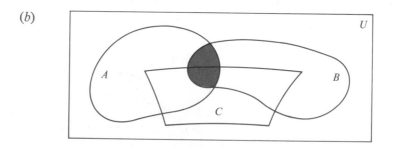

(c)

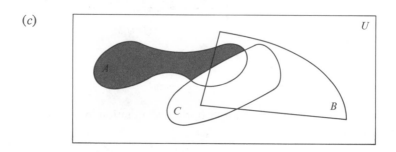

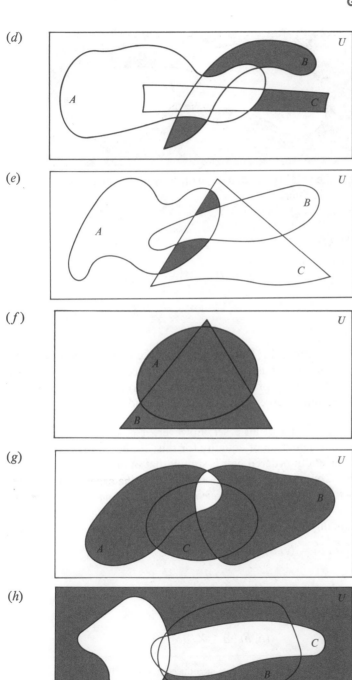

(i)

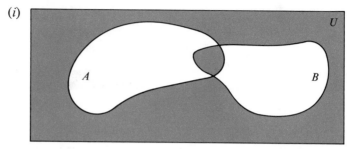

23. Illustrate the validity of each of the following by means of appropriate geometric figures.

(*a*) $L = (L - M) \cup (L \cap M)$

(*b*) $L = L \cap (L \cup M)$

(*c*) $(L \cup M)' = L' \cap M'$

(*d*) $(L \cup M')' = L' \cap M$

(*e*) $(L - M) \cup (M - L) = (L \cup M) - (L \cap M)$

1.4
Relations and Functions

In the previous section, the operations union, intersection, and difference of two sets were introduced. We now introduce another method of describing a set in terms of two given ones. In order to define this new concept, the notion of *ordered pair* must be considered.

An ordered pair is first of all a set; however, it is a set with an added property. By our definition of equality for sets, $\{a, b\} = \{b, a\}$. The idea behind the concept of ordered pair is simply that of imposing an order on the set $\{a, b\}$. Consequently, we denote by (a, b), the ordered pair having a as *first element* and b as *second element*. Now we *no longer have the equality* $(a, b) = (b, a)$ *as a general law*. Rather, we have the following rule governing equality of ordered pairs:

$$(a, b) = (c, d) \text{ if and only if } a = c \text{ and } b = d.^*$$

Definition

If X, Y are sets, the *Cartesian product* of X, Y is

$$\{(a, b) \mid a \in X, b \in Y\}.$$

---

*The notion of ordered pair, as we have introduced it, obviously lacks precision. This can be easily remedied by *defining* (a, b) to be the set $\{\{a, b\}, \{b\}\}$. It is then an easy matter to prove the above law of equality. The reader may find the proof of this fact a somewhat interesting exercise.

The Cartesian product is commonly denoted by $X \times Y$; *i.e.*,

$$X \times Y = \{(a, b) \mid a \in X, b \in Y\}.$$

If $X = \{4, 3, 8\}$ and $Y = \{1, a\}$, we have the Cartesian products given below:

Example
1–12

$$X \times Y = \{(4, a), (4, 1), (3, a), (3, 1), (8, a), (8, 1)\}$$
$$Y \times X = \{(a, 4), (1, 4), (a, 3), (1, 3), (a, 8), (1, 8)\}$$
$$X \times X = \{(4, 4), (4, 3), (4, 8), (3, 4), (3, 3), (3, 8),$$
$$(8, 4), (8, 3), (8, 8)\}$$
$$Y \times Y = \{(a, a), (a, 1), (1, a), (1, 1)\}$$

A useful geometric interpretation of $X \times Y$ may be obtained by representing the members of X and Y on intersecting lines (see Fig. 1–9), say L_1 and L_2 respectively. Then through each point of L_1 which represents an element of X we take a line parallel to L_2; and through each point of L_2 representing a member of Y we draw a line parallel to L_1. The intersections of these various lines represent the members of $X \times Y$ as indicated in Fig. 1–9.

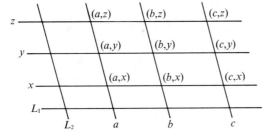

Figure
1–9

If X, Y are sets, a subset G of $X \times Y$ is called a *relation* from X to Y. The *domain* of G is the set

$$\mathcal{D}(G) = \{x \mid (x, y) \in G \text{ for some } y \in Y\}$$

and the *range* of G is the set

$$\mathcal{R}(G) = \{z \mid (t, z) \in G \text{ for some } t \in X\}.$$

It is a rather simple matter to convince oneself that this definition of relation corresponds to our common use of the word. For example, if $X = \{1, 2, 3\}$, then

$$X \times X = \{(1, 1), (1, 2), (1, 3), (2, 1), (2, 2), (2, 3), (3, 1), (3, 2), (3, 3)\}$$

Consider now the property "is the same as." This property we ordinarily consider a relation. We may define a set using this property:

$$\{(x, y) \mid x \in X, y \in X, x \text{ is the same as } y\}.$$

This is the set $\{(1, 1), (2, 2), (3, 3)\}$, a subset of $X \times X$, and therefore a relation by our definition. The domain as well as the range of this relation is X.

Again, consider the property "is less than." With $X = \{1, 2, 3\}$ as before, define $L = \{(x, y) \mid x \in X, y \in X, x \text{ is less than } y\}$. Then L is the subset $\{(1, 2), (1, 3), (2, 3)\}$ of $X \times X$. One should not conclude from these examples that all relations are as familiar as these; the set $\{(1, 1), (2, 1), (2, 2), (3, 3)\}$ is a relation by our definition, but does not appear to be as familiar as the two examples we have just seen.

One of our major concerns here will be the study of certain kinds of relations which we call *functions*.

Definition

Let G be a relation having the property that if $(a, b) \in G$, and $(a, c) \in G$, then $b = c$. Then G is called a *function*.

This definition states that *a relation is a function provided it contains no two distinct ordered pairs with the same first element*. Thinking of a relation as a pairing of elements in its domain with elements in its range, we can describe a function as a relation which pairs with each element in its domain *one and only one* element in its range.

Example 1–13

The relation $F = \{(1, 1), (2, 2), (3, 3)\}$ is a function, while $G = \{(1, 1), (1, 2)\}$ is not. The second relation is not a function by our definition since 1 is paired with two (not just one) elements in its range — namely both 1 and 2. We illustrate these two relations in Fig. 1–10 in two different ways: the first emphasizes that they are subsets of a Cartesian product, and the second emphasizes the pairing.

Example 1–14

Let $X = \{a, b, c, d\}$ and $Y = \{7, 4, 9\}$. Then each of the following relations is a function with domain X.

$$F = \{(a, 7), (b, 4), (c, 9), (d, 9)\}.$$
$$G = \{(a, 7), (b, 7), (c, 7), (d, 7)\}.$$
$$H = \{(a, 9), (b, 7), (c, 9), (d, 4)\}.$$

Figure
1–10

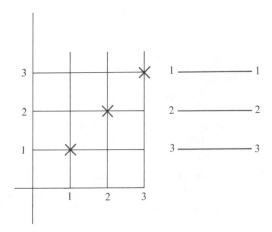

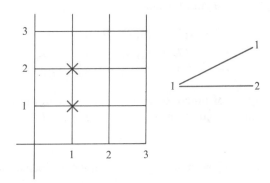

The relation $J = \{(a, 9), (b, 9), (a, 7), (c, 4), (d, 4)\}$ is not a function since $(a, 9) \in J$ and $(a, 7) \in J$ but $9 \neq 7$. We illustrate F, G, H, and J in Fig. 1–11 in two ways.

Note the fact that J is not a function is reflected on the grid-type "picture" (Fig. 1–11) in that the line through a parallel to L_2 has two ×-marks, and that in the other picture of J, two lines go *from a*. This will clearly always be the case for such representations when the relation represented is not a function.

If F is a function and the pair $(x, y) \in F$, it is customary to write $y = F(x)$. In other words, $F(x)$ is used to denote the second member of the ordered pair whose first member is x. $F(x)$ is said to be a *functional value* of F or the *value* of F at x. In Example 1–15 we have

$$7 = F(a), \ 4 = F(b), \ 9 = F(c) \text{ and } 9 = F(d).$$
$$7 = G(a), \ 7 = G(b), \ 7 = G(c) \text{ and } 7 = G(d).$$
$$9 = H(a), \ 7 = H(b), \ 9 = H(c) \text{ and } 4 = H(d).$$

Note that this notation is inappropriate for J since we would have $9 = J(a)$ and also $7 = J(a)$.

Suppose that X, Y are sets and that the relation R is a subset of $X \times Y$. Then we may associate with R a relation — called the *inverse* of R and denoted by R^{-1} — which is a subset of $Y \times X$. R^{-1} is defined to be the set

$$\{(w, z) \mid (z, w) \in R\}.$$

In other words, $(w, z) \in R^{-1}$ if and only if $(z, w) \in R$.

From Example 1–15 again, we have

$$F^{-1} = \{(7, a), (4, b), (9, c), (9, d)\}.$$
$$G^{-1} = \{(7, a), (7, b), (7, c), (7, d)\}.$$
$$H^{-1} = \{(9, a), (7, b), (9, c), (4, d)\}.$$

From the definition of function, we see that F^{-1} is not a function unless $c = d$; G^{-1} is not a function unless $a = b = c = d$; H^{-1} is not a function unless $a = c$.

In case R is a function, we may use the notation described above and write

$$(w, z) \in R^{-1} \text{ if and only if } w = R(z).$$

Also if R and R^{-1} are both functions,

$$z = R^{-1}(w) \text{ if and only if } w = R(z).$$

If R is a function with domain X and range Y, let us consider the question of whether R^{-1} is a function. According to the definition, R^{-1} is a function if and only if the following is true:

$$(w, z_1) \in R^{-1} \text{ and } (w, z_2) \in R^{-1} \text{ implies } z_1 = z_2.$$

Using the definition of R^{-1} we may state this condition as:

(1-2) $(z_1, w) \in R \text{ and } (z_2, w) \in R \text{ implies } z_1 = z_2.$

Figure
1–11

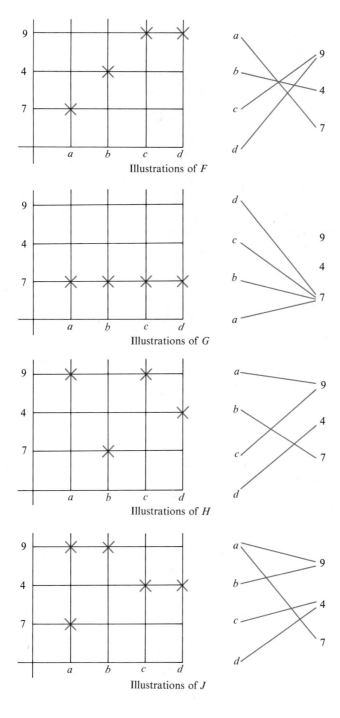

Illustrations of F

Illustrations of G

Illustrations of H

Illustrations of J

Definition

A function R is a *one-one* function if and only if it has the property expressed in **(1-2)**.

In terms of pairing we may state the condition that a function R be *one-one* as follows: *A function* R *is one-one if and only if each element in the range of* R *is paired with only one element in the domain of* R.

Returning to the question of whether R^{-1} is a function, it is evident that R^{-1} is a function if and only if R is a *one-one* function.

Exercises

24. Make the following computations for the sets $V = \{\frac{1}{2}, 3, 7\}$, $W = \{8, 3\}$, $X = \{2, 7, 4\}$. For example, $V \times W = \{(\frac{1}{2}, 8), (\frac{1}{2}, 3), (3, 8), (3, 3), (7, 8), (7, 3)\}$.

(a) $V \times X$ (b) $V \times V$ (c) $W \times V$
(d) $X \times V$ (e) $W \times X$ (f) $(V \times W)^{-1}$
(g) $(X \times V)^{-1}$ (h) $(W \times X)^{-1}$

25. For each of the relations R below compute $\mathcal{D}(R)$, $\mathcal{R}(R)$ and R^{-1}. For example, let $R = \{(1, 2), (0, \frac{1}{3})\}$. Then $\mathcal{D}(R) = \{1, 0\}$, $\mathcal{R}(R) = \{2, \frac{1}{3}\}$, $R^{-1} = \{(2, 1), (\frac{1}{3}, 0)\}$.

(a) $R = \{(10, 7), (a, 5), (a, b)\}$
(b) $R = \{(14, -1), (\frac{3}{8}, \frac{7}{6}), (x, 3), (-10, 1)\}$
(c) $R = \{(x, 2) \mid x$ is a whole number$\}$
(d) $R = \{(t, p) \mid t$ is a triangle and p is the area of $t\}$
(e) $R = \{(a, a + 2) \mid a$ is a whole number$\}$
(f) $R = \{(A, n) \mid A \subseteq \{1, 2, 3\}$ and n is the number of elements of $A\}$

26. Decide whether each of the following relations is a function.

(a) $\{(1, 1), (1, 2)\}$
(b) $\{(1, 2), (7, 6)\}$
(c) $\{(1, 4), (2, 4)\}$
(d) $\{(2, 5), (2, 6)\}$
(e) $\{(0, b), (0.5, 7), (9, 0.7)\}$
(f) $\{(a, b) \mid b = 2$ and a is a whole number$\}$
(g) $\{(a, b) \mid a = 2$ and b is a whole number$\}$

27. (a) If $g = \{(1, 4), (2, 4), (5, 6)\}$, what is $g(1)$, $g(2)$, $g(5)$?
    (b) If $h = \{(0, b), (0.5, 7), (9, 0.7)\}$, what is $h(0)$, $h(0.5)$, $h(9)$?
    (c) If $i = \{(a, b) \mid b = 2$ and a is a whole number$\}$, what is $i(x)$ for any whole number x?
    (d) If $k = \{(n, n + 1) \mid n$ is a whole number$\}$, what is $k(y)$ for any whole number y?

The reader is likely to be familiar with the practice of expressing functions by equations or formulas. For example, one may say: let f be the function defined as follows,

$$f(x) = 2x + 1$$

In such a case as this a set of pairs $(x, 2x + 1)$ is under consideration. Here, the domain of the function is not explicitly given and the reader is to understand that the domain is to be taken as *the set of all those numbers for which the formula will apply*. For the function $f(x) = 2x + 1$ the domain is the set of all numbers since we may multiply any number by 2 and add 1.

The formula $f(x) = 2x + 1$ shows precisely how the number $f(x)$ "depends" on the number x. For this reason it is customary to say that x is the *independent variable* and $f(x)$ is the *dependent variable*.

When a function is given in the formula notation such as $f(x) = 2x + 1$, the idea of pairing is being stressed: for each number x there is associated or paired one and only one number $f(x)$. In fact, this idea is quite often used as the definition of the function concept: *if* X, Y *are sets*, f *is said to be a function from* X *to* Y *if for each* x $\in$ X, f *is a rule which associates one and only one* y $\in$ Y.

This way of thinking of functions is very convenient in certain applications where we wish to show exactly how one quantity depends upon another. The area of a square, for example, depends on the length of one side. This dependence is explicitly shown by the formula

$$A = s^2$$

where s is the length of a side and A is the area. This is also expressed as

$$A(s) = s^2.$$

Functional Notation

Example
1–15

An auto is traveling at a constant speed of 60 miles per hour. At the end of the first hour the auto has traveled 60 miles; at the end of the second hour it has traveled 120 miles, etc. We may construct a table as we have done below to indicate the distance the auto has traveled after certain periods of time (in hours). In the table t represents time and d represents distance traveled (in miles).

t	0	$\frac{1}{2}$	1	$1\frac{1}{2}$	2	$2\frac{1}{2}$	3	$3\frac{1}{2}$	4	$4\frac{1}{2}$	5
d	0	30	60	90	120	150	180	210	240	270	300

The table above is just another way of writing the ordered pairs $(0, 0)$, $(\frac{1}{2}, 30)$, $(1, 60)$, $(1\frac{1}{2}, 90)$, etc. All the information given in the table (and more) is more conveniently expressed by the formula

$$d(t) = 60t$$

i.e., we express the distance, $d(t)$, traveled at the end of time t *as a function of time.*

Example
1–16

Suppose that a farmer wishes to enclose a rectangular area of land with 100 feet of wire. Express the area A of the enclosed field as a function of the length of one side.

Solution. Let w be the width and h the length of the field. Then $A = wh$, expressing A as a function of w and h which is not a solution to the problem. Since only 100 feet of wire is available, $2w + 2h = 100$ or $w + h = 50$. Therefore, we may write w as a function of h or h as a function of w:

$$w = 50 - h \quad \text{and} \quad h = 50 - w.$$

By substituting into the formula $A = wh$ we have

$$A = (50 - h)h \quad \text{or} \quad A = w(50 - w).$$

Both of these formulas express A as a function of the length of one of the sides of the field.

28. Given the function $h(x) = x^2 + 1$.
    (a) What is the domain of h?
    (b) Compute $h(x)$ for $x = 0, 1, -1, \frac{1}{2}, 5, t + 1$.

Answer. (*a*) The domain of h is the set of all numbers since we may square any number and add 1.

(*b*) $h(0) = 0^2 + 1 = 1, h(1) = 1^2 + 1 = 2, h(-1) = (-1)^2 + 1 = 2,$
$h(\frac{1}{2}) = (\frac{1}{2})^2 + 1 = \frac{5}{4}, h(5) = 5^2 + 1 = 25 + 1 = 26, h(t + 1) =$
$(t + 1)^2 + 1 = t^2 + 2t + 2.$

29. Given the function $f(x) = 2/x$.
    (*a*) What is the domain of f?
    (*b*) Compute $f(x)$ for $x = 1, -1, \frac{1}{2}, 5, t + 1$ (for $t \neq -1$).

30. Given the function $g(x) = x/(x - 1)$.
    (*a*) What is the domain of g?
    (*b*) Compute $g(x)$ for $x = 0, 2, -1, \frac{1}{2}, 5, t + 1$ ($t \neq 0$).

31. Given the function $F(m) = \sqrt{m}$.
    (*a*) What is the domain of F?
    (*b*) Compute $F(m)$ for $m = 0, 4, \frac{1}{4}, 16, 100$.

32. Given the function $k(t) = 4/(t + 1)(t - 2)$.
    (*a*) What is the domain of k?
    (*b*) Compute $k(t)$ for $t = 0, 1, \frac{1}{2}, 5, 3, m - 1$ (where $m \neq 0, m \neq 3$).

33. Express the area A of a circle as a function of the radius r.

34. Express the area A of an isosceles right triangle as a function of the altitude h.

35. Express the volume V of a cube as a function of the length of one edge w.

36. Express the area A of a rectangle which is twice as long as it is wide as a function of the width w.

If we are given the function $f(x) = x^2 + x$, it is natural to think of f as a sum of the functions x^2 and x. It is therefore convenient and proper to define what we mean by the *sum* of two functions.

1.6
The Algebra of
Functions

Let g, h be functions having a common domain X. Then $g + h, g - h, g \cdot h$ and g/h are functions defined as follows:

Definition

$$(g + h)(x) = g(x) + h(x) \text{ for all } x \in X;$$
$$(g - h)(x) = g(x) - h(x) \text{ for all } x \in X;$$
$$(g \cdot h)(x) = g(x) \cdot h(x) \text{ for all } x \in X;$$
$$(g/h)(x) = g(x)/h(x) \text{ for all } x \in X \text{ such that } h(x) \neq 0.$$

The functions $g + h$, $g - h$, $g \cdot h$, g/h are called the *sum, difference, product* and *quotient* respectively of the functions g, h.

Example
1–17

If $g(x) = x^2$ and $h(x) = x$, then $f(x) = x^2 + x = g(x) + h(x)$ and f is the sum of g and h.

Example
1–18

Let g, h be functions with the domain $\{0, 1, 2, 3, 4\}$ and suppose these are defined as follows:

$$g(x) = x + 1 \qquad h(x) = 1 - x.$$

We may exhibit the functions $g + h$, $g - h$, $g \cdot h$ and g/h in the table below.

x	0	1	2	3	4
$g(x)$	1	2	3	4	5
$h(x)$	1	0	-1	-2	-3
$(g + h)(x)$	2	2	2	2	2
$(g - h)(x)$	0	2	4	6	8
$(g \cdot h)(x)$	1	0	-3	-8	-15
$(g/h)(x)$	1	not defined	-3	-2	$-5/3$

Example
1–19

If $g(x) = 2x + 5$ and $h(x) = 2 - x$, describe the functions $g + h$ and $g \cdot h$.

Solution. $(g + h)(x) = g(x) + h(x) = (2x + 5) + (2 - x)$
$$= x + 7$$

$(g \cdot h)(x) = g(x) \cdot h(x) = (2x + 5)(2 - x) = 10 - x - 2x^2.$

Let g, h be functions such that $\mathcal{R}(h) \subseteq \mathcal{D}(g)$. Then the *composite function of g by* h is

$$\{(x, y) \mid y = g(h(x))\}.$$

The composite of g by h is denoted by $g \circ h$.

It should be carefully noted that the order in which the functions g, h occur in the definition of $g \circ h$ is very important. Also note that the notation and definition imply that if x is in the domain of $g \circ h$, then

$$g \circ h(x) = g(h(x)).$$

If $h(x) = x^2$ and $g(x) = x + 2$, then the domain of h is the set of all numbers so that the range of g is included in the domain of h. The function $h \circ g$ is given by the following formula:

$$h \circ g(x) = h(g(x)) = h(x + 2) = (x + 2)^2 = x^2 + 4x + 4.$$

Example 1–20

Let $S = \{1, 2, 3\}$ and define functions h, g with domain S as follows:

$$
\begin{array}{ll}
h(1) = 2 & g(1) = 1 \\
h(2) = 3 & g(2) = 3 \\
h(3) = 1 & g(3) = 2
\end{array}
$$

Example 1–21

Then

$$
\begin{aligned}
h \circ g(1) &= h(g(1)) = h(1) = 2 \\
h \circ g(2) &= h(g(2)) = h(3) = 1 \\
h \circ g(3) &= h(g(3)) = h(2) = 3
\end{aligned}
$$

Let $T = \{1, 3, 5\}$ and let h, g be functions with domain T defined as follows:

$$
\begin{array}{ll}
h(1) = 5 & g(1) = 7 \\
h(3) = 3 & g(3) = 8 \\
h(5) = 1 & g(5) = 9
\end{array}
$$

Example 1–22

Then $\mathcal{R}(h) = \{1, 3, 5\}$, $\mathcal{R}(g) = \{7, 8, 9\}$. Since $\mathcal{R}(g) = \{7, 8, 9\}$ and $\{7, 8, 9\}$ is *not* a subset of T, $h \circ g$ is *not* defined. But it is true that $\mathcal{R}(h) = \{1, 3, 5\} \subseteq T = \mathcal{D}(g)$. Hence, $g \circ h$ is defined and may be given by

$$
\begin{aligned}
g \circ h(1) &= g(h(1)) = g(5) = 9 \\
g \circ h(3) &= g(h(3)) = g(3) = 8 \\
g \circ h(5) &= g(h(5)) = g(1) = 7
\end{aligned}
$$

If we are given a relation T and wish to construct its inverse, we only need to interchange first and second members of the ordered pairs of T. If $T = \{(x, y) \mid y = 2x + 3\}$, then by definition

$$T^{-1} = \{(y, x) \mid y = 2x + 3\}.$$

This is an adequate description of T^{-1}, but it is customary to express the second members of ordered pairs — when possible — in terms of the first members. This is not done in this description of T^{-1}. But this may be accomplished from

$$y = 2x + 3$$

by solving for x in terms of y:

$$x = \tfrac{1}{2}(y - 3).$$

Then,

$$T^{-1} = \{(y, x) \mid x = \tfrac{1}{2}(y - 3)\}.$$

Everything is fine now except that we have violated the convention of calling first coordinates x and second coordinates y. Since it makes no difference what names we give these elements, we conform and write

$$T^{-1} = \{(x, y) \mid y = \tfrac{1}{2}(x - 3)\}.$$

In the $T(x)$-notation,

$$T(x) = 2x + 3$$

and

$$T^{-1}(x) = \tfrac{1}{2}(x - 3).$$

Let us compute the composites $T \circ T^{-1}$ and $T^{-1} \circ T$:

$$\begin{aligned}
T \circ T^{-1}(x) = T\big(T^{-1}(x)\big) &= T\big(\tfrac{1}{2}(x - 3)\big) \\
&= 2\big(\tfrac{1}{2}(x - 3)\big) + 3 \\
&= (x - 3) + 3 = x \\
T^{-1} \circ T(x) = T^{-1}\big(T(x)\big) &= T^{-1}(2x + 3) \\
&= \tfrac{1}{2}(2x + 3 - 3) \\
&= x.
\end{aligned}$$

Thus,

$$T \circ T^{-1}(x) = x$$

and

$$T^{-1} \circ T(x) = x$$

This shows that in this case both $T \circ T^{-1}$ and $T^{-1} \circ T$ are functions which associate with each number x the number x itself. A function J which associates the number x with each number x in the domain of J is called an *identity function*. In other words, if J is an identity function, then

$$J(x) = x$$

for all $x \in \mathfrak{D}(J)$.

The fact that $T \circ T^{-1}$ and $T^{-1} \circ T$ are both identity functions is not peculiar to the above example. If T is any function such that T^{-1} is a function, we have $y = T(x)$ if and only if $x = T^{-1}(y)$. Therefore, by substitution,

$$y = T(T^{-1}(y)) \text{ and } x = T^{-1}(T(x)).$$

The above shows that $T \circ T^{-1}$ and $T^{-1} \circ T$ *are identity functions for any* **T** *which is 1-1.*

This property may be used to compute inverses when only functions are involved. For the function

$$T(x) = 2x + 3,$$

for example, we know that $T(T^{-1}(x)) = x$ so that

$$x = T(T^{-1}(x)) = 2(T^{-1}(x)) + 3.$$

Then solving for $T^{-1}(x)$:

$$T^{-1}(x) = \tfrac{1}{2}(x - 3).$$

If $h(x) = 3 - 5x$, show that h is 1-1 and compute $h^{-1}(x)$ by using the composite function $h \circ h^{-1}$.

Example
1–23

Solution. According to the definition of 1-1 function we must show that if $h(x) = h(y)$, then $x = y$. But if $h(x) = h(y)$, we have $3 - 5x = 3 - 5y$. Then $-5x = -5y$ and $x = y$. Therefore h is a 1-1 function. To compute the inverse, h^{-1}, we use the fact that $h \circ h^{-1}(x) = x$. We have $x = h \circ h^{-1}(x) = h(h^{-1}(x)) = 3 - 5h^{-1}(x)$. If we now solve the equation $x = 3 - 5h^{-1}(x)$ for $h^{-1}(x)$, we obtain

$$h^{-1}(x) = (1/5)(3 - x).$$

If $g(x) = \sqrt{x}$ for all non-negative x, compute $g^{-1}(x)$ (assuming that g is 1-1).

Example
1–24

Solution. $x = g \circ g^{-1}(x) = g(g^{-1}(x)) = \sqrt{g^{-1}(x)}$. Then $g^{-1}(x) = x^2$ for all non-negative x.

Let $f(x) = 1/(1 - x)$ for all $x \neq 1$. Assuming that f is 1-1, compute $f^{-1}(x)$.

Example
1–25

Solution. $x = f \circ f^{-1}(x) = f(f^{-1}(x)) = 1/(1 - f^{-1}(x))$. Then $1 - f^{-1}(x) = 1/x$ or $f^{-1}(x) = 1 - 1/x$.

37. For each pair of functions h, g below compute $h + g$ and $g \cdot h$.
    (a) $h(x) = 2x$, $g(x) = 5 + x$
    (b) $h(x) = -x$, $g(x) = x^2 + 1$
    (c) $h(x) = 1 - x$, $g(x) = 1 + x$
    (d) $h(x) = 7x^2 - 1$, $g(x) = 3 - x$
    (e) $h(x) = x^2 + x + 1$, $g(x) = 4 - x + x^2$

38. For each pair of functions h, g of 37 above compute $g - h$.

39. For each pair of functions h, g of 37 above (except (e)) compute h/g.

40. For each pair of functions h, g of 37 above compute $h \circ g$.

41. For each pair of functions h, g of 37 above compute $g \circ h$.

42. For each function, f, below find functions h, g such that $f = h \circ g$.
    For example, if $f(x) = (x^3 + 1)^7$, we may choose $h(x) = x^7$ and
    $g(x) = x^3 + 1$. Then $h \circ g(x) = h(g(x)) = (g(x))^7 = (x^3 + 1)^7 = f(x)$.
    (a) $f(x) = \sqrt{7x + 5}$
    (b) $f(x) = x^3 + 1$
    (c) $f(x) = x^3 + 2x^6 + 7x^9$
    (d) $f(x) = 1/x + 2/x^2 + 3/x^3$, $x \neq 0$.

43. Show that each of the functions below is 1-1.
    (a) $h(x) = 2x$
    (b) $h(x) = 1 - x$
    (c) $h(x) = 5 + x$
    (d) $h(x) = -x$
    (e) $h(x) = 7(4 - 5x)$
    (f) $h(x) = 1/x$, $x \neq 0$
    (g) $h(x) = 4/(x + 2)$, $x \neq -2$

44. For each function, h, of 43 above compute h^{-1}.

2

THE
REAL
NUMBERS

The object of the present chapter is a brief description of those parts of the real number system which the student will need for later work. Much of this description will not be new to him, for he will have encountered a considerable amount of information about numbers in his earlier studies.

The set of *integers* consists of the numbers

$$\ldots, -4, -3, -2, -1, 0, 1, 2, 3, 4, \ldots$$

This set of integers will be denoted by Z. The set $\{1, 2, 3, 4, \ldots\}$ is called the set of *whole numbers*, *natural numbers* or *positive integers* and is denoted by N. It is clear that we have

$$N \subseteq Z.$$

The familiar operations of *addition*, *subtraction*, and *multiplication* may be carried out on pairs of elements of Z and the result will always be an element of Z; this property of Z is generally described by saying that Z is *closed* with respect to addition, subtraction and multiplication. In symbols this may be expressed as follows: if $a, b \in Z$, then $a + b \in Z$, $a - b \in Z$, $a \cdot b \in Z$. As an example, we have $-8, 10 \in Z$ and $-8 + 10 = 2 \in Z$, $-8 - 10 = -18 \in Z$ and $(-8) \cdot 10 = -80 \in Z$. In contrast to this, Z is not closed with respect to division. For example, $1 \in Z$ and $2 \in Z$ but $1 \div 2 = \frac{1}{2} \notin Z$. For this reason we need to "expand" or "broaden" Z to a new set of numbers which will be closed not only with respect to addition, subtraction and multiplication but will also allow division by non-zero numbers. We are thus led to the set Q of rational numbers which is defined as follows:

$$Q = \{a/b \mid b \neq 0, a \in Z \text{ and } b \in Z\}.$$

At this point the definition of addition, subtraction, multiplication and division on Q should be recalled.

Addition in Q: $a/b + c/d = (ad + bc)/bd$ for all a, b, c, $d \in Z$, $b \neq 0, d \neq 0$

Subtraction in Q: $a/b - c/d = (ad - bc)/bd$ for all a, b, c, $d \in Z$, $b \neq 0, d \neq 0$

Multiplication in Q: $(a/b)\cdot(c/d) = (ac)/bd$ for all a, b, c, $d \in Z$, $b \neq 0, d \neq 0$

Division in Q: $(a/b) \div (c/d) = (ad)/(bc)$ for all a, b, c, $d \in Z$, $b \neq 0, c \neq 0, d \neq 0$

If n is any number, $n/1 = n$. In particular, if $n \in Z$, $n/1 \in Q$ and this implies that $Z \subseteq Q$. It then follows that

$$N \subseteq Z \subseteq Q.$$

In the system Q of rational numbers it is always possible to solve certain kinds of equations. If it is desired to find a number $x \in Q$ such that $2x + 3 = 0$, this is possible. We simply solve the equation $2x + 3 = 0$ by the usual method: $2x + 3 = 0$ implies $2x = -3$ and $x = -3/2$. To show that this is actually a solution we substitute: $2x + 3 = 2(-3/2) + 3 = -3 + 3 = 0$. Hence, $-3/2$ is a solution of the equation and is a rational number.

Suppose now that we wish to find a number $x \in Q$ such that

$$x^2 - 2 = 0$$

or, which is the same thing,

$$x^2 = 2.$$

In this case the task is impossible, that is to say, there is no rational number x such that $x^2 = 2$. It will now be shown that this is indeed the case by using the classical *indirect method* of proof. This method is based upon the so-called *law of the excluded middle* which states that a sentence or proposition cannot be both true and false simultaneously. We may find, for example, that it is impossible for a certain statement to be true; it should then be concluded, on the basis of the law of the excluded middle, that the statement must be false. If a statement is assumed to be true and this assumption results in a contradiction, then it is clearly impossible for the statement to be true. It is just this kind of reasoning which will be employed below to prove our result.

There is no number $x \in Q$ such that $x^2 = 2$.

Proof. Suppose there is a number $x \in Q$ such that $x^2 = 2$. We will show that this assumption leads to a contradiction. By definition of Q, if $x \in Q$, there are elements $a, b \in Z, b \neq 0$ such that $x = a/b$. At this point we may assume that a, b have no common factors (other than ± 1), for a common factor could be cancelled and a/b reduced to lowest terms. Then since $x^2 = 2$ and $x = a/b$

$$(a/b)^2 = 2 \qquad \text{or} \qquad \frac{a^2}{b^2} = 2.$$

From the last equation, $a^2 = 2b^2$ which implies that a^2 is an even integer. But if a^2 is even, a must be even since the square of an odd integer is odd. Therefore, a is even and there is some integer, k say, such that $a = 2k$. Then $a^2 = 4k^2$ so that

$$4k^2 = 2b^2 \qquad \text{or} \qquad 2k^2 = b^2.$$

From this we conclude that b^2 is even and, hence, that b is even. We have now arrived at the conclusion that both a and b are even. This means that a, b have a common factor of 2. But this contradicts the fact that a/b is in lowest terms. The assumption made at the beginning of the proof cannot, therefore, hold since it leads to a contradiction. The proof of the theorem is thus complete.

Theorem 1 shows that a solution for the equation

$$x^2 - 2 = 0$$

must be sought outside the set Q of rational numbers. Fortunately, the set of *real* numbers, denoted by R, contains a solution to this equation. The definition of real number given below depends upon *decimal* representations of numbers and for this reason such representations will be briefly reviewed.

The number 1.152 is an example of a decimal number and is simply a convenient way of writing

$$1 + \frac{1}{10} + \frac{5}{100} + \frac{2}{1000}$$

i.e., by definition

$$1.152 = 1 + \frac{1}{10} + \frac{5}{100} + \frac{2}{1000}.$$

This number is also an example of a *terminating* decimal; this means that the digits to the right of the decimal point terminate or do not continue indefinitely. Other examples of terminating decimals are 32.1, −3.25, 0.3, 0.00053425. In contrast to the terminating decimals, there are the *non-terminating* ones. For instance, 8.292929 . . . is a non-terminating decimal with the digits 2, 9 (in that order) continuing indefinitely. This is also an example of a *repeating* decimal (since the digits 2, 9 repeat without end). A convenient notational device for expressing repeating decimals is a "line" or "bar" directly over the repeating part. By employing this notation, the number above may be expressed as

$$8.292929 \ldots = 8.\overline{29}$$

Some other repeating decimals are $262.17\overline{15}$, $-94.\overline{112}$, $0.005\overline{244}$, $1.1259\overline{4}$.

Besides the repeating, non-terminating decimals there are the non-repeating, non-terminating ones. Such a number obviously cannot display all digits to the right of the decimal point. It is possible, however, to indicate that $-3.76152538 \ldots$ is such a number if it is understood that no indefinite repetition occurs as in a repeating decimal.

All of the terminating and repeating decimals are rational numbers. For example, 1.23 is the same as

$$1 + 2/10 + 3/100 = 100/100 + 20/100 + 3/100 = 123/100.$$

Hence, $1.23 \in Q$. Now consider a repeating decimal: $8.\overline{29}$. To see that this number is rational, let

$$x = 8.\overline{29}$$

Then

$$100x = 829.\overline{29}$$

and

$$99x = 100x - x = 829.\overline{29} - 8.\overline{29} = 821.000 \ldots = 821.$$

Solving for x gives $x = 821/99$ and this implies that $8.\overline{29} \in Q$. It should be evident that every terminating and repeating decimal may be shown to be rational in this way. Also, every rational number is either a terminating or repeating decimal. For example, $\frac{1}{2} = 0.5$, $-(3/2) = -1.5$, and $1/3 = 0.\overline{3}$. To express a given rational number as a decimal, we only have to perform the usual long division. For the rational number 456/37, we have

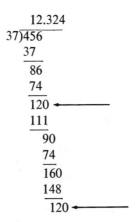

The arrows above indicate that in carrying out this division 120 has been repeated; a continuation of the division will, for this reason, result in a repetition of the digits 3, 2, 4 (in that order). This implies

$$456/37 = 12.\overline{324}.$$

If we now define the *real numbers, R, to be the set of all decimal numbers,* the following relations hold:

$$N \subseteq Z \subseteq Q \subseteq R.$$

1. Express each of the following numbers in the form a/b where a, $b \in Z$ and $b \neq 0$. For example, given $1/2 + 1/3$, we have $1/2 + 1/3 = (3 \cdot 1 + 2 \cdot 1)/2 \cdot 3 = 5/6$.
   (a) $1/2 - 1/4$
   (b) $(2/3) \cdot (5/8 + 7/6)$
   (c) $(2/3) \div (7/8)$
   (d) $(21/10) \cdot (4/7)$
   (e) $(3/2 - 8/5) \div (1/5)$
   (f) 2.5
   (g) 0.01
   (h) $1.12 \div 10^3$
   (i) $(1/2 + 0.25)^2$

Integers, Rationals and Reals

2. Express each of the following rational numbers as terminating or repeating decimals. For example, given 3/8,

$$
\begin{array}{r}
.375 \\
8)\overline{3.0} \\
24 \\
\hline
60 \\
56 \\
\hline
40 \\
40 \\
\hline
\end{array}
$$

Therefore, $3/8 = 0.375$.

(a) 21/10   (b) 5/3   (c) 27/3
(d) 14/99   (e) 111/23   (f) −5132/999
(g) 1/438   (h) 231/2   (i) −756/11

3. Express each of the following repeating decimals in the form a/b for
   $a, b \in Z, b \neq 0$. For example, given $41.7\bar{2}$. We let $x = 41.7\bar{2}$. Then
   $10x = 417.\bar{2}$. Also $417.\bar{2} = 417.2\bar{2}$. Then we subtract as indicated
   below:

$$
\begin{array}{r}
10x = 417.2\bar{2} \\
x = 41.7\bar{2} \\
\hline
9x = 375.5
\end{array}
$$

Solving the equation $9x = 375.5$ for x gives $x = 375.5/9 = 3755/90$.
Hence, $41.7\bar{2} = 3755/90$.

(a) $0.\overline{12}$ (b) $3.\overline{512}$ (c) $5.\overline{9405}$ (d) $-1.3\bar{4}$
(e) $7.23\overline{29}$ (f) $-21.657\bar{3}$ (g) $101.098\overline{12}$ (h) $0.00541\bar{8}$
(i) $0.\bar{9}$ (j) $2.76\bar{9}$ (k) $349.\bar{9}$

4. Decide whether the following decimals are repeating and whether
   they represent rational numbers. For instance, the digits in the
   decimal 0.020020002000020000020000002 . . . are to continue as
   indicated with an extra 0 between 2's at each successive stage. This
   is not a repeating decimal and obviously does not terminate. Hence,
   this decimal is an irrational number.
   (a) 1.127012700127000127000012700000127000000127 . . .
   (b) 2.0101101110111101111101111110111111110 . . .
   (c) 13.98598559855598555598555559855555598 . . .

5.* For this exercise recall that an integer n is *even* if there is some $k \in Z$
   such that $n = 2k$; n is *odd* if there is some $k \in Z$ such that $n = 2k + 1$.
   (a) Show that the sum of two even integers is even.
   (b) Show that the sum of two odd integers is even.

(*c*) Show that the sum of an odd integer and an even integer is odd.

(*d*) Show that the square of an even integer is even.

(*e*) Show that the square of an odd integer is odd.

In this section we wish to review the order properties of the real numbers. For this purpose the following axiom is introduced.

The Order Axiom. There is a non-empty subset of real numbers called the *positive numbers* and denoted by P with the following properties:

(*a*) If $a \in P$ and $b \in P$, then $(a + b) \in P$ and $(ab) \in P$.

(*b*) If a is a real number, then one and only one of the following is true:

$$a \in P, \quad \text{or} \quad -a \in P, \quad \text{or} \quad a = 0.$$

We emphasize here that no information is available concerning *positive* numbers except that which is contained in the above axiom.

Statement (*a*) of the axiom may be expressed by saying: *The sum and product of positive numbers is positive.*

Statement (*b*) may be rendered: *If a number is not zero, then either it is positive or its negative inverse is positive — but not both.*

The Order Axiom enables us to define precisely what we mean when we say that one real number is less than another. This has the effect of "arranging" the reals in a fixed order; hence, the name for the axiom.

The real number a is *less than* the real number b, denoted by $a < b$, if and only if $(b - a) \in P$.

Definition

A consequence of this definition is that for any two real numbers a and b, exactly one of the statements

$$a < b; \quad b < a; \quad a = b$$

is true. To see this, consider the number $a - b$. By the Order Axiom, one and only one of the following is true:

$$(a - b) \in P; \quad -(a - b) \in P; \quad a - b = 0.$$

These statements lead immediately to the result by definition of "$<$." By definition we also have $a - 0 = a \in P$ if and only if $0 < a$ and $0 - a = -a \in P$ if and only if $a < 0$.

Questions such as "For what numbers x is the number $2x - 7$ negative?" arise in the study of calculus. Expressions of the type "$2x - 7 < 0$" are called inequalities, and they can be "solved" by using properties of real numbers together with three special properties of the "less than" relation. Stated informally, the three properties are: (1) Any real number may be added to both sides of an inequality which is a true statement, and the resulting inequality is a true statement. (2) If both sides of an inequality which is a true statement are multiplied by the same *positive* number, the resulting inequality is a true statement. (3) If both sides of an inequality which is a true statement are multiplied by the same negative number and the "sense" or "direction" of the inequality is reversed, then the resulting inequality is a true statement. We now state these formally as theorems, giving proofs for two of them.

Theorem 2

If $a < b$, then $a + c < b + c$.

Proof: If $a < b$ then $b - a \in P$, by the definition of "less than," and, since $c - c = 0$, we have

$$b - a = b - a + c - c = b + c - a - c = (b + c) - (a + c) \in P.$$

Again, by the definition of "less than," we see that this last statement means that $a + c < b + c$.

Theorem 3

If $a < b$ and $0 < c$, then $ac < bc$.

Proof: Since $a < b$ and $0 < c$ we know from the definition of "less than" that $b - a \in P$ and $c \in P$. Then by the order axiom $(b - a)c \in P$, that is, $bc - ac \in P$ and thus $ac < bc$.

Theorem 4

If $a < b$ and $c < 0$, then $bc < ac$.

Proof: Left as an exercise.

If $a < b$ and $b < c$, then $a < c$.

Proof: Left as an exercise.

Some examples illustrating the use of these ideas follow after a few words about notation.

The expression "$a > b$" is read "a is greater than b" and it means that b is less than a; i.e., $a > b$ if and only if $b < a$. The expression "$a \leq b$" is read "a is less than or equal to b" and is an abbreviated way of writing "$a < b$ or $a = b$." All the results established for "$<$" hold for "$\leq$" and can be easily proved by considering separately the cases where equality holds.

Find all real numbers x for which $2x - 7 < 0$.

Example
2–1

Solution. If x is a real number such that $2x - 7 < 0$, by Theorem 2 we have $(2x - 7) + 7 < 0 + 7$ or $2x < 7$. From this, by Theorem 3 since $0 < \frac{1}{2}$, $\frac{1}{2} \cdot (2x) < \frac{1}{2} \cdot 7$ or $x < \frac{7}{2}$. This shows that *if there are any numbers x such that $2x - 7 < 0$, then $x < \frac{7}{2}$*. On the other hand, if $x < \frac{7}{2}$, $2x < 7$ (Why?) and $2x - 7 < 0$ (Why?). And this shows that *every* number x which is less than $\frac{7}{2}$ works; i.e., the *solution set* for this problem is $\{x \mid x < \frac{7}{2}\}$.

Notice in this example that there are two important parts: First, we assumed that there was a number x such that $2x - 7 < 0$, and from this we deduced that $x < \frac{7}{2}$. Then we reversed the steps to show that if $x < \frac{7}{2}$, then $2x - 7 < 0$. This shows, of course, that

$$\{x \mid 2x - 7 < 0\} = \{x \mid x < \tfrac{7}{2}\}.$$

Both of these steps are important, and should be observed in solving any inequalities of this type. In the following examples, we do not supply all these details, but the reader is urged to supply them for himself.

Find the solution set of the inequality $5x + 7 < 8x - 3$.

Example
2–2

Solution. If x is a real number such that $5x + 7 < 8x - 3$, then

$$5x - 8x < -3 - 7$$
$$-3x < -10$$
$$(-\tfrac{1}{3})(-3x) > -\tfrac{1}{3}(-10) \qquad \text{Theorem 4}$$
$$x > \tfrac{10}{3}.$$

Since these steps are reversible, we conclude that the solution set for the inequality $5x + 7 < 8x - 3$ is the set $\{x \mid x > \frac{10}{3}\}$.

Example 2–3

Solve the inequality $(3x - 5)/x < 2$; *i.e.*, find the solution set.

Solution: We may write $(3x - 5)/x$ as $3x/x - 5/x$ or $3 - 5/x$ so that the inequality

$$(3x - 5)/x < 2$$

is the same as the inequality

$$3 - 5/x < 2.$$

Then

$$1 - 5/x < 0$$

and

$$1 < 5/x.$$

This last inequality cannot hold if $x = 0$ or $x < 0$. Therefore, $x > 0$ and from $1 < 5/x$ we see that $x < 5$. This shows that if x is a number which satisfies the inequality

$$(3x - 5)/x < 2$$

then $0 < x$ and $x < 5$. Now if $0 < x$ and $x < 5$, the following steps may be carried out

$$x < 5$$
$$1 < 5/x$$
$$1 - 5/x < 0$$
$$3 - 5/x < 2$$
$$(3x - 5)/x < 2.$$

Hence, the solution set of $(3x - 5)/x < 2$ is $\{x \mid 0 < x \text{ and } x < 5\}$.

The solution set for Example 2–2 is $\{x \mid 0 < x \text{ and } x < 5\}$. The two inequalities

$$0 < x \qquad \text{and} \qquad x < 5$$

may be more conveniently written as

$$0 < x < 5$$

so that the solution set may be expressed as $\{x \mid 0 < x < 5\}$. If a, b are any real numbers, the inequalities

$$a < x \quad \text{and} \quad x < b$$

may be expressed as

$$a < x < b.$$

The *absolute value* of the real number a, denoted $|a|$, is defined as follows:

    (1) $|a| = a$ if $a \geq 0$.
    (2) $|a| = -a$ if $a < 0$.

Definition

Some examples will help illustrate the meaning and use of this definition.

Since $\frac{1}{2} > 0$, $|\frac{1}{2}| = \frac{1}{2}$.

Example
2–4

$|-3| = -(-3)$ since $-3 < 0$. Hence $|-3| = 3$.

Example
2–5

For what numbers x is it true that $|x| = 3$?

Example
2–6

Solution. The definition gives rise to two cases. If $x \geq 0$, then $|x| = x$, and the equation $|x| = 3$ becomes $x = 3$. If $x < 0$, then $|x| = -x$, hence $|x| = 3$ becomes $-x = 3$ or $x = -3$. We also see by the definition that $|\pm 3| = 3$ and we obtain $\{x \mid |x| = 3\} = \{3, -3\}$.

If a is a real number, then $|a| \geq 0$.

Theorem
6

Proof: We know that $a \geq 0$ or $a < 0$. If $a \geq 0$, $|a| = a$ by definition. Thus, $|a| \geq 0$. If $a < 0$, then by definition $|a| = -a$. But in this case $-a > 0$. Hence in either case $|a| \geq 0$.

Let a, x be real numbers such that $a > 0$. Then

$$|x| < a \quad \text{if and only if} \quad -a < x < a.$$

Theorem
7

Proof: First, we prove that if $|x| < a$, then $-a < x < a$.

Case 1. If $x \geq 0$, then $|x| = x$ by definition. So if $|x| < a$, $x < a$.

Hence, $0 \leq x < a$. But since $a > 0$, $-a < 0$ giving

$$-a < x < a.$$

Case 2. If $x < 0$, $|x| = -x$ by definition. So if $|x| < a$, $-x < a$, or, by Theorem 4, $-a < x$. Since $x < 0$ and $0 < a$, $x < a$ so that

$$-a < x < a.$$

Now we prove that if $-a < x < a$, then $|x| < a$.

Case 1. If $x \geq 0$, then $|x| = x$ by definition. So if $-a < x < a$, then $x = |x| < a$.

Case 2. If $x < 0$, then $|x| = -x$ by definition. So if $-a < x$, then $-x < a$ by Theorem 4, or, $-x = |x| < a$.

In both cases we have that if $-a < x < a$, then $|x| < a$.

<table>
<tr><td>

**Example
2–7**

</td><td>

For which real numbers x is it true that $|2x + 3| < 4$?

Solution. We use Theorem 6 with $2x + 3$ playing the role of x. According to the theorem there is a number x such that $|2x + 3| < 4$ if and only if $-4 < 2x + 3 < 4$; *i.e.*, if and only if $-7 < 2x < 1$ or $-\frac{7}{2} < x < \frac{1}{2}$.

</td></tr>
<tr><td>

**Example
2–8**

</td><td>

Find all real numbers x for which $|x - 3| \geq 7$.

Solution. If we first observe that $|x - 3| \geq 7$ is true for all numbers x *except* those for which $|x - 3| < 7$, we may again use Theorem 7. Now $|x - 3| < 7$ is true if and only if $-7 < x - 3 < 7$, and thus, if and only if $-4 < x < 10$. The solution to the original inequality is, then, all those numbers x except those satisfying $-4 < x < 10$. The solution thus must be those numbers x such that $x \leq -4$ or $x \geq 10$.

</td></tr>
</table>

We now wish to represent the set of real numbers as points on a line. On any line L choose two distinct points, say A and B, with B to the right of A. Let 0 be represented by A and let 1 be represented by the point B, called the *unit point*. (See Fig. 2–1.) Now to the right and left of A mark off seg-

**Figure
2–1**

ments equal in length to segment *AB*. Let the resulting points represent the positive and negative integers as indicated in Fig. 2–2. It is evident in

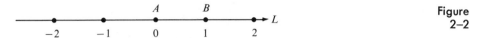

Figure
2–2

Fig. 2–2 that if *n* is any integer the point on *L* which represents *n* is a distance $|n|$ from the point *A*. If *a* is any real number, we mark two points on line *L*, one to the right of *A* and one to the left of *A*. Both of these points will be taken at a distance $|a|$ from the point *A*. (See Fig. 2–3.) Then if

Figure
2–3

$a > 0$, the point to the right of *A* is taken to represent *a*; if $a < 0$, the point to the left of *A* is taken to represent *a*. (See Fig. 2–4.)

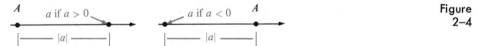

Figure
2–4

This discussion indicates how each real number is represented by a point on the line *L*. Conversely, we could reverse this process and indicate how each point of line *L* represents some number. Such a representation of the real numbers is called a *one-dimensional coordinate system*. Let it be stressed that *if* P *is a point of the line and the number* a *is represented by* P, *then the distance between* A *and* P *is* |a|.

This particular way of representing real numbers on a line has the following convenient connection with "less than": a $<$ b *if and only if the point* P *representing* a *is to the left of the point* M *representing* b. Hence, $\{x \mid x > a\}$ is represented on the line *L* as the set of all points to the right of the point representing *a*. (See Fig. 2–5.) The relative position of the

$$\{x \mid x > a\} \longrightarrow$$
$$a$$

Figure
2–5

sets $\{x \mid x < a\}$, $\{x \mid a < x < b\}$, $\{x \mid x > b\}$ on line *L* is indicated in Fig. 2–6.

$$\longleftarrow \{x \mid x < a\} \qquad \{x \mid a < x < b\} \qquad \{x \mid x > b\} \longrightarrow$$

Figure
2–6

We now return to a consideration of $\sqrt{2}$ in order to illustrate some of the ideas under discussion. First, some elementary results will be established.

Theorem 8

If $a, b \in R$ and $0 < a \leq b$, then $a^2 \leq b^2$.

Proof. Since $0 < a$ and $a \leq b$ we have

$$a^2 \leq ab$$

by multiplying by a. Also, multiplying the inequality $a \leq b$ by b gives

$$ab \leq b^2.$$

Then from $a^2 \leq ab$ and $ab \leq b^2$ it follows that $a^2 \leq b^2$ so that the theorem is proved.

Theorem 9

If $a, b \in R$, $b > 0$ and $a^2 < b^2$, then $a < b$.

Proof. Either $a < b$ or $b \leq a$. If $b \leq a$, then $b^2 \leq a^2$ by Theorem 8 and this contradicts the hypothesis $a^2 < b^2$. Hence, $a < b$.

From Theorem 9 it follows that $1 < \sqrt{2}$ and $\sqrt{2} < 2$ since $1 < 2$ and $2 < 4$. This indicates that $\sqrt{2}$ lies between 1 and 2 as shown below:

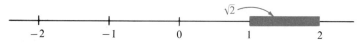

Now we divide the interval from 1 to 2 into 10 equal parts:

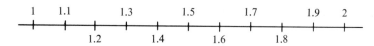

Then

$$(1.1)^2 = 1.21 \text{ shows that } 1.1 < \sqrt{2},$$
$$(1.2)^2 = 1.44 \text{ shows that } 1.2 < \sqrt{2},$$
$$(1.3)^2 = 1.69 \text{ shows that } 1.3 < \sqrt{2},$$
$$(1.4)^2 = 1.96 \text{ shows that } 1.4 < \sqrt{2},$$
$$(1.5)^2 = 2.25 \text{ shows that } 1.5 > \sqrt{2}.$$

Therefore, $1.4 < \sqrt{2} < 1.5$ and $\sqrt{2}$ lies in the shaded area shown below.

Next, the interval from 1.4 to 1.5 may be divided into 10 equal parts:

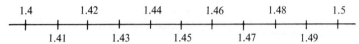

Then

$$(1.41)^2 = 1.9881 \text{ shows that } 1.41 < \sqrt{2},$$
$$(1.42)^2 = 2.0164 \text{ shows that } 1.42 > \sqrt{2}.$$

Therefore, $1.41 < \sqrt{2} < 1.42$ and $\sqrt{2}$ lies in the shaded area shown.

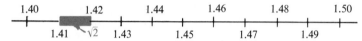

Continuing, the interval from 1.41 to 1.42 is divided into 10 parts:

$$(1.411)^2 = 1.990921 \text{ shows that } 1.411 < \sqrt{2},$$
$$(1.412)^2 = 1.993744 \text{ shows that } 1.412 < \sqrt{2},$$
$$(1.413)^2 = 1.996569 \text{ shows that } 1.413 < \sqrt{2},$$
$$(1.414)^2 = 1.999396 \text{ shows that } 1.414 < \sqrt{2},$$
$$(1.415)^2 = 2.002225 \text{ shows that } 1.415 > \sqrt{2}.$$

Therefore, $1.414 < \sqrt{2} < 1.415$ and $\sqrt{2}$ lies in the shaded area.

We could continue in this manner to achieve any desired degree of approximation to $\sqrt{2}$ as a terminating decimal. Since $\sqrt{2}$ is irrational, none of our terminating decimals would be $\sqrt{2}$, but at each stage a better estimate would be obtained.

6. Find the absolute value of each of the following numbers: 4, 0.5, -12, -1.2, $1 + 99$, $1 - 99$, $-21\frac{1}{2}$.

7. In each of the following problems you are to draw the best conclusion possible from the given information.

(a) Given: $a > 0$. Draw a conclusion about $-a$.
(b) Given: $-a > 0$. Draw a conclusion about a.
(c) Given: $a < 0$. Draw a conclusion about $-a$.
(d) Given: $x = 5$. Draw a conclusion about $-x$.
(e) Given: $y = -2$. Draw a conclusion about $-y$.
(f) Given: $x > 2$. Draw a conclusion about $x - 2$.
(g) Given: $w < 1$. Draw a conclusion about $1 - w$.
(h) Given: $u > 3$. Draw a conclusion about $u + 1$.
(i) Given: $u < 3$. Draw a conclusion about $2u$.
(j) Given: $k > 0$. Draw a conclusion about $|k|$.
(k) Given: $k < 0$. Draw a conclusion about $|k|$.
(l) Given: $-k > 0$. Draw a conclusion about $|-k|$.
(m) Given: $-k < 0$. Draw a conclusion about $|-k|$.
(n) Given: $\frac{1}{2} > w$. Draw a conclusion about $|\frac{1}{2} - w|$.
(o) Given: $x > 0$ and $y < 0$. Draw a conclusion about $|xy|$.

8. Solve the following inequalities:

(a) $4x + 7/8 \geq 0$
(b) $3 - (5x)/4 < 0$
(c) $(1 + 3x)/2 - 9 > 0$
(d) $(8 + x)/4 - x/8 \leq 0$
(e) $(3x + 7)/9 \leq 12$
(f) $0.1\,x - 5 \leq 1/10$

9. Describe the set of all real numbers x that satisfies the following conditions. For example, given $x + 1 > 0$ and $x - 1 > 0$, if $x + 1 > 0$, then $x > -1$ and if $x - 1 > 0$, $x > 1$. If $x > 1$, then $x > -1$ since $1 > -1$. Thus the numbers x for which $x + 1 > 0$ and $x - 1 > 0$ are those numbers x such that $x > 1$. The set description is $\{x \mid x > 1\}$.

(a) $x + 1 < 0$ and $x - 1 < 0$.
(b) $x + 1 > 0$ and $x - 1 < 0$.
(c) $x + 1 < 0$ and $x - 1 > 0$.
(d) $1 \leq 3x - 1 \leq 7$
(e) $-6 < 1 - x \leq 0$
(f) $x + 8 \leq 3x - 19 < 5$
(g) $x + 1 \leq 2x < 12$

10. Solve these equations in the following manner:

Given $|2x| = 4$, by the definition of absolute value we have

$$|2x| = 2x \qquad \text{if } 2x \geq 0$$
$$|2x| = -2x \qquad \text{if } 2x < 0.$$

Therefore, the equation is

$$2x = 4 \qquad \text{if } 2x \geq 0$$

and

$$-2x = 4 \qquad \text{if } 2x < 0.$$

If $2x = 4$, $x = 2$ and if $-2x = 4$, $x = -2$. Thus the only possible solutions are 2, -2. Both of these numbers are solutions since $|2 \cdot 2| = |4| = 4$ and $|2 \cdot (-2)| = |-4| = 4$. The solution set is then $\{-2, 2\}$.

(a) $|x| = 0$ (b) $|x - 1| = 0$ (c) $|x/2| = 5$
(d) $|7x + 2| = 5$ (e) $|y - 2| = -0.9$ (f) $1 - 4|x| = 0$

11. Find the solution set for each pair of equations below. For example, take $|x| = 1$ and $|3 - x| = 2$. If $|x| = 1$, then $x = 1$ or $x = -1$ by definition of $|x|$. If $|3 - x| = 2$, then

$$2 = |3 - x| = 3 - x \qquad \text{for } 3 - x \geq 0$$

and

$$2 = |3 - x| = -(3 - x) = x - 3 \qquad \text{for } 3 - x < 0.$$

From $2 = 3 - x$ we obtain $x = 1$; from $2 = x - 3$, $x = 5$. Therefore, the only number satisfying both equations is $x = 1$.
(a) $|x| = 1$ and $|x + 2| = 1$
(b) $|x + 7| = 10$ and $|x - 3| = 20$
(c) $|1 + x| = 13$ and $|18 - x| = 32$
(d) $|3x - 2| = 4$ and $|x| = x$

12. Find the solution set for each equation:

Example: $|x| = |x + 2|$. By definition

(i) $|x| = x$ for $x \geq 0$
(ii) $|x| = -x$ for $x < 0$
(iii) $|x + 2| = x + 2$ for $x + 2 \geq 0$ or $x \geq -2$
(iv) $|x + 2| = -(x + 2) = -x - 2$ for $x + 2 < 0$ or $x < -2$

On the line below the sets $A = \{x \mid x < -2\}$, $B = \{x \mid -2 \leq x < 0\}$ and $C = \{x \mid x \geq 0\}$ have been indicated. Every number x is in one

$A = \{x \mid x < -2\}$ $B = \{x \mid -2 \leq x < 0\}$ $C = \{x \mid x \geq 0\}$

of these three regions. If $x \in A$, then $x < -2$ and since $-2 < 0$, $x < 0$ as well. Hence, if $x \in A$, (ii) and (iv) hold. The original equation $|x| = |x + 2|$ is, in this case, the same as

$$-x = -x - 2$$

and this equation has no solution. This implies that $|x| = |x + 2|$ has no solution x if $x < -2$.

If $x \in B$, then $-2 \leq x < 0$ which means that (ii) and (iii) hold. In this case the equation $|x| = |x + 2|$ is

$$-x = x + 2 \qquad \text{or} \qquad x = -1.$$

The only possible solution in B is -1.

If $x \in C$, $x \geq 0$ and $x > -2$ so that (i) and (iii) hold. The equation $|x| = |x + 2|$ is the same as

$$x = x + 2$$

which has no solution.

From these three cases one may conclude that the only possible solution for the equation $|x| = |x + 2|$ is the number -1. The number -1 is actually a solution since

$$|-1| = |-1 + 2|$$

is a true statement.

(a) $|x - 1| = |x + 2|$
(b) $|2x| = |x - 10|$
(c) $|2x - 3| = |x + 1|$
(d) $|x| = |x + 1| + 1$

13. Find the solution set for each of the following inequalities.

Example: $|5 - x| \leq 1$. *Solution:* By Theorem 6, $|5 - x| \leq 1$ if and only if $-1 \leq 5 - x \leq 1$. Subtracting 5 from the second pair of inequalities gives

$$-6 \leq -x \leq -4.$$

Then multiplying by -1: $6 \geq x \geq 4$. Hence, the solution set is $\{x \mid 4 \leq x \leq 6\}$.

(a) $|x + 2| < 4$ (b) $|4x - 1| \leq 12$ (c) $|1 - 5x| \leq 0.5$
(d) $|x + 1| > 1$ (e) $|2x + 3| > 2$ (f) $|10x - 7| \geq \frac{1}{2}$

14. Solve the following inequalities.

Example: $|x - 1| \leq |x - 2|$. By Theorem 6, $|x - 1| \leq |x - 2|$ if and only if

$$-|x - 2| \leq x - 1 \leq |x - 2|$$

or

$$-(x - 2) \leq x - 1 \leq x - 2 \qquad \text{for } x - 2 \geq 0.$$

Then

$$-x + 2 \leq x - 1 \leq x - 2$$

and

$$-2x + 2 \leq -1 \leq -2.$$

Since it is not true that $-1 \leq -2$, there can be no solution for $x - 2 \geq 0$ or $x \geq 2$. If $x - 2 < 0$ or $x < 2$, the above inequalities become

$$-[-(x - 2)] \leq x - 1 \leq -(x - 2)$$

or

$$x - 2 \leq x - 1 \leq -x + 2$$

and

$$-2 \leq -1 \leq -2x + 2.$$

From the inequality $-1 \leq -2x + 2$ we obtain $-3 \leq -2x$ or

$$x \leq 3/2.$$

If x is any number such that $x \leq 3/2$, the preceding steps may be reversed to show that

$$-|x - 2| \leq x - 1 \leq |x - 2|$$

and this shows that $\{x \mid x \leq 3/2\}$ is the solution set of the inequality $|x - 1| \leq |x - 2|$.

(a) $|x + 2| < |5 - x|$
(b) $|2x + 1| \leq |2x - 1|$
(c) $|x - 1| < |x + 1| + 1$

15. (a) Show that if x is any number, then $(x + |x|)/2$ is x or 0.
    (b) Show that if x is any number, then $(x - |x|)/2$ is x or 0.
    (c) Can you conclude from (a) and (b) that for all numbers x

$$(x + |x|)/2 = (x - |x|)/2?$$

16. Show that for a real number x

   (a) $-|x| \leq x \leq |x|$.

   (b) $|x|^2 = x^2$.

17. If a is a real number, show that
$$\sqrt{a^2} = |a|.$$

   [*Hint:* If $x \geq 0$ is a real number, $\sqrt{x}$ *means* that *non-negative* real number whose square is x.]

18. Prove each of the following:

   (a) If $x \neq 0$, then $x^2 > 0$.

   (b) There is no number x such that $x^2 + 1 = 0$.

   (c) If $x \leq y$, then $-y \leq -x$.

   (d) If $0 < x < y$, then $0 < \dfrac{1}{y} < \dfrac{1}{x}$.

   (e) If $x < y$ and $w < z$, then $x + w < y + z$.

   (f) If $x < y$, then $x < (\frac{1}{2})(x + y) < y$.

   (g) If $x < y < z$, then $x < (\frac{1}{3})(x + y + z) < z$.

   (h) Prove that $(\frac{1}{2})(x + y + |x - y|)$ is the larger number of x, y.

---

*2.3
The Completeness Axiom

We now take up a discussion of the Completeness Axiom for the real numbers. This axiom in a sense "completes" the system of real numbers. We say "completes" because it is this axiom which allows us to show that we have all the numbers which we desire. For example, without the completeness axiom we could not show that there is a real number x such that $x^2 = 2$; *i.e.*, we would not know that $\sqrt{2}$ exists. We must go beyond the collection of rational numbers to obtain such a number as Theorem 1 shows.

Definition

If S is a non-empty set of real numbers, then a real number x is an *upper bound* of S if and only if

$$s \leq x \qquad \text{for all} \qquad s \in S.$$

Some sets have upper bounds and some do not, as the following examples show.

Example 2–9

The set of real numbers has no upper bound since if x is any real number, then $x < x + 1$.

The set $\{\frac{1}{2}, -8, 2, \frac{7}{8}, \sqrt{2}\}$ has the numbers $2, 3, \frac{11}{2}, 100, 4086$ as some of its upper bounds. Of course, these are not all the upper bounds of this set, since *any* number larger than the given upper bounds is also an upper bound.

Example
2–10

Let $S = \{t \mid t$ is a real number and $t^3 < 10\}$. The number 1 is not an upper bound for S since $1 < 2$ and $2 \in S(2^3 < 10)$. However, we claim that 4 is an upper bound for S. Suppose that $t \geq 4$. Then $t^3 \geq 64 > 10$ so that $t^3 > 10$. Hence $t \notin S$. Therefore, if $t \in S$, $t < 4$.

Example
2–11

Let S be a non-empty set of real numbers and let t be an upper bound of S such that if u is any upper bound of S, then

$$t \leq u.$$

Then t is called the *least upper bound* of S.

We may describe a least upper bound of a set (if it has one) as the *smallest* of all its upper bounds.

Of all the upper bounds for the set $\{\frac{1}{2}, -8, 2, \frac{7}{8}, \sqrt{2}\}$ we see that 2 is the smallest. Hence, 2 is the least upper bound of this set.

Example
2–12

Let $S = \{x \mid x < 1\}$. Then 1 is clearly an upper bound for S. Also, 1 is the smallest of all the upper bounds for S, for we can show that no number smaller than 1 can be an upper bound for S. To do this, suppose that t is a number such that $t < 1$. Then $t < (\frac{1}{2})(t + 1) < 1$. [See Exercise 18($f$)]. Therefore $(\frac{1}{2})(t + 1)$ is an element of S which is larger than t so that t cannot be an upper bound for S.

Example
2–13

Observe that the least upper bound of the set $\{\frac{1}{2}, -8, 2, \frac{7}{8}, \sqrt{2}\}$ is a member of the set, while the least upper bound of $\{x \mid x < 1\}$ is not a member since $1 \notin \{x \mid x < 1\}$.

Suppose that u is the least upper bound of some set S. Then, in particular, u is an upper bound for S; i.e., $x \leq u$ for all $x \in S$. Also, if d is *any* positive number, there is an element $y \in S$ such that $u - d < y$. This is true because if there is no such $y \in S$, then $x \leq u - d$ for *all* $x \in S$ and this would imply that $u - d$ is an upper bound for S. But since d is positive,

$u - d < u$, and u was supposed to be the *least* upper bound for S. Therefore, if u is the least upper bound of S, we may say the following:

(*a*) $x \leq u$ for all $x \in S$.
(*b*) There is an element $y \in S$ for each positive number d such that

$$u - d < y.$$

The reader may be familiar with the fact that not every set of real numbers contains a least or smallest element. The set $D = \{t \mid 0 < t < 10\}$, for example, contains no smallest element. Suppose to the contrary that $d \in D$ and $d \leq t$ for *all* $t \in D$. Then since $d \in D$, $0 < d < 10$. But $0 < (\frac{1}{2})d < d$ or $0 < (\frac{1}{2})d < 10$, which shows that $(\frac{1}{2})d \in D$ and is smaller than the element which was supposed to be smallest. Thus, the set D contains no smallest element.

The content of the Completeness Axiom is that certain kinds of sets of real numbers *always* have least members. We start with a set of real numbers $S \neq \emptyset$ which has upper bounds. Then we consider the set $\triangle$ of *all* upper bounds of S. The completeness axiom asserts that $\triangle$ contains a smallest or least member.

Completeness Axiom (or Least Upper Bound Axiom). Let S be a set of real numbers such that $S \neq \emptyset$ and S has upper bounds. Then S has a least upper bound.

If a non-empty set S of real numbers has upper bounds, then we say that S is *bounded above*. Using this terminology, the Completeness Axiom may be expressed as follows: *Every non-empty set of real numbers which is bounded above has a least upper bound.*

The use of this axiom will now be illustrated in proving some results.

Theorem 10

The set N of positive integers is not bounded above.

Proof: Suppose to the contrary that there is a real number y such that $n \leq y$ for all $n \in N$. Then, by the Completeness Axiom N has a least upper bound, say x. Since $x - 1 < x$, there is an integer $m \in N$ such that $x - 1 < m$ and $x < m + 1$. But $m + 1 \in N$. Hence x is *not* an upper bound for N. This contradiction thus proves the theorem.

If d is any positive real number, then there is an integer n such that $(1/n) < d$.

Proof: By Theorem 10 the set of positive integers is not bounded above. Hence there is an integer n such that $n > 1/d$. Then $1/n < d$.

Let x, y be real numbers such that $y > 0$. Then there is a positive integer m such that $x < my$.

Proof: If x is negative or zero, then, since $y > 0$, $x < 1 \cdot y$ and we may take $m = 1$. If x is positive, then y/x is positive and by Theorem 11 there is an integer $m > 0$ such that $(1/m) < y/x$. Hence $x < my$.

It was pointed out earlier that without the completeness property of the real numbers, we do not know that there is a number x such that $x^2 = 2$. With the completeness axiom we may prove that there is such a number x. First, we define

$$S = \{y \mid y \text{ is a positive rational number and } y^2 < 2\}.$$

Then $1 \in S$ so that $S \neq \varnothing$. Also, 3 is an upper bound for S (for if $y \geq 3$, then $y^2 \geq 9$ so that $y \notin S$). Therefore, by the Completeness Axiom, S has a least upper bound, say x. It is now possible to show that $x^2 = 2$. ($\sqrt{2}$ is by definition this number x.) We shall not give the proof here; the reader may find this carried further in *Introduction to Analysis*, Vol. I, by N. B. Haaser, J. P. LaSalle, J. A. Sullivan, Ginn & Co., 1959, or in *The Number System* by B. K. Youse, Dickenson Publishing Co., 1965.

19. Formulate definitions for the following terms: *lower bound, greatest lower bound, largest element.*

20. For each of the following sets, find some lower bounds and upper bounds when these exist.
    (*a*) $\{\frac{1}{10}, -\sqrt{2}, \frac{1}{2}, 0.78\}$
    (*b*) $\{n^2 \mid n \text{ is an integer}\}$
    (*c*) $\{m^2 \mid m \text{ is an integer and } m < 101\}$

    (d) $\{(\tfrac{1}{2})n \mid n$ is an integer and $n \geq 0\}$

    (e) $\{1/m \mid m$ is an integer and $m > 0\}$

    (f) $\{1 - 1/n \mid n$ is a positive integer$\}$

21. For each of the sets in 20,
    (a) find the least upper bound if it exists;
    (b) find the greatest lower bound when it exists.

22. Use the Completeness Axiom to prove that a non-empty set S of real numbers which has lower bounds has a greatest lower bound. [*Hint:* Consider the set $N = \{-x \mid x \in S\}$.]

23. Show that the set $\{1 - 1/n \mid n$ is a positive integer$\}$ has no largest element.

24. (a) Show that if a set $S \neq \varnothing$ of real numbers has a largest member q, then q is the least upper bound of S.
    (b) Show that if a set $S \neq \varnothing$ has a smallest member p, then p is the greatest lower bound of S.

25. Prove that every real number x is between two integers; *i.e.*, prove that there are integers n, m such that $n < x < m$.

26. As indicated above, $\sqrt{2}$ is defined to be the least upper bound of the set

$$\{y \mid y \text{ is a positive rational number and } y^2 < 2\}.$$

In a similar fashion formulate definitions for each of the following:

    (a) $\sqrt[3]{2}$ (b) $\sqrt[3]{7}$ (c) $3^{3/5}$

Be sure in each case to show that the set which you use in your definition is not empty and that it has an upper bound.

---

*2.4

The Natural Numbers and Mathematical Induction

The reader will recall that the set N of natural numbers or positive integers consists of the numbers $1, 1 + 1, 1 + 1 + 1, 1 + 1 + 1 + 1,$ etc., and that these are denoted by

$$1, 2, 3, 4, \ldots .$$

A basic characteristic of the set N of natural numbers is the so-called well ordering property which we assume to be true.

Well-Ordering Property of the Natural Numbers. Let S be a subset of the natural numbers such that $S \neq \varnothing$. Then there is a natural number $t \in S$ such that $t \leq x$ for all $x \in S$.

The number t of this statement is called the *least* member of S. In this language we may say that *every non-void set of natural numbers contains a least element.*

Example
2–14

The least natural number in the set N of all natural numbers is 1; 2 is the least element of the set of even natural numbers.

Example
2–15

Show that there are no natural numbers x such that $0 < x < 1$.

Solution. Suppose there is at least one natural number x such that $0 < x < 1$. Then if

$$S = \{x \mid x \text{ is a natural number and } 0 < x < 1\},$$

$S \neq \emptyset$. By the well-ordering property there is a least member, say t, of S. Hence, $0 < t < 1$. If we multiply by t, we have

$$0 < t^2 < t < 1.$$

But since the product of natural numbers is a natural number, $t^2 \in S$ with $t^2 < t$. This is a contradiction and, therefore, proves the statement.

Example
2–16

Prove that if m is any natural number, then $m < 2^m$.

Solution. Suppose that there is some natural number n such that $n \geq 2^n$. Then the set

$$S = \{k \mid k \text{ is a natural number and } k \geq 2^k\}$$

is not empty. So S has a least member, say t. Thus t is the smallest natural number such that

$$t \geq 2^t.$$

We note that $t > 1$ since $1 < 2$. Therefore, $t - 1 \geq 1$ and is a natural number smaller than t. Since t is the smallest natural number such that $t \geq 2^t$, we conclude that $t - 1 < 2^{t-1}$. Then, by adding 1, we have $t < 2^{t-1} + 1$. But since $1 < 2^{t-1}$,

$$2^{t-1} + 1 < 2^{t-1} + 2^{t-1} = 2 \cdot 2^{t-1} = 2^t.$$

Hence $t < 2^t$, contrary to the fact that t was assumed to be the smallest natural number such that $t \geq 2^t$. We conclude that our original supposition that there was at least one natural number n such that $n \geq 2^n$ is incorrect, and that $m < 2^m$ for all natural numbers m.

In Example 2–16 we have the statement

$$m < 2^m.$$

In fact, we have many statements here — one for each natural number m — and we wish to prove that they are all true. It quite often happens that we have a function P which associates with each natural number n a statement $P(n)$ and we wish to show that the set

$$\{P(n) \mid n \text{ is a natural number}\}$$

contains only true statements. In the example above the statement $P(n)$ for each natural number n was

$$n < 2^n,$$

and we were able to prove that these are all true by using the well-ordering property of the natural numbers. Another method of doing this, which is based on the well-ordering property, is called mathematical induction.

Principle of Mathematical Induction. Let P be a function which associates with each natural number n a statement $P(n)$ and suppose that the following are true:
   (*a*) $P(1)$ is true.
   (*b*) If, for any natural number k, $P(k)$ is true, then $P(k + 1)$ is true.
Then $P(n)$ is true for all natural numbers n.

Proof: Suppose that there is at least one natural number s such that $P(s)$ is not true. Then the set

$$S = \{s \mid s \text{ is a natural number and } P(s) \text{ is false}\}$$

is not empty. Let t be the least element of S. Then $1 \notin S$ since by (*a*) $P(1)$ is true. Thus $t > 1$ and $t - 1$ is a natural number not in S (the smallest element of S is t and $t - 1 < t$). Hence, $P(t - 1)$ is true. By (*b*), then, $P(t - 1 + 1)$ is true; *i.e.*, $P(t)$ is true. But $t \in S$ so that $P(t)$ is false — a contradiction. This implies that our supposition that there was at least one natural number s such that $P(s)$ is false is incorrect. Therefore $P(n)$ is true for all natural numbers n.

Let us apply the principle of mathematical induction to the statement

$$n < 2^n.$$

The principle states that we can conclude that this statement is true for all natural numbers if we can show that it has properties (*a*) and (*b*). It has property (*a*) since $P(1)$ is the statement

$$1 < 2,$$

and this is true. If we know that $P(k)$ is a true statement for a natural number k; *i.e.*, that

$$k < 2^k,$$

then,

$$k + 1 < 2^k + 1 < 2^k + 2^k = 2^{k+1}$$

or $P(k + 1)$ is true. This shows that (*b*) also is satisfied by P. Hence, we conclude that $P(n)$ is true for all natural numbers.

The procedure we have just used in applying the principle of mathematical induction must be observed generally; *i.e.*, to prove that a statement is true for all natural numbers by mathematical induction we must:

(*a*) Show that the statement is true for 1.

(*b*) Show that if the statement is true for a natural number k, then it is true for $k + 1$.

Example
2–17

Let $P(n)$ be the statement

$$n = n + 1.$$

Show that P has property (*b*) but does not have property (*a*).

Solution. $P(1)$ is the statement

$$1 = 1 + 1$$

which is not true, so that P does not have property (*a*). To show that P has property (*b*) we must show that *if $P(k)$ is true, then $P(k + 1)$ is true.* If $P(k)$ is true, then

$$k = k + 1,$$

and adding 1 gives

$$k + 1 = (k + 1) + 1.$$

Hence, if $k = k + 1$, then $(k + 1) = (k + 1) + 1$. Thus (*b*) is satisfied. (The reader has no doubt decided that $n = n + 1$ is false for every natural number n.)

Example 2–18

Show that the statement

$$I(n): n \neq n + 1$$

satisfies both conditions (*a*) and (*b*) of the principle of mathematical induction.

Solution. The statement $I(1)$ is

$$1 \neq 1 + 1$$

and this is true, so (*a*) is satisfied. Now suppose that $k \neq k + 1$. We need to demonstrate that this assumption implies that $k + 1 \neq (k + 1) + 1$. But if we did not have $k + 1 \neq (k + 1) + 1$, we would have $k + 1 = (k + 1) + 1$, and subtracting 1 would give $k = k + 1$. Hence, (*b*) is satisfied and we may conclude that $n \neq n + 1$ for all natural numbers n.

Example 2–19

Prove that $[n(n + 1)(n + 2)]/3$ is a natural number for all natural numbers n.

Solution. We apply the principle of mathematical induction to the statement $P(n)$: $[n(n + 1)(n + 2)]/3$ is a natural number. For $n = 1$ the statement $P(n)$ is

$$[1(1 + 1)(1 + 2)]/3 \text{ is a natural number;}$$

i.e., "2 is a natural number," and this is true, so $P(n)$ satisfies condition (*a*) of the induction principle. To show that condition (*b*) is also satisfied, suppose that $P(k)$ is true:

$$[k(k + 1)(k + 2)]/3 \text{ is a natural number.}$$

We wish to show that $P(k + 1)$ is true if $P(k)$ is true. The statement $P(k + 1)$ is

$$[(k + 1)(k + 2)(k + 3)]/3 \text{ is a natural number.}$$

For the numerator of the above we may write

$$(k + 1)(k + 2)(k + 3) = k(k + 1)(k + 2) + 3(k + 1)(k + 2)$$

and therefore,

$$[(k + 1)(k + 2)(k + 3)]/3 = [k(k + 1)(k + 2)]/3 + [(k + 1)(k + 2)].$$

Since $(k + 1)(k + 2)$ is a natural number, the sum on the right above is a natural number if $[k(k + 1)(k + 2)]/3$ is a natural number. In other words, $[(k + 1)(k + 2)(k + 3)]/3$ is a natural number if $[k(k + 1)(k + 2)]/3$ is a natural number. This shows that $P(n)$ satisfies condition (*b*) and concludes the proof that $[n(n + 1)(n + 2)]/3$ is a natural number for all natural numbers n.

Prove that for all natural numbers n

Example
2–20

$$2 + 4 + 6 + \cdots + 2n = n(n + 1).$$

Solution. We are to prove that the sum of the first n even natural numbers is $n(n + 1)$. The induction principle may be applied to the statement

$$P(n): 2 + 4 + 6 + \cdots + 2n = n(n + 1).$$

$P(1)$ is the statement: "$2 \cdot 1 = 1(1 + 1)$." Thus, $P(1)$ is true. The statements $P(k)$ and $P(k + 1)$ are

$$P(k): 2 + 4 + 6 + \cdots + 2k = k(k + 1),$$

and

$$P(k + 1): 2 + 4 + 6 + \cdots + 2k + 2(k + 1) = (k + 1)(k + 2).$$

Now suppose that $P(k)$ is true. Then,

$$2 + 4 + 6 + \cdots + 2k + 2(k + 1)$$
$$= (2 + 4 + 6 + \cdots + 2k) + 2(k + 1)$$
$$= k(k + 1) + 2(k + 1)$$
$$= (k + 1)(k + 2).$$

This shows that if $P(k)$ is true, then so is $P(k + 1)$.

27. Prove that for all natural numbers n each of the following is a natural number.
    (a) $[n(n + 1)]/2$
    (b) $[n(n + 1)(n + 2)]/6$
    (c) $[n(n + 1)(n + 2)(n + 3)]/24$

28. Prove the following generalization of the induction principle: Let $P(n)$ be a statement that is associated with every natural number n such that $n \geq a$ where a is a natural number. Suppose the following are true:
    (a) $P(a)$ is true.
    (b) If $P(k)$ is true for any natural number $k \geq a$, $P(k + 1)$ is true.
        Then $P(n)$ is true for all natural numbers $n \geq a$.

29. Prove each of the following is true for all natural numbers n:
    (a) $1 + 2 + 3 + \cdots + n = [n(n + 1)]/2$.
    (b) $1 + 3 + 5 + \cdots + (2n - 1) = n^2$.

(c) $1 + 5 + 9 + \cdots + (4n - 3) = n(2n - 1)$.

(d) $1 + 4 + 9 + \cdots + n^2 = [n(n + 1)(2n + 1)]/6$.

(e) $1 + 8 + 27 + \cdots + n^3 = [n^2(n + 1)^2]/4$.

(f) $1/(1 \cdot 2) + 1/(2 \cdot 3) + 1/(3 \cdot 4) + \cdots + 1/[n(n + 1)] = n/(n + 1)$.

(g) $2 + 2(2^2) + 3(2^3) + \cdots + n(2^n) = 2 + (n - 1)2^{n+1}$.

(h) $a + ar + ar^2 + \cdots + ar^{n-1} = (ar^n - a)/(r - 1)$, where a, r are real numbers and $r \neq 1$.

(i) $2 \cdot 3 + 2 \cdot 3 + 2 \cdot 3^2 + \cdots + 2 \cdot 3^{n-1} = 3^n - 1$.

(j) $1 + \dfrac{1}{\sqrt{2}} + \dfrac{1}{\sqrt{3}} + \cdots + \dfrac{1}{\sqrt{n}} \geq \sqrt{n}$.

30. Prove each of the following:

(a) If n is a positive integer and if x is a number such that $0 < x < 1$, then $0 < x^n < 1$.

(b) If n is a natural number such that $n \geq 4$, then $2^n < n!$ ($n!$ means $1 \cdot 2 \cdot 3 \ldots n$). [*Hint:* Use the induction principle of Exercise 28.]

(c) If a line segment of unit length is given, then a line segment of length $\sqrt{n}$ can be constructed with ruler and compass for any natural number n.

(d) If x, y are real numbers and n is a natural number, then
$x^n - y^n = (x - y)(x^{n-1} + x^{n-2}y + x^{n-3}y^2 + \cdots + xy^{n-2} + y^{n-1})$.

(e) The sum of the interior angles of a convex polygon of n sides is $180° \cdot (n - 2)$, where $n \geq 3$.

(f) Let $a, b, x_1, x_2, x_3, \ldots, x_n$ be real numbers and n a natural number. If $a < x_1 < b, a < x_2 < b, a < x_3 < b, \ldots, a < x_n < b$, then,

$$a < (x_1 + x_2 + x_3 + \cdots x_n)/n < b.$$

(g) If x is a real number such that $x \geq -1$ and n is a natural number, then

$$(1 + x)^n \geq 1 + nx.$$

31. Use the well-ordering property of the natural numbers to prove the following.

Let S be a subset of the natural numbers N which satisfies the following two conditions:

(a) $1 \in S$

(b) If $k \in S$ then $(k + 1) \in S$

Then $S = N$.

REAL RELATIONS, FUNCTIONS AND THEIR GRAPHS

3

Now that we have described a method of associating a number with each point of a line, we may discuss *rectangular coordinate systems*. A rectangular two-dimensional coordinate system is established by taking two lines perpendicular to each other and, on each of these lines, a one-dimensional coordinate system is constructed as described above. We take one of the lines horizontal and the other vertical, and choose their point of intersection to correspond to 0, while the unit points are taken as indicated in Fig. 3–1.

Figure
3–1

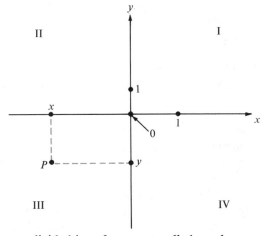

The plane is now divided into four parts called *quadrants* as indicated in Fig. 3–1. Notice that the unit point on the vertical line is above the point corresponding to zero, and on the horizontal line it is to the right; also the unit lengths are not the same in the figure. In most of the work in this chapter the unit lengths chosen will be the same, although this is by no means necessary. It is now possible to associate with each point P of the plane an ordered pair (x, y) of real numbers. The method for doing this was described in Chapter 1 but for emphasis the procedure which establishes this association will now be reviewed.

First of all, if the point P in the plane is given, how is the ordered pair (x, y) which goes with P found? As usual, two lines are drawn through P: one line parallel to the horizontal line and the other one parallel to the vertical line. (See Fig. 3–1.) The two numbers obtained in this way, as indicated in Fig. 3–1, determine the ordered pair (x, y) that is associated with P. It should be stressed that the number x of the horizontal line is the first member of the ordered pair (x, y), and that the number y of the vertical line is the second member of the ordered pair.

It has become standard practice to use the letter "x" to name numbers corresponding to points of the horizontal line, and to use the letter "y" to name numbers corresponding to points of the vertical line. Hence, these lines are referred to as the "x-axis" and the "y-axis" respectively. (A more appropriate name for them would be "axis of first coordinates" and "axis of second coordinates.") Of course other names would do just as well, but the x, y-terminology is well established. If the ordered pair (a, b) corresponds to the point Q, we say that a is the *first* or *x-coordinate* of Q and that b is the *second* or *y-coordinate* of Q.

Now, suppose that the ordered pair (x, y) is given and we wish to determine the point P associated with (x, y). Choose the point on the x-axis which corresponds to the number x and construct a line on this point parallel to the y-axis; on the y-axis construct a line parallel to the x-axis which passes through the point corresponding to the number y. These two lines intersect (Why?) and their intersection point is the point P which we seek.

Several consequences of this particular method for establishing a coordinate system are immediate. For example, *if a point* Q *is on the* x-axis *and* (a, b) *corresponds to* Q, *then* b = 0. Remember that b is obtained from the y-axis as the intersection of the y-axis and a line on Q drawn parallel to the x-axis. But since Q is on the x-axis, the line on Q parallel to the x-axis is the x-axis itself, and this line meets the y-axis at 0. Hence, $b = 0$. We may therefore conclude that *every point of the* x-axis *corresponds to an ordered pair of the form* (x, 0); *i.e.*, the second coordinate of every such point is 0.

By a similar argument we obtain: *if* (a, b) *corresponds to a point of the* y-axis, a = 0; *every point of the* y-axis *corresponds to an ordered pair of the form* (0, b).

We might refer to our two-dimensional coordinate system as being "distance oriented" for the following reason: *If the point* Q *corresponds to the ordered pair* (a, b), *then the distance between* Q *and the* y-*axis is* |a| *and the distance between* Q *and the* x-*axis is* |b|. This can be seen very easily by recalling the method for establishing the coordinates on the x- and y-axis. For example, let Q be a point in the second quadrant. Then our result is illustrated by Fig. 3–2.

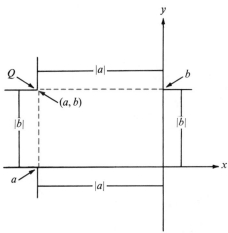

Figure
3–2

Because of the way our coordinate system was constructed, we may easily verify the following:

Example
3–1

   (i) If F is a point in the first quadrant with coordinates (a, b), then $a \geq 0$ and $b \geq 0$.
  (ii) If S is a point in the second quadrant with coordinates (a, b), then $a \leq 0$ and $b \geq 0$.
 (iii) If T is a point in the third quadrant with coordinates (a, b), then $a \leq 0$ and $b \leq 0$.
 (iv) If F is a point in the fourth quadrant with coordinates (a, b), then $a \geq 0$ and $b \leq 0$.

Let k be a line which is 3.5 units below the x-axis and parallel to the x-axis. Show that if a point Q of k has coordinates (a, b) then $b = -3.5$.

Example
3–2

Solution. By the above result, the point Q is a distance $|b|$ from the x-axis. But Q is on k, which is 3.5 units below the x-axis. Hence, $|b| = 3.5$ and $b = 3.5$ or $b = -3.5$. Since Q is *below* the x-axis, it is in either quadrant III or IV; from Example 3–1 (iii) and (iv), $b < 0$ so $b = -3.5$.

Let m be the line which passes through the point with coordinates $(0, 0)$ (this point is called the *origin*) and bisects the second and fourth quad-

Example
3–3

rants as indicated in Fig. 3–3. We mean by this that every point P of m is either in quadrant II or IV and P is the same distance from the x-axis as from the y-axis. Let us prove that if a point P of m has coordinates (a, b), then $b = -a$.

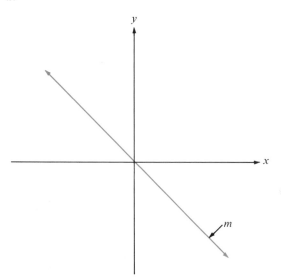

Solution. We are given that P is the same distance from the x-axis as from the y-axis; hence, $|a| = |b|$. If P is in quadrant II, $a \leq 0$ and $b \geq 0$ by Example 3–1 (ii). Then $|a| = -a$ and $|b| = b$, giving $b = -a$. If P is in quadrant IV, $a \geq 0$, $b \leq 0$ by Example 3–1 (iv). In this case, $|a| = a$, $|b| = -b$, and again $b = -a$. We may state this result by saying that *the sum of the coordinates of every point on this line is zero.*

1. Construct a rectangular coordinate system with equal unit lengths on the two axes and plot the points having the following coordinates:

   (a) $(0, 0)$ (b) $(0, 1)$ (c) $(0, -1)$ (d) $(-1, 0)$
   (e) $(\frac{1}{2}, -4)$ (f) $(-3, 3)$ (g) $(2.75, -8)$ (h) $(0, \frac{5}{8})$

2. Construct a rectangular coordinate system with different unit lengths on the axes and plot the points having the coordinates of Exercise 1 above. In the rest of these problems take equal unit lengths.

3. Prove each of the following statements:
   (a) If n is a line 2 units to the right and parallel to the y-axis and (a, b) are coordinates of a point on n, then $a = 2$.

(b) If m is a line d units to the left and parallel to the y-axis and (a, b) are coordinates of a point on m, then $a = -d$.

(c) If k is a line t units above and parallel to the x-axis and (a, b) are coordinates of a point on k, then $b = t$.

4. Let W, X be two points on the x-axis with a corresponding to W and b corresponding to X. Show that the distance between W and X is $|a - b|$.

5. If P is a point with coordinates (a, b), we say that the point with coordinates $(a, 0)$ is the *projection of P on the x-axis* and that the point having coordinates $(0, b)$ is the *projection of P on the y-axis*. Prove the following:

   (a) If A and B are points of a line parallel to the x-axis whose projections on the x-axis are $(a, 0)$ and $(b, 0)$ respectively, then the distance between A and B is $|a - b|$.

   (b) If C and D are points of a line parallel to the y-axis whose projections on the y-axis are $(0, c)$ and $(0, d)$ respectively, then the distance between C and D is $|c - d|$.

6. Let m be the line which passes through the origin and bisects the first and third quadrants. Show that if a point P of m has coordinates (a, b), then $b = a$. (See Example 3–3.)

7. Let t be the line on the two points with coordinates $(0, 1)$ and $(1, 0)$ respectively. Show that if Q is a point of t having coordinates (c, d), then $c + d = 1$.

8. On a coordinate system, plot each of the following sets:

   (a) $S_1 = \{(x, y) \mid x = 2\}$.

   (b) $S_2 = \{(x, y) \mid y = 0\}$.

   (c) $S_3 = \{(x, y) \mid x = 1 \text{ and } y \geq 0\}$.

   (d) $S_4 = \{(x, y) \mid x = -\frac{3}{2} \text{ and } |y| \leq 1\}$.

9. On a coordinate system, plot each of the following sets:

   (a) $T_1 = \{(x, y) \mid y = -1 \text{ and } x \leq 3\}$.

   (b) $T_2 = \{(x, y) \mid |x| \leq 1 \text{ and } |y| = 2\}$.

   (c) $T_3 = \{(x, y) \mid |x| > 1 \text{ and } y = -3\}$.

We wish now to consider the following problem. Suppose that K and L are points of the plane having coordinates (a, b) and (c, d), respectively; can we express the distance between K and L in terms of the coordinates of the given points, *i.e.*, in terms of the numbers a, b, c, d? We refer to Fig. 3–4 in the following consideration, but direct the reader's attention

3.2
The Distance Formula
and Circles

The Distance Formula and Circles

Figure
3–4

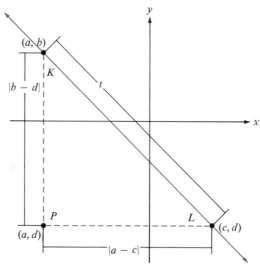

to the fact that only one of several possibilities for the location of the two points K and L with respect to the quadrants is pictured here. If a line on K is drawn parallel to the y-axis and a line on L is drawn parallel to the x-axis, these lines intersect at right angles in a point P having coordinates (a, d) as shown. Now, by Exercise 5, the lengths of the short sides of the right triangle formed are $|b - d|$ and $|a - c|$. Since the object here is to determine the distance t between K and L or the length of the line segment KL, we may use the pythagorean theorem (which says the sum of the squares of the lengths of the short sides of a right triangle is the square of the length of the long side) to obtain

$$|a - c|^2 + |b - d|^2 = t^2.$$

By a previous result (see Exercise 16, Chapter 2) $|a - c|^2 = (a - c)^2$ and $|b - d|^2 = (b - d)^2$ so that we may express the above equation as

$$t^2 = (a - c)^2 + (b - d)^2.$$

Since t is a length and must therefore be non-negative

$$t = \sqrt{(a - c)^2 + (b - d)^2}.$$

Thus, the question we started with has been answered: *If K and L are points having coordinates* (a, b) *and* (c, d) *respectively, then the distance between K and L is the number*

$$\sqrt{(a - c)^2 + (b - d)^2}.$$

This expression for the distance between points is known as the *distance formula.*

Example
3–4

If two points have coordinates $(-1, \sqrt{2})$ and $(\frac{1}{2}, 1)$, the distance between them is given by the distance formula as

$$\sqrt{(-1-\tfrac{1}{2})^2 + (\sqrt{2}-1)^2} = \sqrt{\tfrac{9}{4} + (3 - 2\sqrt{2})}$$
$$= \sqrt{\tfrac{21}{4} - 2\sqrt{2}}$$
$$= \sqrt{(\tfrac{1}{4})(21 - 8\sqrt{2})} = \tfrac{1}{2}\sqrt{21 - 8\sqrt{2}}.$$

Example
3–5

Show that a point having coordinates (a, b) is at a distance

$$\sqrt{a^2 + b^2}$$

from the origin.

Solution. The origin has coordinates $(0, 0)$ so that the distance between the point having coordinates (a, b) and the origin is given by the distance formula as

$$\sqrt{(a - 0)^2 + (b - 0)^2} = \sqrt{a^2 + b^2}.$$

Example
3–6

Let S be the set of coordinates (x, y) of points which are equidistant from the points having coordinates $(1, 1)$ and $(-1, -1)$, respectively. Now let

$$W = \{(x, y) \mid y = -x\}.$$

We will show that $W = S$. (See Fig. 3–5 and compare this with Example 3–3.)

Figure
3–5

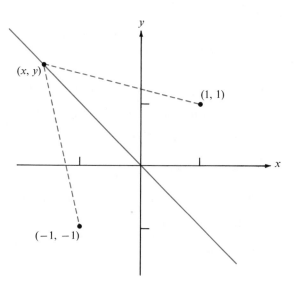

Solution. Let $(x, y) \in W$. Then $y = -x$ and the distance formula gives

$$\sqrt{(x - 1)^2 + (y - 1)^2} = \sqrt{(x - 1)^2 + (-x - 1)^2}$$
$$= \sqrt{(x - 1)^2 + (x + 1)^2}$$

and

$$\sqrt{(x + 1)^2 + (y + 1)^2} = \sqrt{(x + 1)^2 + (-x + 1)^2}$$
$$= \sqrt{(x + 1)^2 + (x - 1)^2}$$

as the distance between (x, y) and $(1, 1)$ and (x, y) and $(-1, -1)$ respectively. This shows that the distance between (x, y) and $(1, 1)$ is the same as the distance between (x, y) and $(-1, -1)$. Therefore, by definition of S, $(x, y) \in S$, which shows that every member of W is a member of S or $W \subseteq S$. Now, if (x, y) is a member of S we must have

$$\sqrt{(x - 1)^2 + (y - 1)^2} = \sqrt{(x + 1)^2 + (y + 1)^2}$$

by definition of S and the distance formula. This equation reduces to

$$(x - 1)^2 + (y - 1)^2 = (x + 1)^2 + (y + 1)^2$$
$$x^2 - 2x + 1 + y^2 - 2y + 1 = x^2 + 2x + 1 + y^2 + 2y + 1$$
$$-2x - 2y = 2x + 2y$$
$$-4y = 4x$$
$$y = -x.$$

But this shows that $(x, y) \in W$ and that $S \subseteq W$. It then follows that $S = W$.

The distance formula is extremely useful for describing circles in terms of coordinates of points. The reader will recall that a circle is a set of points in a plane having a center D and radius r, and that each of the points is a distance $r > 0$ from the point D. In other words, *a point P is on the circle of radius* r *and center* D *if and only if the distance between* P *and* D *is* r. If P has coordinates (x, y) and D has coordinates (a, b), then the distance between P and D is, according to the distance formula,

$$\sqrt{(x - a)^2 + (y - b)^2}.$$

Therefore, P is a point of the circle if and only if

$$\sqrt{(x - a)^2 + (y - b)^2} = r$$

or if and only if

$$(x - a)^2 + (y - b)^2 = r^2.$$

The circle of radius r *and center* D (a, b)* *may then be described as the set*

$$\{(x, y) \mid (x - a)^2 + (y - b)^2 = r^2\}.$$

---

*This is a common notation that means D is a point whose coordinates are (a, b).

The circle with center at the origin and radius 2 is

$$\{(x, y) \mid (x - 0)^2 + (y - 0)^2 = 2^2\}, \quad \text{or} \quad \{(x, y) \mid x^2 + y^2 = 4\}.$$

Example
3–7

The set $\{(x, y) \mid x^2 + y^2 + 2x + 4y = 0\}$ describes the circle with center at $(-1, -2)$ and radius $\sqrt{5}$ since

Example
3–8

$$x^2 + y^2 + 2x + 4y = 0$$

is equivalent to

$$[x - (-1)]^2 + [y - (-2)]^2 = (\sqrt{5})^2.$$

This is obtained by completing the squares in the equation above as follows:

$$\begin{aligned}
x^2 + y^2 + 2x + 4y &= x^2 + 2x + y^2 + 4y \\
&= (x^2 + 2x + 1) + (y^2 + 4y + 4) - 5 \\
&= (x + 1)^2 + (y + 2)^2 - 5 \\
&= (x - (-1))^2 + (y - (-2))^2 - (\sqrt{5})^2.
\end{aligned}$$

Hence,

$$\begin{aligned}
\{(x, y) \mid x^2 + y^2 + 2x + 4y = 0\} \\
= \{(x, y) \mid (x - (-1))^2 + (y - (-2))^2 = (\sqrt{5})^2\}.
\end{aligned}$$

(See Fig. 3–6.)

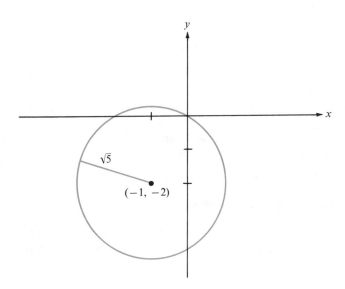

Figure
3–6

10. Calculate the distance between the following pairs of points.

   (a) $(1, 1); (1, 2)$ (b) $(0, 0); (7, 9)$

   (c) $(\frac{1}{4}, -5); (-1, 0)$ (d) $(-\sqrt{2}, 1); (1, -\frac{1}{8})$

11. Compute the distance between the following pairs of points.

   (a) $(a, 1)$ and $(1, 1)$ (b) $(a, 1)$ and $(2, 2)$

   (c) $(a, 2)$ and $(2, b)$ (d) $(b, -a)$ and $(-b, a)$

12. Describe each of the following in set terminology:
   (a) The circle having center at $(0, 0)$ and radius 1.
   (b) The circle having center at $(0, 1)$ and radius 1.
   (c) The circle having center at $(-1, 0)$ and radius 1.
   (d) The circle having center at $(-1, -1)$ and radius 2.
   (e) The circle having center at $(3, -7)$ and radius $2/3$.
   (f) The circle having center at $(-1, -2)$ and radius $3/8$.
   (g) The circle having center at $(\sqrt{2}, -4)$ and radius $\frac{1}{2}$.
   (h) The circle having center at $(-8/9, 0.6)$ and radius $\sqrt{6}$.

13. (a) In the proof of the distance formula, only the case in which one point is in quadrant II and the other is in quadrant IV was considered. List all the other possibilities and draw an appropriate figure for each.
   (b) The results of Exercise 5 give the distance between points which lie on lines parallel to one of the axes. Show that the distance formula gives the same result.

14. Plot the following points and determine the area of the triangle obtained.

$$(-2, 0), (0, 2), (5, 0)$$

15. Determine whether the following sets are circles; if they are, give the center and radius.
   (a) $\{(x, y) \mid x^2 + y^2 = 1\}$
   (b) $\{(x, y) \mid x^2 + y^2 = 0\}$
   (c) $\{(x, y) \mid (x - 1)^2 + y^2 = -1\}$
   (d) $\{(s, t) \mid (s - 1)^2 + (t - 1)^2 = 4\}$
   (e) $\{(a, b) \mid a^2 - 4 = b^2\}$
   (f) $\{(p, m) \mid (p + 8)^2 + (m + 5)^2 = \sqrt{2}\}$
   (g) $\{(p, m) \mid 4(p^2 + p) + 4(m^2 - 2m) = -1\}$

16. Show that the triangle with vertices $(-3, -1)$, $(2, 5)$ and $(4, \frac{10}{3})$ is a right triangle. [*Hint:* Use the converse of the Pythagorean theorem.]

17. (a) What conclusion about the numbers x, y may be reached if we are given that the distance between (x, y) and $(2, 2)$ is the same as the distance between (x, y) and $(-1, -3)$? *Solution:* The distance between (x, y) and $(2, 2)$ is $\sqrt{(x - 2)^2 + (y - 2)^2}$, and the distance between (x, y) and $(-1, -3)$ is $\sqrt{(x + 1)^2 + (y + 3)^2}$. Therefore, if we are given that these distances are the same, we may conclude that

$$\sqrt{(x - 2)^2 + (y - 2)^2} = \sqrt{(x + 1)^2 + (y + 3)^2},$$

or squaring, $(x - 2)^2 + (y - 2)^2 = (x + 1)^2 + (y + 3)^2$. This last equation may be written as follows by squaring the indicated terms:

$$x^2 - 4x + 4 + y^2 - 4y + 4 = x^2 + 2x + 1 + y^2 + 6y + 9$$

or $6x + 10y = -2$. Dividing this last equation by 2 gives $3x + 5y = -1$. Therefore, a conclusion we may reach about the numbers x, y is that $3x + 5y = -1$.

(b) If A is the set of all points (x, y) whose distance from $(2, 2)$ is the same as the distance between $(-1, -3)$ and (x, y), and $B = \{(x, y) \mid 3x + 5y = -1\}$, what conclusion about A, B can be drawn?

(c) If A, B are defined as in (b) above, show that $B \subseteq A$.

(d) Show that $A = B$ and interpret the sets A and B on a coordinate system.

18. (a) What conclusion about numbers x, y may be reached if we are given that the distance between (x, y) and $(4, 5)$ is the same as the distance between (x, y) and $(0, 0)$?

(b) If X is the set of all points (x, y) equidistant from $(4, 5)$ and $(0, 0)$ and $Y = \{(x, y) \mid 8x + 10y = 41\}$, what conclusion about X, Y may be drawn?

(c) Show that $Y \subseteq X$ and interpret the sets X, Y on a coordinate system.

By a *graph* of a relation we mean a set of points, and only those points, in the plane corresponding to the ordered pairs of the relation. For example, the circle of Fig. 3–7 is a graph of

$$\{(x, y) \mid x^2 + (y - 1)^2 = 4\}.$$

3.3
Graphs of Real
Relations

Figure
3–7

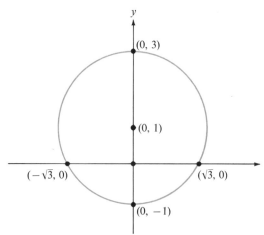

It will be recalled from Section 2.2 that on a one-dimensional coordinate system a set of the form $\{x \mid a \leq x \leq b\}$ is represented by a line segment

Figure
3–8

$$\{x \mid a \leq x \leq b\}$$

as in Fig. 3–8. This set is called a *closed interval* and is denoted by $[a, b]$; *i.e.*, by definition

$$[a, b] = \{x \mid a \leq x \leq b\}.$$

Variations are:

$$(a, b) = \{x \mid a < x < b\},$$

called an *open interval;*

$$[a, b) = \{x \mid a \leq x < b\},$$

called a *right half-open interval;*

$$(a, b] = \{x \mid a < x \leq b\},$$

called a *left half-open interval.*

Example
3–9

Construct the graph of the relation $\{(x, y) \mid x \in [-1, 1]\}$.

Solution. We claim that the shaded area of Fig. 3–9 is the graph of this relation, since any (x, y) of this relation must have $x \in [-1, 1]$ or $-1 \leq x \leq 1$. On the other hand, if (x, y) is a point of the shaded area, then $-1 \leq x \leq 1$ and (x, y) is a member of the relation. Observe that this relation is the Cartesian product $[-1, 1] \times R$, where R is the set of reals.

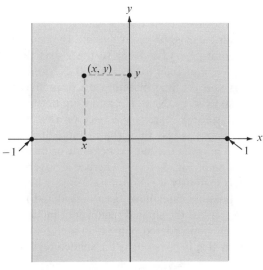

Figure
3–9

Construct the graph of the relation $\{(x, y) \mid |x| \leq 1\}$.

Example
3–10

Solution. It is clear that $\{(x, y) \mid |x| \leq 1\} = \{(x, y) \mid -1 \leq x \leq 1\}$ (see Theorem 7, page 41) since $|x| \leq 1$ if and only if $-1 \leq x \leq 1$. Hence, the graph of $\{(x, y) \mid |x| \leq 1\}$ is given by Fig. 3–9.

Construct the graph of the relation $\{(x, y) \mid y = -x\}$.

Example
3–11

Solution. In Example 3–3, it is shown that if a point (a, b) is on the line which bisects the quadrants II and IV, then $b = -a$; *i.e.*, every point on this line represents an element of the relation $\{(x, y) \mid y = -x\}$. Conversely, suppose the ordered pair $(a, b) \in \{(x, y) \mid y = -x\}$. Then $b = -a$ and $|b| = |-a| = |a|$. But $|a|$ is the distance that (a, b) is from the *y*-axis and $|b|$ is the distance that (a, b) is from the *x*-axis. Therefore, the point (a, b) is the same distance from the *x*-axis as from the *y*-axis. We cannot conclude at this point that (a, b) is a point of the line which bisects quadrants II and IV because points of the line bisecting quadrants I and III are also the same distance from one axis as from the other. But in this case the condition that $b = -a$ requires a and b to have *different signs* unless $a = b = 0$; and this is *not* true for the points of the line bisecting quadrants I and III. Hence, we conclude that if $b = -a$, (a, b) represents a point of the line bisecting quadrants II and IV. (See Fig. 3–5.)

Construct the graph of the relation $\{(x, y) \mid x < y\}$.

Example
3–12

Solution. If x is negative and $y \geq 0$, the condition $x < y$ is obviously met, and thus it is easy to see that every point to the left of the *y*-axis in quadrant II represents an element (x, y) of our relation. [See Fig. 3–10(a).]

If both x and y are positive, then $x = |x|$ and $y = |y|$ [the distance (x, y) is from the y-axis and x-axis, respectively] and the condition that $x < y$ requires that the point (x, y) be closer to the y-axis than to the x-axis. [See Fig. 3–10(b).] We conclude from this that every point in quadrant I between the y-axis and the line bisecting quadrant I represents an element (x, y) of $\{(x, y) \mid x < y\}$. Next, if x and y are both negative $|x| = -x$ and $|y| = -y$ and since $x < y$, $-y < -x$ or the distance the point (x, y) is from the x-axis is less than the distance it is from the y-axis. [See Fig. 3–10(c).] This tells us that every point of quadrant III between the x-axis and the line of Fig. 3–10(c) (where $x = y$) represents a point of the given relation. (Obviously we cannot have $x < y$ if x is positive and y is negative, so all possible cases have been considered.) Conversely, it is evident that every point of the shaded region (not including the line L) of Fig. 3–10(d) represents an ordered pair (x, y) such that $x < y$. Thus, the graph of the relation $\{(x, y) \mid x < y\}$ is the shaded region of Fig. 3–10(d).

**Figure
3–10**

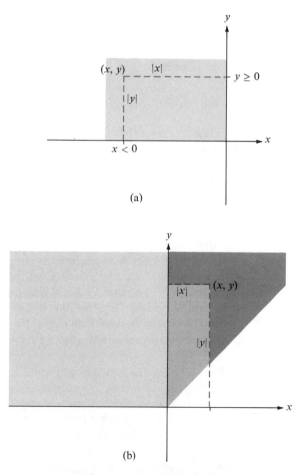

(a)

(b)

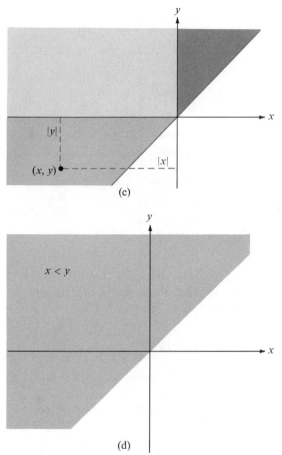

Of particular importance is the so-called *linear* relation

$$\{(x, y) \mid Ax + By = C\},$$

where A, B, C are real numbers and not both of A and B are zero. We now investigate the graphs of linear relations. If $B = 0$, then (since not both of A and B are zero) $A \neq 0$ and

$$\{(x, y) \mid Ax + By = C\} = \{(x, y) \mid Ax = C\}$$
$$= \{(x, y) \mid x = C/A\}.$$

If $C/A > 0$, every (x, y) of the linear relation is $|x| = |C/A| = C/A$ units from the y-axis and to the right of the y-axis. Thus, every (x, y) lies on the line parallel to, and C/A units to the right of, the y-axis. Conversely, every point of this line is C/A units from the y-axis and to the right, and hence represents a pair (x, y) where $x = C/A$. This line is the graph of the linear relation in this case. We have a similar situation if $C/A < 0$, except that the line is to the left of the y-axis. (See Fig. 3–11.)

**Figure
3–11**

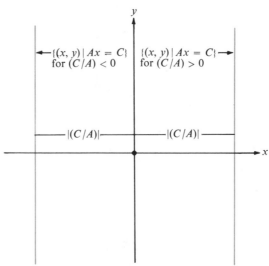

In case $B \neq 0$ and $A = 0$, $\{(x, y) \mid Ax + By = C\} = \{(x, y) \mid y = C/B\}$, and this case is similar to that above except that the line is parallel to the x-axis instead of to the y-axis. (See Fig. 3–12.)

**Figure
3–12**

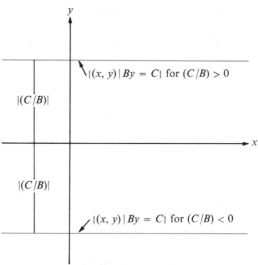

If $B \neq 0$, then the pair $(0, C/B)$ satisfies the equation $Ax + By = C$, i.e.,

$$A \cdot 0 + B \cdot (C/B) = C.$$

This implies that $(0, C/B) \in \{(x, y) \mid Ax + By = C\}$. In the same way we can see that if $A \neq 0$, then $(C/A, 0) \in \{(x, y) \mid Ax + By = C\}$. The point

$(0, C/B)$ is called the *y-intercept* and the point $(C/A, 0)$ is called the *x-intercept* of the linear relation. We have indicated these points in Fig. 3–13 in the case $C/B > 0$ and $C/A < 0$. It turns out that if $C \neq 0$, the line on these two points is the graph of $\{(x, y) \mid Ax + By = C\}$. To demonstrate this fact let (x, y) be any point on this line different from X, Y as in Fig. 3–13. From the figure we have two similar triangles: $\triangle YOX$ is similar to $\triangle YP'P$. Thus from geometry we have that the ratios of corresponding sides are equal; in particular

$$\frac{YO}{OX} = \frac{YP'}{P'P}.$$

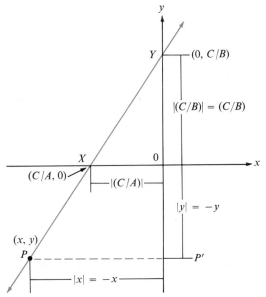

Figure 3–13

But from the figure $YO = C/B$, $YP' = (C/B) + (-y)$, $OX = -C/A$ and $P'P = -x$. Therefore, substitution into the above equation gives

$$(C/B)/(-C/A) = [C/B + (-y)]/(-x)$$

or

$$(A/B)x = (C/B) - y.$$

Now, multiplying by B gives

$$Ax = C - By \qquad \text{or} \qquad Ax + By = C.$$

Hence, if (x, y) is on the line connecting X and Y, then $Ax + By = C$ and so

$$(x, y) \in \{(x, y) \mid Ax + By = C\}.$$

If (p, q) is a point *not* on the line connecting X and Y, then we may show that $(p, q) \notin \{(x, y) \mid Ax + By = C\}$. For we may find a point of the line, say R, whose first coordinate is p. (See Fig. 3–14.) If t is the second coordinate of R, it has just been shown above that

$$Ap + Bt = C \quad \text{or} \quad t = C/B - (A/B)p.$$

Figure 3–14

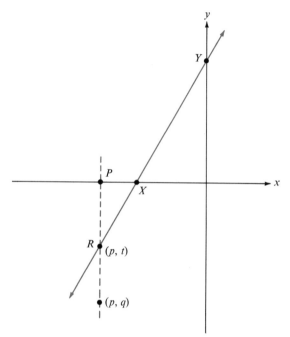

Now since the point (p, t) is different from the point (p, q), $q \neq t$, and since $t = C/B - (A/B)p$, therefore we have

$$Ap + Bq \neq C,$$

which proves that $(p, q) \notin \{(x, y) \mid Ax + By = C\}$. *We may now conclude that a point* (a, b) *is on the line connecting* X, Y *if, and only if,* (a, b) $\in$ {(x, y) $\mid$ Ax + By = C}.

The above argument shows that in this particular case ($C/B > 0$ and $C/A < 0$) *the graph of the linear relation* {(x, y) $\mid$ Ax + By = C} *is a line.* The other cases arising when $C \neq 0$ present no new problems and may be handled just as has been done above. However, if $C = 0$, $C/B = 0$ and $C/A = 0$ (assuming $B \neq 0$ and $A \neq 0$) making the *x*- and *y*-intercepts the same point. Thus the above argument cannot be used. (See Exercise 24.)

Graph the relation $\{(x, y) \mid y = x + 1\}$.

Example
3–13

Solution. The equation $y = x + 1$ may be written as $x + (-1)y = -1$. This relation is therefore a linear relation in which $A = 1$, $B = -1$ and $C = -1$. We have just seen from the above discussion that the graph of a linear relation is a line; to obtain the graph only two points will be needed since a line is determined by any two points on it. It is easy to see that the points $(0, 1)$ and $(-1, 0)$ both satisfy the equation $y = x + 1$ and are therefore points on the graph of the relation. The graph is given in Fig. 3–15.

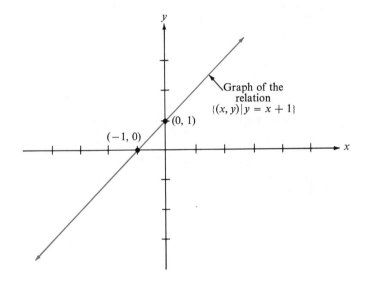

Figure
3–15

19. Construct the graph of each of the following relations.

   (a) $\{(x, y) \mid y = x + 2\}$
   (b) $\{(x, y) \mid y = 2x\}$
   (c) $\{(x, y) \mid y = -2x\}$
   (d) $\{(x, y) \mid y = 2x - 1\}$

20. Construct the graph of each of the following relations:

   (a) $\{(x, y) \mid y = x + 2, x \geq 0\}$
   (b) $\{(x, y) \mid y = 2x, x < 0\}$
   (c) $\{(x, y) \mid y = -2x, x > 0\}$
   (d) $\{(x, y) \mid y = 2x - 1, x \leq 0\}$

21. Construct the graph of each of the following relations:
    (a) $\{(x, y) \mid x + y = 1\}$
    (b) $\{(x, y) \mid x + y > 1\}$
    (c) $\{(x, y) \mid x + y < 1\}$.

22. Construct the graph of the following relations:
    (a) $\{(x, y) \mid y \leq x + 1, x \geq 0, y \geq 0\}$
    (b) $\{(x, y) \mid y > 2x - 1, x \leq 0\}$

Figure 3–16

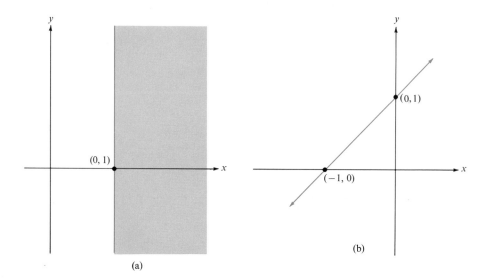

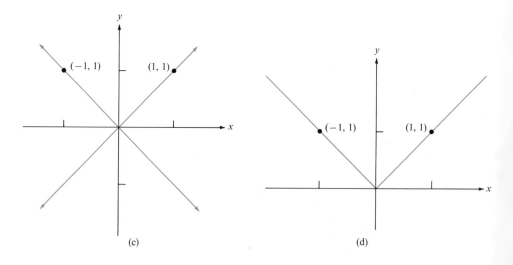

23. Construct the graph of each of the following relations.
    (a) $\{(x, y) \mid y \le |2.5|\}$
    (b) $\{(x, y) \mid x^2 + y^2 = 1\}$
    (c) $\{(x, y) \mid x^2 + y^2 < 1\}$
    (d) $\{(x, y) \mid x^2 + y^2 > 1\}$
    (e) $\{(x, y) \mid (x + 1)^2 + (y - 1)^2 = 1 \text{ and } |y| < 1\}$
    (f) $\{(x, y) \mid (x + 1)^2 + (y - 1)^2 < 1 \text{ and } |y| = 1\}$
    (g) $\{(x, y) \mid (x + 1)^2 + (y - 1)^2 > 1 \text{ and } |y| > 1\}$

24. Show that the graph of the linear relation $\{(x, y) \mid Ax + By = 0\}$ is a line. [*Hint:* Consider the two points $(0, 0)$ and $(B, -A)$.]

25. What are the relations which have the graphs of Fig. 3–16?

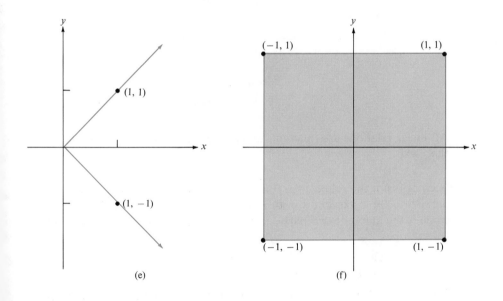

Figure 3–16 (cont.)

(e) (f)

We now consider those real relations which are functions; *i.e.*, those real relations R having the property:

$$(a, b) \in R \text{ and } (a, c) \in R \text{ imply } b = c.$$

As before, let us rephrase this definition: R *contains no two distinct ordered pairs with the same first element;* R *pairs no more than one element in the range of* R *with each element of the domain.*

3.4
Real Functions

In terms of the graph of R, if $(a, b) \in R$ and $(a, c) \in R$, we have a situation as indicated in Fig. 3–17 if $b \neq c$. The important aspect of Fig. 3–17 is that if $b \neq c$, then $(a, b) \neq (a, c)$ and the two points (a, b), (a, c) are the same distance from the y-axis and hence on a line parallel to the y-axis. The conclusion to be drawn from this is that *the relation* R *is not a function if its graph is cut more than once by any line parallel to the* y-*axis.*

Figure 3–17

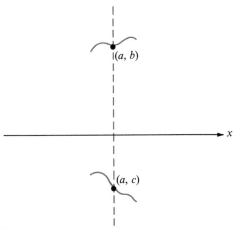

Example 3–14

All linear relations are functions except those having a graph parallel to the y-axis.

Example 3–15

A circle is not the graph of a function since an infinite number of lines parallel to the y-axis cut it in two points. (See Fig. 3–18(a).) However, a circle is the union of two parts, each of which is the graph of a function; for example,

$$\{(x, y) \mid x^2 + y^2 = 1\} = \{(x, y) \mid x^2 + y^2 = 1, y \geq 0\} \ \cup$$
$$\{(x, y) \mid x^2 + y^2 = 1, y \leq 0\}.$$

The first member of the above union is the top half and the second member is the lower half of the circle as indicated in Fig. 3–18(b). For a pair (x, y) of the top half we have $y \geq 0$ and $x^2 + y^2 = 1$ or $y^2 = 1 - x^2$. Hence, y is the non-negative number whose square is $1 - x^2$, i.e., $y = \sqrt{1 - x^2}$. If (x, y) is an ordered pair belonging to the lower half, $x^2 + y^2 = 1$ and $y \leq 0$ or $y^2 = 1 - x^2$. In this instance y is the non-positive number whose square is $1 - x^2$ or $y = -\sqrt{1 - x^2}$. We may then write

$$\{(x, y) \mid x^2 + y^2 = 1\} = \{(x, y) \mid y = \sqrt{1 - x^2}\} \ \cup$$
$$\{(x, y) \mid y = -\sqrt{1 - x^2}\}.$$

(See Fig. 3–18(c), (d).)

**Figure
3–18**

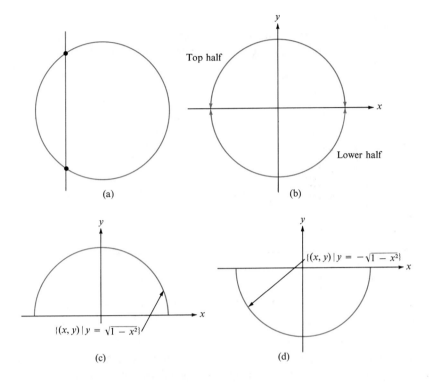

(a)

(b)

Top half

Lower half

(c)

$\{(x, y) \mid y = \sqrt{1 - x^2}\}$

(d)

$\{(x, y) \mid y = -\sqrt{1 - x^2}\}$

The relation $F = \{(x, y) \mid y = x^2\}$ is a function since if $(a, b) \in F$ and $(a, c) \in F$, $b = a^2$ and $c = a^2$ so that $b = c$. The graph of F is indicated in Fig. 3–19.

**Example
3–16**

**Figure
3–19**

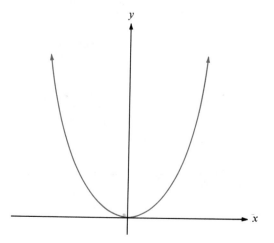

**Example
3–17**

The relation $G = \{(x, y) \mid x = y^2\}$ is not a function since $(1, 1) \in G$ and $(1, -1) \in G$. Observe that G is the inverse relation of the function in Example 3–16 above; *i.e.*, $G = F^{-1}$. The graph of G is indicated in Fig. 3–20.

**Figure
3–20**

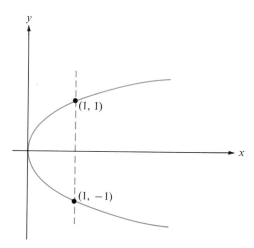

**Example
3–18**

The relation $A = \{(a, b) \mid b = |a|\}$ is a function whose graph is indicated in Fig. 3–21. The graph is easily constructed by using the definition of absolute value:

$$b = |a| = a \text{ if } a \geq 0$$

and

$$b = |a| = -a \text{ if } a < 0$$

**Figure
3–21**

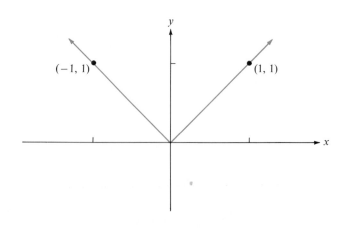

so that

$$A = \{(a, b) \mid b = a \text{ and } a \geq 0\} \cup \{(a, b) \mid b = -a \text{ and } a < 0\}.$$

Hence, A is the union of two linear functions.

Example
3–19

Let $C = \{(w, z) \mid z = (-1)^w \cdot w \text{ and } w \text{ is an integer} \geq 0\}$. Since w is an integer, $(-1)^w = 1$ for even w and $(-1)^w = -1$ for odd w. Therefore,

$$C = \{(w, z) \mid z = w \text{ and } w \text{ is an even integer} \geq 0\} \cup$$
$$\{(w, z) \mid z = -w \text{ and } w \text{ is an odd integer} \geq 0\}.$$

(See Fig. 3–22 for the graph of C.)

Figure
3–22

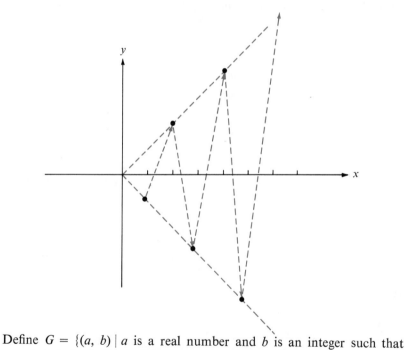

Example
3–20

Define $G = \{(a, b) \mid a \text{ is a real number and } b \text{ is an integer such that } a - 1 < b \leq a\}$. G is called the *greatest integer* function; this is a quite natural name for this function since it will be noted that b — in the definition of G — is the *largest integer* which is not greater than a. Some ordered pairs in G are $(1, 1)$, $(\frac{1}{2}, 0)$, $(\sqrt{2}, 1)$, $(-\frac{1}{2}, -1)$. Suppose that $(x, b) \in G$ and $x \in [0, 1)$, *i.e.*, $0 \leq x < 1$. Then, evidently 0 is the largest integer which is not larger than x. Hence, $b = 0$. Similar reasoning shows that if $(x, b) \in G$ and $x \in [1, 2)$, $b = 1$; generally if $(x, b) \in G$ and

$x \in [n, n + 1)$, where n is an integer, then $b = n$. A standard notation for this function is "[ ]", *i.e.*, instead of writing $b = G(a)$ we write

$$b = [a].$$

In this notation the greatest integer function on the half-open interval $[-2, 3)$ may be described as follows:

$$\begin{aligned}
[x] &= -2 && \text{for all } x \in [-2, -1) \\
[x] &= -1 && \text{for all } x \in [-1, 0) \\
[x] &= 0 && \text{for all } x \in [0, 1) \\
[x] &= 1 && \text{for all } x \in [1, 2) \\
[x] &= 2 && \text{for all } x \in [2, 3)
\end{aligned}$$

This function is an example of a kind of function known as a *step function;* this term is clearly justified as the graph given in Fig. 3–23 shows.

Figure 3–23

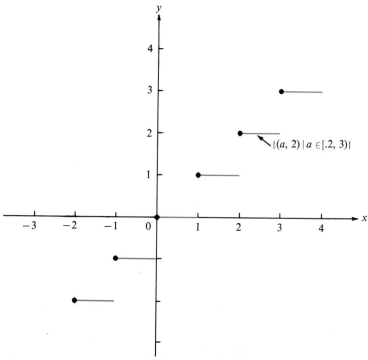

$\{(a, 2) \mid a \in [.2, 3)\}$

In some cases the domain and range of a function can easily be seen from the graph of the function. Remembering that the domain is the set of all first coordinates and the range is the set of all second coordinates, we may determine these as indicated in Fig. 3–24.

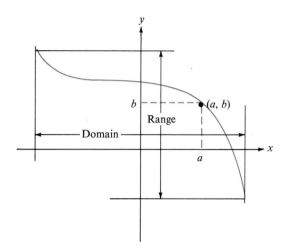

Figure
3–24

26. The graphs given in Fig. 3–25 are graphs of real relations; which are graphs of functions?

27. Construct the graph of each of the following functions and give their domain and range.
    (a) $\{(a, b) \mid b = a^2 + 1\}$
    (b) $\{(x, y) \mid y = x^2 - 5, \ x \in [-6, -5]\}$
    (c) $\{(c, d) \mid d = \sqrt{c}\}$
    (d) $\{(p, q) \mid q = \sqrt{8 - p}\}$

28. Write each of the functions of Exercise 27 above in the $f(x)$-notation.

29. Sketch the graph of each of the functions given below:

    (a) $f(x) = 5x$ (b) $f(x) = 4x - 2$
    (c) $f(x) = x^2$ (d) $g(x) = -x^2$
    (e) $g(x) = x^2 - 1$

30. Sketch the graph of each function below:

    (a) $f(x) = \sqrt{x}$ (b) $f(x) = \sqrt{1 - x}$
    (c) $f(x) = \sqrt{1 - x^2}$ (d) $f(x) = \sqrt{1 + x^2}$

31. Sketch the graph of each relation given below and decide which ones are functions.

    (a) $\{(x, y) \mid x = \sqrt{y}\}$ (b) $\{(x, y) \mid x = |y|\}$
    (c) $\{(x, y) \mid y = |x|\}$ (d) $\{(x, y) \mid x = \sqrt{-y}\}$
    (e) $\{(x, y) \mid y = \sqrt{-x}\}$ (f) $\{(x, y) \mid |y| = x^2\}$

**Figure
3–25**

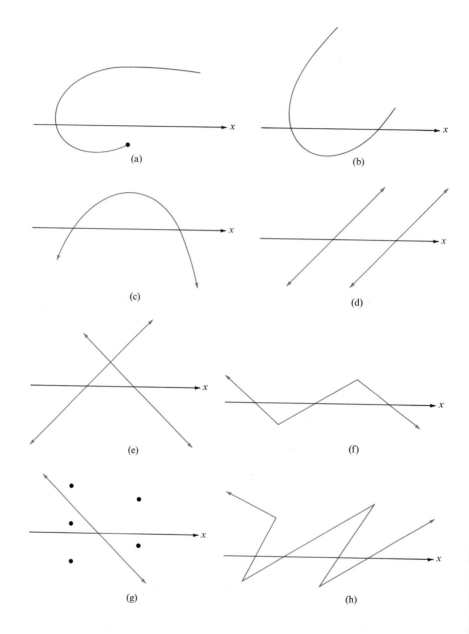

32. Solve each of the following:

For example, $[2x] = 0$. By definition of the greatest integer function $[2x]$ means the largest integer which is not greater than $2x$, $i.e.$, $[2x]$ is an integer and

$$[2x] \leq 2x < [2x] + 1.$$

In this case $[2x] = 0$ so that

$$0 \leq 2x < 1 \qquad \text{or} \qquad 0 \leq x < \tfrac{1}{2}.$$

Hence, if x is a solution for the equation $[2x] = 0$, then $x \in [0, \tfrac{1}{2})$. Conversely, if $x \in [0, \tfrac{1}{2})$, then $0 \leq x < \tfrac{1}{2}$ and $0 \leq 2x < 1$. Since $2x \in [0, 1)$ we see that 0 is the largest integer which is not larger than $2x$; this implies that $[2x] = 0$. Thus, the solution set for the equation $[2x] = 0$ is $[0, \tfrac{1}{2})$.

(a) $[2x] = 1$ (b) $[x + 1] = 0$ (c) $[x/5] = 2$

(d) $[x] < 0$ (e) $[x] = x$ (f) $[x] = -x$

33. Sketch the graph of each of the following functions:

Example: $f(x) = [-x]$, $x \in [-3, 3]$. If $x = -3$, then $-x = 3$ and $f(x) = f(-3) = [3] = 3$. If $x \in (-3, -2]$, then $-3 < x \leq -2$ and $2 \leq -x < 3$ which implies that $[-x] = 2$. Therefore, $f(x) = [-x] = 2$ for all $x \in (-3, -2]$. In a similar way we have

$$f(x) = [-x] = 1 \qquad \text{for } x \in (-2, -1]$$
$$f(x) = [-x] = 0 \qquad \text{for } x \in (-1, 0]$$
$$f(x) = [-x] = -1 \quad \text{for } x \in (0, 1]$$
$$f(x) = [-x] = -2 \quad \text{for } x \in (1, 2]$$
$$f(x) = [-x] = -3 \quad \text{for } x \in (2, 3]$$

From this information the graph of f may be sketched as has been done in Fig. 3–26.

(a) $f(x) = [x + 1]$, $x \in (-\tfrac{1}{2}, 1\tfrac{1}{2})$

(b) $f(x) = [x - 1]$, $x \in (-1.5, 2.3)$

(c) $g(x) = [2x]$, $x \in (-3, 2]$

(d) $g(x) = x + [x]$, $x \in [-4, 4]$

(e) $h(x) = |x| - [x]$, $x \in [-4, 4]$

**Figure
3–26**

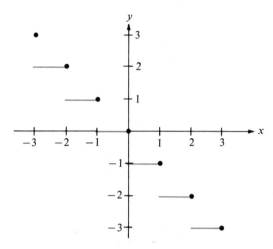

34. Sketch the graph of each of the following:

(a) $f(x) = \begin{cases} x & \text{if } x \le 0 \\ 2x & \text{if } x > 0 \end{cases}$

(b) $f(x) = \begin{cases} 1 - x & \text{for } x < 0 \\ 1 & \text{for } x \ge 0 \end{cases}$

(c) $f(x) = \begin{cases} x/2 & \text{for } x \in [-5, -1] \\ x^2 & \text{for } x \in [0, 5] \end{cases}$

(d) $f(x) = \begin{cases} \sqrt{1 - x^2} & \text{for } |x| < 1 \\ x - [x] & \text{for } x \ge 1 \end{cases}$

35. Prove the following properties of the greatest integer function.

(a) $[\sqrt{1 - x^2}] = 0$ if $|x| < 1$, $x \ne 0$
  $= 1$ if $x = 0$

(b) $[1/x] = 0$ for $x > 1$
  $[1/x] \ge 1$ for $0 < x \le 1$

(c) $[|x|/x] = 1$ for $x > 0$
  $[|x|/x] = -1$ for $x < 0$

(d) $[x + 1] = [x] + 1$

36. Let $a = 0.\bar{2}$ and let n be a positive integer.

(a) Show that $[10^n a] = 22222 \ldots 2$, where the number of 2's is equal to n.

(b) Show that $10^n a - [10^n a] = a$.

POLYNOMIAL FUNCTIONS

4

We have defined in the previous chapter the linear relation

We have defined in the previous chapter the linear relation

$$f = \{(x, y) \mid Ax + By = C\},$$

where A, B, C are real numbers and not both A, B are zero. Since the graphs of these relations are straight lines, the only linear relations that are not functions are those having graphs which are lines parallel to the y-axis. The graphs are parallel to the y-axis if and only if $B = 0$ (see page 75). Thus, if $B \neq 0$, f is a function and may be written

$$\{(x, y) \mid y = C/B - (A/B)x\}$$

or in the $f(x)$-notation $y = f(x) = C/B - (A/B)x$. In this case, f is an example of what is called a polynomial function of degree one.

P is a *polynomial* function *of degree n* where $n \geq 0$ is an integer if and only if there are real numbers (called the *coefficients* of P) $a_0, a_1, a_2, \ldots, a_n$ with $a_n \neq 0$ such that

Definition

$$P(x) = a_0 + a_1x + a_2x^2 + \cdots + a_nx^n$$

for all x in the domain of P.

In the linear function

**Example
4–1**

$$f(x) = C/B - (A/B)x,$$
$$n = 1, a_0 = C/B, \text{ and } a_1 = -A/B.$$

The polynomial functions

**Example
4–2**

$$P(x) = a_0, a_0 \neq 0$$

of degree 0 are usually called *constant functions*. It should be noted that our definition of polynomial function does not include the constant function

$$Z(x) = 0.$$

We shall refer to this as a *zero function* or *polynomial function having no degree*.

Example 4–3

The function

$$f(x) = x^3 - 8x^2 - \sqrt{2}x + 9, x \in [1, 3]$$

is a polynomial function of degree 3.

Example 4–4

The absolute value function

$$A(x) = |x|, x \geq 0$$

is a polynomial function of degree 1, since in this case

$$A(x) = x.$$

However, if the domain is $[-2.5, 1]$ for

$$A(x) = |x|,$$

then this is *not* a polynomial function. On the other hand, since

$$A(x) = |x| = -x \quad \text{for} \quad x \in [-2.5, 0]$$
$$A(x) = |x| = x \quad \text{for} \quad x \in [0, 1]$$

we might view this function as being composed of *pieces* of polynomial functions.

Definition

Let q be a function and c a number in the domain of q. Then c is called a *zero of q* if and only if $q(c) = 0$.

Example 4–5

The non-zero constant polynomial functions

$$P(x) = a, a \neq 0$$

have no zeros $\big($since $P(c) = 0$ for any number c implies $a = 0\big)$. However, every number in the domain of a zero function is a zero of that function.

If $Q(x) = x^2 - 2$, then

Example
4–6

$$Q(\sqrt{2}) = (\sqrt{2})^2 - 2 = 2 - 2 = 0,$$

so that $\sqrt{2}$ is a zero of Q. Also $-\sqrt{2}$ is a zero of Q.

Let

Example
4–7

$$G(x) = [x], \ x \in [-2, 8].$$

Then for all x such that $0 \leq x < 1$,

$$G(x) = 0.$$

Conversely, if $x \notin [0, 1)$, then $G(x) \neq 0$. Hence, every number in the interval $[0, 1)$ is a zero of G, and there are no others.

The zeros (if any) of a function can be helpful when one is graphing the function; for each zero c of a function f determines a point $(c, f(c)) = (c, 0)$ where the graph of f crosses the x-axis. In the case of a linear function

$$f(x) = a_0 + a_1 x,$$

the x-intercept determines the only zero of the function. Since $a_1 \neq 0$, we have

$$f\left(-\frac{a_0}{a_1}\right) = a_0 + a_1\left(-\frac{a_0}{a_1}\right)$$

$$= a_0 - a_0 = 0.$$

Then the x-intercept is the point $\left(-\dfrac{a_0}{a_1}, 0\right)$.

The polynomial functions of degree 2 are called quadratic functions:

$$Q(x) = a_0 + a_1 x + a_2 x^2.$$

We shall consider these polynomials in some detail in this chapter.

We have seen the quadratic

$$f(x) = x^2$$

in the previous chapter; this has a graph as in Fig. 4–1 and only one zero, which is 0: $f(0) = 0^2 = 0$. The reader will have observed that the general appearance of this graph has not been justified; the tools at our disposal

**Figure
4–1**

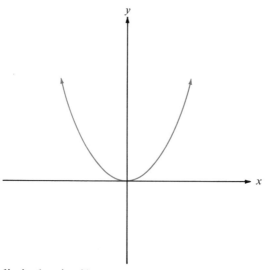

now are too limited to justify some aspects of the graphs of polynomials. For example, how do we know that the graph of $f(x) = x^2$ is *smooth* — as in Fig. 4–1 — and has no *holes* or *gaps?* Generally, the graphs of polynomial functions are smooth with no jumps *as long as the domain is without gaps.* The graph of $f(x) = x^2$, for all real x, is as in Fig. 4–1; for $f(x) = x^2$, $x \in [-1, 1]$, the graph is as in Fig. 4–2. For the first of these,

**Figure
4–2**

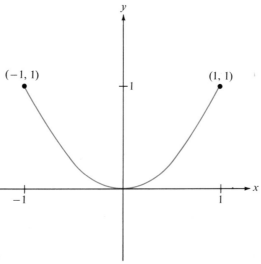

the domain is the set of all real numbers; and for the second, the set $[-1, 1]$. Neither of these domains has breaks or gaps. However, we get breaks and gaps in the graph if we make breaks or gaps in the domain. Fig. 4–3 indicates this situation for the function

$$f(x) = x^2, \ x \in [-1, 0] \cup [\tfrac{1}{2}, 2].$$

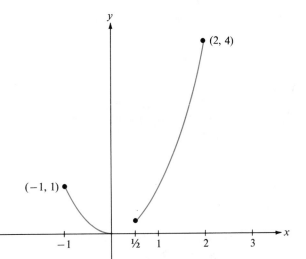

**Figure
4–3**

The break in the domain of this function between 0 and $\tfrac{1}{2}$ produces the break in the curve. To construct the graph of a polynomial function, we may proceed by calculating enough points on the curve until we get a good idea of its general shape. We then connect the points thus obtained with a smooth line and hope for the best.

Sketch the graph of

$$P(x) = 1 - 5x + x^5.$$

**Example
4–8**

Solution. We calculate the points of the graph corresponding to $-2, -1,$ $-\tfrac{1}{2}, 0, \tfrac{1}{2}, 1, 2$. We have

$$
\begin{aligned}
P(-2) &= 1 - 5(-2) + (-2)^5 = 1 + 10 - 32 = -21 \\
P(-1) &= 1 - 5(-1) + (-1)^5 = 1 + 5 - 1 = 5 \\
P(-\tfrac{1}{2}) &= 1 - 5(-\tfrac{1}{2}) + (-\tfrac{1}{2})^5 = 1 + \tfrac{5}{2} - \tfrac{1}{32} = 3 + \tfrac{15}{32} \\
P(0) &= 1 - 5(0) + (0)^5 = 1 \\
P(\tfrac{1}{2}) &= 1 - 5(\tfrac{1}{2}) + (\tfrac{1}{2})^5 = 1 - \tfrac{5}{2} + \tfrac{1}{32} = -\tfrac{47}{32} \\
P(1) &= 1 - 5(1) + (1)^5 = 1 - 5 + 1 = -3 \\
P(2) &= 1 - 5(2) + (2)^5 = 1 - 10 + 32 = 23.
\end{aligned}
$$

We now plot the points

$$(-2, P(-2)) = (-2, -21)$$
$$(-1, P(-1)) = (-1, 5)$$
$$(-\tfrac{1}{2}, P(-\tfrac{1}{2})) = (-\tfrac{1}{2}, 3 + \tfrac{15}{32})$$
$$(0, P(0)) = (0, 1)$$
$$(\tfrac{1}{2}, P(\tfrac{1}{2})) = (\tfrac{1}{2}, -\tfrac{47}{32})$$
$$(1, P(1)) = (1, -3)$$
$$(2, P(2)) = (2, 23)$$

as shown in Fig. 4–4. In this figure we have indicated the general shape of the graph with the line through the plotted points.

1. Give the degree of each of the following polynomial functions.
   (a) $f(x) = x^2 + x - 18$, $x \in [-9, 8]$
   (b) $g(x) = 12x + 9x^2 - x^5$, $x \in [0, 12]$
   (c) $h(x) = 13 + 2x$, $x \geq 0$
   (d) $p(x) = 12$, $x < -1$
   (e) $f(x) = (x - 1)(x - 2)(x - 3)$
   (f) $b(x) = 1 + 2x + 3x^2 + \cdots + 100x^{99}$
   (g) $q(x) = 1 + x + x^2 + x^3 + \cdots + x^k$

2. Determine whether the following are polynomial functions. If a function is a polynomial function, give its degree.
   (a) $f(x) = \sqrt{2x}$
   (b) $g(x) = (x + 2)(x - 2)$
   (c) $h(x) = [x]$
   (d) $p(x) = [x]$, $x \in (0, 1)$
   (e) $R(x) = 7x^5 - 4x + \sqrt{7}x^3 - x^9 - 8$
   (f) $f(x) = \begin{cases} x^2 + 2x + 5 & \text{if } x \in [-1, 0] \\ -x^2 + 2x + 5 & \text{if } x \in [0, 1] \end{cases}$
   (g) $g(x) = \begin{cases} x^2 + 2 & \text{if } x \leq 0 \\ x + 2 & \text{if } x > 0 \end{cases}$

3. Determine the zeros (if any) of each of the following functions.

   Example: $f(x) = x^2 - 1$. By definition, a zero of this function is a number x such that $f(x) = 0$ or such that $x^2 - 1 = 0$. If we factor $x^2 - 1$ as $(x - 1)(x + 1)$ we see that $x^2 - 1 = 0$ implies that

$x - 1 = 0$ or $x + 1 = 0$. Therefore, $x = 1$ or $x = -1$. Thus 1 and -1 are the only possible zeros of f. These two numbers are zeros since

$$f(1) = 1^2 - 1 = 1 - 1 = 0$$

and

$$f(-1) = (-1)^2 - 1 = 1 - 1 = 0.$$

(a) $g(x) = [x] + 1, x \in (-5, 10)$
(b) $h(x) = [x] + 1, x \geq 0$
(c) $k(x) = [x - 2]$
(d) $k(x) = 5 - 4x$
(e) $m(x) = x/2 - 3/2$
(f) $w(x) = 4x^2 - 1$
(g) $f(x) = x + [x]$
(h) $g(x) = 2x - [x]$
(i) $a(x) = 2|x| + 5$
(j) $b(x) = 2|x| - 5$

4. For each function f below compute $f(x)$ for the given values of x, plot the points $(x, f(x))$ and then sketch the graph of f. Also give the zeros of each f.

Example: $f(x) = -x^2 + 3x, x \in [-2, 4]; x = -2, -1\frac{1}{2}, -1, -\frac{1}{2}, 0, 1, 2, 3, 4.$

Then

$$f(-2) = -(-2)^2 + 3(-2) = -10$$
$$f(-1\tfrac{1}{2}) = -(-1\tfrac{1}{2})^2 + 3(-1\tfrac{1}{2}) = -27/4$$
$$f(-1) = -(-1)^2 + 3(-1) = -4$$
$$f(-\tfrac{1}{2}) = -(-\tfrac{1}{2})^2 + 3(-\tfrac{1}{2}) = -7/4$$
$$f(0) = -0^2 + 3 \cdot 0 = 0$$
$$f(1) = -1^2 + 3 \cdot 1 = 2$$
$$f(2) = -2^2 + 3 \cdot 2 = 2$$
$$f(3) = -3^2 + 3 \cdot 3 = 0$$
$$f(4) = -4^2 + 3 \cdot 4 = -4.$$

Now the points $(x, f(x))$ are plotted and the graph of f is sketched as in the figure below. That 0 and 3 are zeros of f may be seen from the equations $f(0) = f(3) = 0$. That 0, 3 are the only zeros of f may be seen by factoring $f(x)$ as $x(-x + 3)$; then $f(x) = 0$ only if $x = 0$ or $x = 3$.

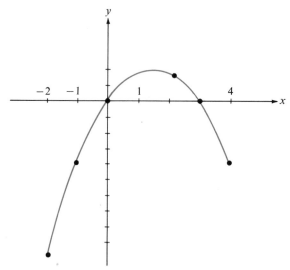

(a) $f(x) = x/2 - 7, x \in [-8, 2]; x = -8, -7, -6, \ldots, 0, 1, 2$
(b) $f(x) = 2x - 7, x \in [-8, 2]; x = -8, -7, -6, \ldots, 0, 1, 2$
(c) $f(x) = x^2 + 1; x = -3, -2, -1, 0, 1, 2, 3$
(d) $f(x) = x^2 + x + 1; x = -3, -2, -1, 0, 1, 2, 3$
(e) $f(x) = x^2 - 1; x = -3, -2, -1, 0, 1, 2, 3$
(f) $f(x) = x^2 - x + 1; x = -3, -2, -1, 0, 1, 2, 3$
(g) $f(x) = -x^2 - 1; x = -3, -2, -1, 0, 1, 2, 3$
(h) $f(x) = -x^2 + x + 1; x = -3, -2, -1, 0, 1, 2, 3$
(i) $f(x) = x^3; x = -3, -2, -1, 0, 1, 2, 3$
(j) $f(x) = -x^3; x = -3, -2, -1, 0, 1, 2, 3$
(k) $f(x) = x^3 - x; x = -2, -1, 0, 1, 2, -3/2, -\frac{1}{2}, \frac{1}{2}, 3/2$
(l) $f(x) = -x^3 + x; x = -2, -1, -\frac{1}{2}, -3/2, 0, \frac{1}{2}, 3/2, 1, 2$

5. Sketch the graph of each function given below.
   (a) $f(x) = 3x, \mathfrak{D}(f) = [-2, 0] \cup [\frac{1}{2}, 2]$
   (b) $g(x) = 1 - x, \mathfrak{D}(g) = [-5, -4] \cup [-3, 1] \cup [2, 4]$
   (c) $f(x) = -x^2, \mathfrak{D}(f) = [-3, 0) \cup (0, 3]$
   (d) $g(x) = |x + 2|$
   (e) $h(x) = |x^2 - 1|$
   (f) $f(x) = |x^3|, x \in [-3, 3]$

   (g) $f(x) = \begin{cases} x + 5, & \text{for } x \leq -5 \\ x/2 + 5/2, & \text{for } x \in (-5, 0) \\ 5/2, & \text{for } x \geq 0 \end{cases}$

   (h) $g(x) = \begin{cases} x^2 + x, & \text{for } x \leq 0 \\ -x^2 + x, & \text{for } x > 0 \end{cases}$

Returning now to the general quadratic function

$$Q(x) = c + bx + ax^2 \qquad a \neq 0$$

we will investigate the graphs of these. We may write

$$
\begin{aligned}
Q(x) &= ax^2 + bx + c \\
&= a\left(x^2 + (b/a)x\right) + c \\
&= a\left(x^2 + (b/a)x + b^2/4a^2\right) + c - b^2/4a \\
&= a(x + b/2a)^2 + \frac{(4ac - b^2)}{4a}.
\end{aligned}
$$

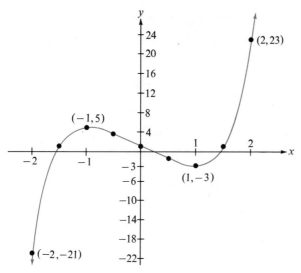

**Figure
4–4**

Now, for all x

$$(x + b/2a)^2 \geq 0$$

and if $a > 0$,

$$a(x + b/2a)^2 \geq 0$$

so that

$$Q(x) = a(x + b/2a)^2 + \frac{4ac - b^2}{4a} \geq \frac{4ac - b^2}{4a}.$$

We thus conclude that if $a > 0$,

$$Q(x) \geq \frac{4ac - b^2}{4a}$$

for all x. In constructing the graph of Q, we will be plotting points $(x, Q(x))$ where the second coordinate, $Q(x)$, is never smaller than $(4ac - b^2)/4a$. This tells us that no part of the graph of Q lies below the line $y = (4ac - b^2)/4a$. (See Fig. 4–5.)

Figure 4–5

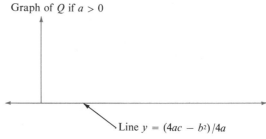

Graph of Q if $a > 0$

Line $y = (4ac - b^2)/4a$

Example 4–9

If $Q(x) = 1 + x + x^2$, we have $a = 1$, $b = 1$, $c = 1$ and

$$\frac{4ac - b^2}{4a} = \frac{4 \cdot 1 \cdot 1 - 1^2}{4 \cdot 1} = \frac{3}{4}.$$

Thus, the graph of Q is nowhere below the line $y = \frac{3}{4}$. We may, of course, conclude from this that Q has no zeros.

In case $a < 0$, the situation is similar:

$$a(x + b/2a)^2 \leq 0$$

and

$$Q(x) = a(x + b/2a)^2 + (4ac - b^2)/4a \leq (4ac - b^2)/4a,$$

or

$$Q(x) \leq (4ac - b^2)/4a$$

for all x. This gives rise to the conclusion that the graph of Q is nowhere above the line $y = (4ac - b^2)/4a$.

Since for all x

$$Q(x) = a(x + b/2a)^2 + \frac{4ac - b^2}{4a},$$

we have

$$Q(-b/2a) = a \cdot 0 + (4ac - b^2)/4a$$
$$= (4ac - b^2)/4a.$$

Therefore, the graph of Q meets the line $y = (4ac - b^2)/4a$ since the point

$$(-b/2a, Q(-b/2a))$$

is on this line. Also, this is the only point of Q which is on this line since

$$a(x + b/2a)^2 = 0$$

only if $x = -b/2a$.

Consider the case when $a > 0$ and the number $(4ac - b^2)/4a < 0$. (See Fig. 4–6.) Let $x_1 < x_2$ be two numbers to the right of $-b/2a$ as shown in

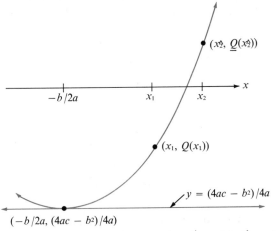

Figure
4–6

Fig. 4–6. Since $a > 0$, we know that both points $(x_1, Q(x_1))$ and $(x_2, Q(x_2))$ are above the line L. We also have

$$-b/2a < x_1 < x_2$$

so that

$$0 < x_1 + b/2a < x_2 + b/2a$$

and so

$$(x_1 + b/2a)^2 < (x_2 + b/2a)^2.$$

Finally,

$$a(x_1 + b/2a)^2 + (4ac - b^2)/4a < a(x_2 + b/2a)^2 + (4ac - b^2)/4a$$

or

$$Q(x_1) < Q(x_2).$$

It thus is seen that the point $(x_1, Q(x_1))$ is *below* the point $(x_2, Q(x_2))$. The relative positions of these points is shown in Fig. 4–6.

Similar considerations show that if $x_1 < x_2 < -b/2a$, then

$$Q(x_1) > Q(x_2),$$

and that $(x_1, Q(x_1))$ is *above* $(x_2, Q(x_2))$. This information may be summarized as follows:

(*a*) If $a > 0$ for the quadratic

$$Q(x) = c + bx + ax^2,$$

the graph of Q rises as x increases to the right of the number $-b/2a$, *and the graph of Q rises as x decreases to the left of* $-b/2a$. *The graph of Q, therefore, has the general appearance given in Fig. 4–7.*

Figure 4–7

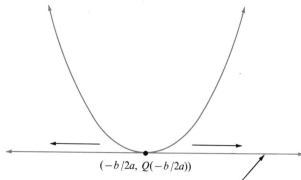

$(-b/2a, Q(-b/2a))$

(*b*) If $a < 0$ for the quadratic $y = (4ac - b^2)/4a$

$$Q(x) = c + bx + ax^2,$$

the graph of Q falls as x increases to the right of the number $-b/2a$, *and the graph of Q falls as x decreases to the left of* $-b/2a$. *The graph of Q therefore has the general appearance given in Fig. 4–8.*

Figure 4–8

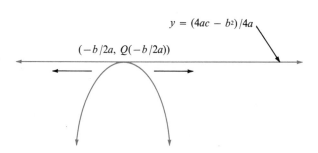

$y = (4ac - b^2)/4a$

$(-b/2a, Q(-b/2a))$

Example
4–10

We have seen above that the graph of

$$Q(x) = 1 + x + x^2$$

is nowhere below the line $y = \frac{3}{4}$.

Also, $-b/2a = -\frac{1}{2}$;

$$Q(-\tfrac{1}{2}) = \tfrac{3}{4}$$
$$Q(-1) = 1$$
$$Q(0) = 1$$
$$Q(1) = 3.$$

(See Fig. 4–9 for the graph of Q.)

Figure
4–9

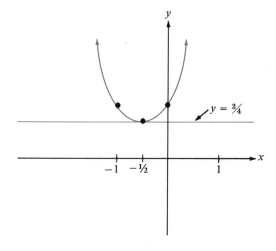

6. Analyze the following quadratics and then sketch their graphs. **Exercises**
Example: $Q(x) = 3x^2$. In comparing this example with the standard quadratic $Q(x) = ax^2 + bx + c$ we see that $a = 3$, $b = 0$, $c = 0$. Now the numbers $-b/2a$ and $(4ac - b^2)/4a$ may be computed:

$$-b/2a = 0 \quad \text{and} \quad (4ac - b^2)/4a = (0 - 0)/12 = 0.$$

Since the number $(4ac - b^2)/4a = 0$, the graph of Q is nowhere below the line $y = 0$ and since the number $-b/2a = 0$, the graph of Q has a low point at $(0, 0)$. Also, the graph of Q opens upward since $a > 0$. The graph is given below.

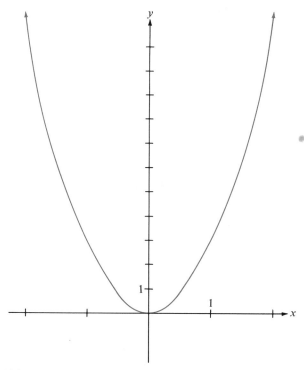

(a) $Q(x) = 2x^2 - 3$
(b) $Q(x) = -2x^2 + x$
(c) $Q(x) = -2x^2 - x + 1$
(d) $Q(x) = x^2 - 10x + 2$
(e) $Q(x) = x^2 - 5x + 7$
(f) $Q(x) = -x^2/4 + 4x - 17$
(g) $Q(x) = 10x^2$
(h) $Q(x) = -3x^2 + x - 5$
(i) $Q(x) = -x^2 + x/2 + 1$

7. *Example:* Find a number m so that the point $(1, 1)$ is on the graph of the quadratic $Q(x) = x^2 + mx - 1$. By definition, $(1, 1)$ is on the graph of $Q(x)$ if and only if $Q(1) = 1$. Since $Q(1) = 1^2 + m \cdot 1 - 1 = m$, $(1, 1)$ is on the graph of $Q(x)$ if and only if $m = 1$.

(a) Find a number n so that the point $(1, 0)$ is on the graph of $Q(x) = 2x^2 + nx$.

(b) Find a number p so that the point $(2, 1)$ is on the graph of $Q(x) = -5x^2 + px + 1$.

8. *Example:* Find a number m so that the graph of the quadratic $f(x) = x^2 - mx + 1$ has its low point on the x-axis. For the

quadratic $f(x) = ax^2 + bx + c$, $a > 0$, we know that the low point is $(-b/2a, f(-b/2a))$. In this case $a = 1$, $b = -m$ so that $-b/2a = m/2$. Hence, the low point is $(m/2, f(m/2))$ and this point is on the x-axis only if $f(m/2) = 0$. Therefore, we wish to find m so that $f(m/2) = 0$ and this is equivalent to finding m so that $m^2/4 - m^2/2 + 1 = 0$ since $f(m/2) = m^2/4 - m^2/2 + 1$. This requires that the equation $-m^2/4 + 1 = 0$ be solved for m. This equation has the solution set $\{-2, 2\}$ and we may conclude that the quadratics $f(x) = x^2 + 2x + 1$ and $f(x) = x^2 - 2x + 1$ each have a low point on the x-axis.

(a) Find a number m such that the graph of $Q(x) = x^2 - mx$ has its low point on the x-axis.

(b) Find a number n such that the graph of $Q(x) = -2x^2 + nx$ has its high point on the line $y = \frac{1}{2}$.

9. *Example:* Suppose that x, y are numbers such that $x + y = 8$. What numbers must x, y be so that xy is as large as possible? Since $x + y = 8$, we may write $y = 8 - x$. Then $xy = 8x - x^2$ so that xy is a function of x. The problem thus reduces to finding the high point on the graph of the quadratic $8x - x^2$. In this case $a = -1$, $b = 8$ and $-b/2a = 8/2 = 4$. This implies that the high point occurs at $x = 4$. Therefore, the numbers x, y must both be 4.

(a) Suppose that x, y are numbers such that $x - y = 8$. What numbers must x, y be so that xy is as small as possible? What is the smallest number xy can be?

(b) Suppose that x, y are numbers such that $2x + y = 4$. What numbers must x, y be so that xy is as large as possible? What is the largest number xy can be?

(c) What is the maximum area of a rectangle which has a perimeter of 40?

(d) What is the maximum area of a right triangle which has a perimeter of 10 and hypotenuse of length 5?

10. (a) Is it possible for the graph of

$$P(x) = 1 + 2kx + x^2$$

to be nowhere above the line $y = 101$ for some number k? If so, find such a k.

(b) Repeat (a) if

$$P(x) = 1 + 2kx - x^2.$$

(c) Is it possible for the graph of

$$P(x) = 4kx + 4x^2$$

to be nowhere above the line $y = 25$ for some number k? If so, find such a k.

(d) Repeat (c) if

$$P(x) = 4kx - 4x^2.$$

4.3

Zeros of Quadratic Functions

If Q is a quadratic given by

$$Q(x) = c + bx + ax^2, a \neq 0,$$

we have seen in Sec. 4–2 that

$$Q(x) = a(x + b/2a)^2 + \frac{4ac - b^2}{4a}.$$

A number $d \in \mathcal{D}(Q)$ is a zero of Q if and only if $Q(d) = 0$ or

$$a(d + b/2a)^2 + \frac{4ac - b^2}{4a} = 0.$$

From this we may conclude

(4-1) $$4a^2(d + b/2a)^2 = b^2 - 4ac.$$

For every real number d,

$$4a^2(d + b/2a)^2 \geq 0;$$

and hence, if $b^2 - 4ac < 0$, there can be no real number d such that **(4-1)** is true. If $b^2 - 4ac < 0$, the quadratic Q has no zeros.

If $b^2 - 4ac \geq 0$, **(4-1)** may be written

$$(d + b/2a)^2 = \frac{b^2 - 4ac}{4a^2}.$$

Therefore, since $\dfrac{b^2 - 4ac}{4a^2} \geq 0$,

$$d + b/2a = \pm\sqrt{\frac{b^2 - 4ac}{4a^2}} = \pm\frac{\sqrt{b^2 - 4ac}}{|2a|}$$

and

$$d = -b/2a \pm \frac{\sqrt{b^2 - 4ac}}{2a} = \frac{-b \pm \sqrt{b^2 - 4ac}}{2a}.$$

[Notice that $\pm(\sqrt{b^2 - 4ac})/|2a| = \pm(\sqrt{b^2 - 4ac})/2a$ by the definition of absolute value.]

In this case, if Q has any zeros at all, they must be one of the numbers

$$\frac{-b + \sqrt{b^2 - 4ac}}{2a}, \quad \text{or} \quad \frac{-b - \sqrt{b^2 - 4ac}}{2a}.$$

It is easy to show that these numbers actually are zeros of Q (if they are in the domain of Q); for example

$$Q\left(\frac{-b + \sqrt{b^2 - 4ac}}{2a}\right)$$

$$= a\left(\frac{-b + \sqrt{b^2 - 4ac}}{2a}\right)^2 + b\left(\frac{-b + \sqrt{b^2 - 4ac}}{2a}\right) + c$$

$$= a \cdot \frac{b^2 - 2b\sqrt{b^2 - 4ac} + (b^2 - 4ac)}{4a^2} + \frac{-b^2 + b\sqrt{b^2 - 4ac}}{2a} + c$$

$$= \frac{2b^2 - 2b\sqrt{b^2 - 4ac} - 4ac}{4a} + \frac{-b^2 + b\sqrt{b^2 - 4ac}}{2a} + c$$

$$= \frac{b^2 - b\sqrt{b^2 - 4ac} - 2ac}{2a} + \frac{-b^2 + b\sqrt{b^2 - 4ac}}{2a} + c$$

$$= -c + c = 0.$$

Thus,

$$Q\left(\frac{-b + \sqrt{b^2 - 4ac}}{2a}\right) = 0, \text{ and similarly, } Q\left(\frac{-b - \sqrt{b^2 - 4ac}}{2a}\right) = 0.$$

It may, of course, happen that the numbers

$$\frac{-b + \sqrt{b^2 - 4ac}}{2a} \quad \text{and} \quad \frac{-b - \sqrt{b^2 - 4ac}}{2a}$$

are not different; *i.e.*,

$$\frac{-b + \sqrt{b^2 - 4ac}}{2a} = \frac{-b - \sqrt{b^2 - 4ac}}{2a}.$$

This is the case if and only if $b^2 - 4ac = 0$ as the following shows:

$$-b + \sqrt{b^2 - 4ac} = -b - \sqrt{b^2 - 4ac},$$

and

$$2\sqrt{b^2 - 4ac} = 0$$

$$b^2 - 4ac = 0.$$

In this case, then, Q can have at most one zero.

These results may be summarized as follows for the quadratic function

$$Q(x) = c + bx + ax^2, a \neq 0:$$

1. *If the number* $b^2 - 4ac < 0$, *then Q has no real zeros.*
2. *If the number* $b^2 - 4ac = 0$, *then Q has no more than one real zero.*
3. *If the number* $b^2 - 4ac > 0$, *then Q has no more than two real zeros.*

Example 4–11

If $Q(x) = 1 + x + x^2$, then $a = 1, b = 1, c = 1$ and

$$b^2 - 4ac = 1 - 4 = -3 < 0$$

so that Q has no real zeros.

Example 4–12

If $Q(x) = 1 + 2x + x^2$, then $a = 1, b = 2, c = 1$ and

$$b^2 - 4ac = 4 - 4 = 0$$

so that Q can have no more than one zero and this must be the number

$$\frac{-b}{2a} = \frac{-2}{2} = -1.$$

Since -1 is in the domain of Q, -1 is a zero and the only zero of Q:
$Q(-1) = 1 + 2(-1) + (-1)^2 = 0.$

Example 4–13

If $Q(x) = 1 + 2x + x^2$ for $x \in [0, 3]$, then Q has no zeros; for the only possible zero is -1 and this is not in the domain of Q.

Example 4–14

If $P(x) = 7 - x - x^2$ for $x \in [-10, 10]$, then $a = -1, b = -1, c = 7$ and $b^2 - 4ac = (-1)^2 - 4(-1)(7) = 1 + 28 = 29$. Therefore, the two numbers

$$\frac{1 + \sqrt{29}}{-2} \quad \text{and} \quad \frac{1 - \sqrt{29}}{-2}$$

are zeros of $P(x)$ if they are in the domain of P. We have

$$5 < \sqrt{29} < 6$$
$$6 < 1 + \sqrt{29} < 7$$

and

$$\frac{7}{-2} < \frac{1 + \sqrt{29}}{-2} < \frac{6}{-2} = -3$$

or

$$-3\tfrac{1}{2} < \frac{1 + \sqrt{29}}{-2} < -3$$

so that $\dfrac{1 + \sqrt{29}}{-2}$ is in the domain of P and is thus a zero. Similarly,

$\dfrac{1 - \sqrt{29}}{-2}$ is a zero, so that P has two zeros.

The possibilities for the zeros of the quadratic

$$Q(x) = c + bx + ax^2, a \neq 0$$

may be seen clearly by considering the graph of Q. Consider the case in which $b^2 - 4ac < 0$ (this implies Q has no real zeros). Then $4ac - b^2 > 0$; and

      (a) if $a > 0$, $(4ac - b^2)/4a > 0$.
      (b) if $a < 0$, $(4ac - b^2)/4a < 0$.

For convenience we let $(4ac - b^2)/4a = L$. In case (a) we know by our previous considerations that the graph of Q can be nowhere below the line $y = L$. But since $L > 0$, we have a situation as indicated in Fig. 4–10.

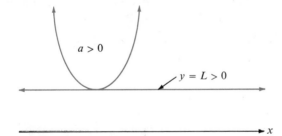

**Figure
4–10**

This clearly shows that the graph of Q cannot cross the x-axis and Q has no zeros. In case (b) $a < 0$ and $L < 0$ giving rise to a graphical situation as indicated in Fig. 4–11. Again, the graph does not cross the x-axis and Q cannot have zeros.

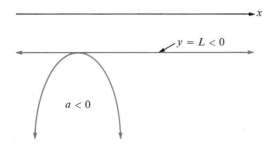

**Figure
4–11**

In case $L = 0$, the line $y = L$ is the x-axis; so the graph of Q may *touch* the x-axis once as indicated in Fig. 4–12, but cannot cross it twice.

Figure 4–12

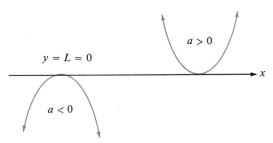

We leave as an exercise for the student the graphical consideration of the case $b^2 - 4ac > 0$.

Example 4–15

Find the zeros of the quadratic $Q(x) = x^2 - 9x + 14$.

Solution: This is a quadratic which can be easily factored:

$$Q(x) = x^2 - 9x + 14 = (x - 7)(x - 2).$$

Then $Q(x) = 0$ only if $x - 7 = 0$ or $x - 2 = 0$. This implies that $x = 7$ or that $x = 2$. To verify that 2, 7 are the zeros:

$$Q(2) = 2^2 - 9 \cdot 2 + 14 = 4 - 18 + 14 = 0,$$
$$Q(7) = 7^2 - 9 \cdot 7 + 14 = 49 - 63 + 14 = 0.$$

Example 4–16

Find the zeros of the quadratic $Q(x) = x^2 - 9x + 15$.

Solution: This quadratic is not as easily factored as the one of Example 4–15. It does have zeros, however, since $b^2 - 4ac = (-9)^2 - 4 \cdot 1 \cdot 15 = 81 - 60 = 21$. Therefore, using the formula for the zeros gives

$$\frac{9 + \sqrt{21}}{2} \quad \text{and} \quad \frac{9 - \sqrt{21}}{2}$$

as the zeros of this quadratic.

Exercises

11. Find the zeros of each of the following quadratics.
   (a) $Q(x) = x^2 - 4x + 4$
   (b) $Q(x) = x^2 - 4x + 4$, $\mathfrak{D}(Q) = [0, 5]$
   (c) $Q(x) = x^2 + 3x - 10$

(d) $Q(x) = x^2 + 3x - 10$, $\mathfrak{D}(Q) = \{x \mid x \geq 0\}$
(e) $Q(x) = x^2 - 12x + 11$
(f) $Q(x) = x^2 + 12x + 11$
(g) $Q(x) = 2x^2 - 5x - 12$
(h) $Q(x) = 3x^2 + 4x - 7$, $x \geq 0$
(i) $Q(x) = 4x^2 + 4x - 3$
(j) $Q(x) = 35x^2 + 45x + 10$

12. Find the zeros of each of the following quadratics.
   (a) $F(x) = x^2 - 2x - 1$
   (b) $F(x) = x^2 + x - 1$
   (c) $F(x) = 2x^2 - x - 2$
   (d) $F(x) = 2x^2 - x - 2$, $x \leq 0$
   (e) $F(x) = x^2 + 5x + 5$
   (f) $F(x) = -x^2 + 5x + 5$
   (g) $F(x) = -2x^2 + 4x + 1$
   (h) $F(x) = -2x^2 + 4x + 1$, $x \geq 1$
   (i) $F(x) = 9x^2 + x - 1$
   (j) $F(x) = -3x^2 + \sqrt{2}x + 1$

13. The number $b^2 - 4ac$ associated with the quadratic function

$$Q(x) = c + bx + ax^2, a \neq 0$$

is called the *discriminant* of the quadratic. Calculate the discriminant of each of the following quadratic functions and use this number to discuss the zeros of the functions.
   (a) $Q(x) = 1 + x^2 + x$
   (b) $Q(x) = \frac{1}{2} + 2x + \sqrt{2}x^2$
   (c) $Q(x) = -3x^2 + (\frac{1}{2})x + 1$
   (d) $Q(x) = (\frac{5}{2})x^2 + \sqrt{5}x + 2$

14. (a) Discuss the zeros of $Q(x) = 1 + x + x^2$ in each of the domains below:

   (i) $[-1, 1]$
   (ii) The set of rational numbers
   (iii) The set of irrational numbers
   (iv) $[10, 101]$
   (v) The set of real numbers.

   (b) Same question as (a) above except that

$$Q(x) = -3x^2 + (\frac{1}{2})x + 1.$$

   (c) Same question as (a) above except that

$$Q(x) = (\frac{5}{2})x^2 + \sqrt{5}x + 2.$$

(d) Same question as (a) above except that

$$Q(x) = \tfrac{1}{2} + 2x + \sqrt{2}x^2.$$

15. If the domain of the quadratic function

$$Q(x) = 3x^2 - 6x + k$$

is the set of all real numbers, what value of k will insure that Q has exactly one real zero?

16. Let r, s, t be rational numbers. Show that the zeros of the quadratic

$$R(x) = x^2 - 2rx + r^2 - s^2 + 2st - t^2$$

must be rational numbers.

17. If the quadratic function

$$F(x) = x^2 + 2(1 + t)x + t^2$$

has exactly one zero, what number is t?

4.4

More About the Zeros of a Quadratic Function

Let the number e be a zero of the quadratic polynomial

$$Q(x) = c + bx + ax^2$$

where $a \neq 0$. Then, by the definition of a zero,

$$Q(e) = c + be + ae^2 = 0.$$

Using the distributive law, we may write the last equation as

$$c + be + ae^2 = [c/a + (b/a)e + e^2]a = 0.$$

It is possible now to conclude that $c/a + (b/a)e + e^2 = 0$ since $a \neq 0$. This shows that the number e is also a zero of the quadratic

$$P(x) = c/a + (b/a)x + x^2$$

and that *the zeros of Q and P are the same*. It is possible, therefore, to study the zeros of Q by studying the zeros of P.

Suppose that $Q(x) = c + bx + ax^2$ has as zeros the real numbers z_1 and z_2. As we have seen already, this implies that the discriminant $d = b^2 - 4ac \geq 0$. We also know that

$$z_1 = \frac{-b + \sqrt{d}}{2a} \quad \text{and} \quad z_2 = \frac{-b - \sqrt{d}}{2a}.$$

Then,

$$z_1 + z_2 = (-2b)/(2a) = -(b/a)$$

and

$$z_1 \cdot z_2 = (b^2 - d)/4a^2 = [b^2 - (b^2 - 4ac)]/4a^2 = c/a.$$

This demonstrates that the sum of the zeros of Q *is the negative of the co-efficient of* x *in* P(x) *and that the product of the zeros of* Q *is the constant term of* P(x).

Without finding the zeros of the quadratic function $Q(x) = 1 + 2x - 5x^2$, determine the sum and product of its zeros.

Example 4–17

Solution. According to the above considerations, we may solve this problem by considering the quadratic

$$P(x) = \frac{1}{-5} + \left(\frac{2}{-5}\right)x + x^2.$$

The sum of the zeros of Q is $\frac{2}{5}$ and their product is $-\frac{1}{5}$.

Find a quadratic which has the numbers $1 + \sqrt{2}$ and $1 - \sqrt{2}$ as zeros.

Example 4–18

Solution. Since $(1 + \sqrt{2}) + (1 - \sqrt{2}) = 2$ and $(1 + \sqrt{2})(1 - \sqrt{2}) = 1 - 2 = -1$, we have that the quadratic

$$-1 - 2x + x^2$$

has $1 + \sqrt{2}$ and $1 - \sqrt{2}$ as zeros. There are, of course, other possibilities; for example, $-3 - 6x + 3x^2$.

The information just obtained concerning the zeros of a quadratic function may be used to factor the quadratic if it can be factored at all. If $Q(x) = c + bx + ax^2$, $a \neq 0$, has zeros z_1, z_2, then $z_1 + z_2 = -(b/a)$ and $z_1 \cdot z_2 = c/a$. Hence,

$$
\begin{aligned}
Q(x) &= ax^2 + bx + c \\
&= a[x^2 + (b/a)x + (c/a)] \\
&= a[x^2 - (z_1 + z_2)x + z_1 \cdot z_2] \\
&= a(x - z_1)(x - z_2).
\end{aligned}
$$

Example 4–19

Factor the quadratic $5x^2 - x - 1$.

Solution. First, we calculate the zeros:

$$z_1 = \frac{1 + \sqrt{1 - (4)(5)(-1)}}{2(5)} = \frac{1 + \sqrt{21}}{10}.$$

$$z_2 = \frac{1 - \sqrt{21}}{10}.$$

Then, the quadratic may be factored:

$$5x^2 - x - 1 = 5[x - (1 + \sqrt{21})/10][x - (1 - \sqrt{21})/10].$$

If the quadratic $Q(x) = c + bx + ax^2$ can be factored as

$$Q(x) = a(x - z_1)(x - z_2),$$

then $Q(z_1) = Q(z_2) = 0$ and consequently z_1, z_2 are zeros of Q. This shows that if Q can be factored, then Q has zeros. If the discriminant of Q is negative ($b^2 - 4ac < 0$), Q has no real zeros. Hence, *if the discriminant of Q is negative, Q cannot be factored as the product of two real linear factors.*

Example 4–20

The quadratic $x^2 + x + 1$ cannot be factored since the discriminant is $1^2 - 4 \cdot 1 \cdot 1 = -3 < 0$.

Example 4–21

The quadratic $3x^2 + 4x + \sqrt{6}$ cannot be factored since the discriminant is $16 - 4 \cdot 3 \cdot \sqrt{6} = 4(4 - 3\sqrt{6}) < 0$.

Example 4–22

Let the function F be defined by

$$F(x) = x/(x^2 - 5x + 9).$$

Determine the range of F.

Solution. Suppose that c is a number in the range of F, i.e., $c \in \mathcal{R}(F)$. Then, by definition of range, there is a number, say t, in the domain of F such that

$$F(t) = t/(t^2 - 5t + 9) = c.$$

Therefore,

$$t = c(t^2 - 5t + 9) = ct^2 - 5ct + 9c$$

and

$$ct^2 - (5c + 1)t + 9c = 0.$$

This equation shows that t is a zero of the quadratic

$$cx^2 - (5c + 1)x + 9c.$$

But the quadratic cannot have real zeros unless its discriminant is positive or zero. Hence,

$$(5c + 1)^2 - 4 \cdot c \cdot 9c \geq 0,$$

or

$$25c^2 + 10c + 1 - 36c^2 \geq 0,$$

or

$$1 + 10c - 11c^2 \geq 0.$$

Thus c is a number for which the quadratic $Q(x) \geq 0$, where

$$Q(x) = 1 + 10x - 11x^2 = (1 + 11x)(1 - x).$$

This quadratic may be shown to be non-negative between its zeros: $-\frac{1}{11}$, 1. We conclude that $\mathcal{R}(F) \subseteq [-\frac{1}{11}, 1]$. This argument may be reversed to show that $[-\frac{1}{11}, 1] \subseteq \mathcal{R}(F)$, and therefore that the range of F is the closed interval $[-\frac{1}{11}, 1]$.

Factor the polynomial function $x^4 - 2x^2 - 3$.

Example
4–23

Solution. If we write $x^4 - 2x^2 - 3 = (x^2)^2 - 2x^2 - 3$, the given polynomial may be considered a quadratic "in x^2." Consider the quadratic $y^2 - 2y - 3$. This has the zeros -1, 3 so that

$$y^2 - 2y - 3 = (y - 3)(y + 1).$$

Then,

$$x^4 - 2x^2 - 3 = (x^2 - 3)(x^2 + 1)$$
$$= (x + \sqrt{3})(x - \sqrt{3})(x^2 + 1).$$

No further factorization is possible since $x^2 + 1$ has no zeros.

18. Without computing the zeros, determine the sum and product of the zeros of the following quadratics:

(a) $x^2 - x + \frac{1}{4}$ (b) $4x^2 - 4x + 1$

(c) $x^2 + \sqrt{2}x + \frac{1}{4}$ (d) $4x^2 + 4\sqrt{2}x + 1$

(e) $x^2 + 3x + 1$ (f) $8x^2 + 24x + 8$

(g) $x^2 - 4x + \frac{1}{2}\sqrt{2}$ (h) $2x^2 - 8x + \sqrt{2}$

19. Find a quadratic which has the given numbers as zeros.

(a) $1, -2$ (b) $8, -1$

(c) $2 + \sqrt{3}, 2 - \sqrt{3}$ (d) $\frac{1}{8}, -\frac{1}{2}$

(e) $\frac{1}{2} + \sqrt{5}, -\frac{1}{2} + \sqrt{5}$

20. Factor the polynomial functions of Exercise 18.

21. Determine all the numbers k so that the quadratic

$$x^2 + 2(k + 2)x + 9k$$

can be factored.

22. Show that if a, b are any real numbers, the quadratic

$$x^2 - 2ax + a^2 - b^2$$

can be factored.

23. If a, b, c are any real numbers, show that the quadratic

$$(a - b + c)x^2 + 2(a - b)x + (a - b - c)$$

can be factored if $c \neq b - a$.

24. Let $a \neq 0$ be a real number. Determine all numbers m such that the quadratic

$$2a^2x^2 + 2amx + m^2 - 2$$

can be factored.

25. Determine the range of each of the following functions.

(a) $F(x) = (x^2 + 1)/x$

(b) $F(x) = x/(x^2 + 1)$

(c) $F(x) = (x^2 + 1)/x^2$

(d) $F(x) = 2/(x^2 + 1)$

(e) $F(x) = (2x^2 - 1)/x$

(f) $F(x) = 2x^2/(x + 1)$

(g) $F(x) = (x^2 - x + 1)/(x^2 + 2x + 1)$

26. Factor each of the following.

(a) $x^4 - 4x^2 + 4$ (b) $x^4 + 3x^2 - 10$
(c) $x^4 - 12x^2 + 11$ (d) $x^4 + 12x^2 + 11$
(e) $2x^4 - 5x^2 - 12$ (f) $3x^4 + 4x^2 - 7$
(g) $2x^4 - 10x^2 - 7$ (h) $3x^4 - 29x^2 + 18$
(i) $x^6 - 1$ (j) $x^6 + 7x^3 - 8$

If we are given a function f and a number c in the domain of f, the determination of the number $f(c)$ can sometimes be very difficult. In case f is a polynomial function, the calculation of $f(c)$ involves only addition and multiplication; but even so, if the degree of f is large, the calculation can be quite tedious. A technique — applicable to polynomial functions only — known as synthetic division can be used to advantage to aid in these computations. To illustrate the process let

$$P(x) = a_0 + a_1x + a_2x^2 + a_3x^3$$

and suppose c is a number in the domain of P. Then

$$\begin{aligned} P(c) &= a_0 + a_1c + a_2c^2 + a_3c^3 \\ &= a_0 + (a_1 + a_2c + a_3c^2)c \\ &= a_0 + (a_1 + [a_2 + a_3c]c)c. \end{aligned}$$

Notice from this that $P(c)$ may be obtained by the following sequence of steps:

Multiply a_3 by c to obtain a_3c; then add a_2 to this to obtain $a_2 + a_3c$. Next multiply $(a_2 + a_3c)$ by c to obtain $(a_2 + a_3c)c$; then add to this a_1 to obtain $a_1 + (a_2 + a_3c)c$. The final step is to multiply by c and add a_0.

It will be obvious to the reader that the pattern is simply one of multiplying by c and adding the coefficients successively. If we wish to arrive at $P(c)$ in this fashion, the problem is one of hardly more than bookkeeping proportions. Consider then the schematic process for keeping track of the appropriate numbers as given in Table (a). In Table (a), the first row simply consists of the coefficients of P written in the appropriate order. It should be evident that this process can be carried out on a polynomial of any degree.

For the polynomial function

Example
4–24

$$f(x) = 3x^7 - x^6 + 31x^4 + 22x + 5$$

we calculate $f(-2)$. By the process just described, we write the coefficients in the first row and proceed as before [see Table (b)]. Notice that zeros are written for missing terms of the polynomial. The -2 is placed at the far right simply to remind us what to multiply by. Thus,

$$f(-2) = 9.$$

Table (a)

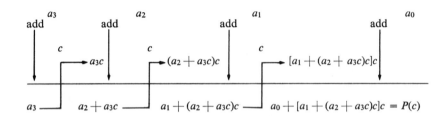

Table (b)

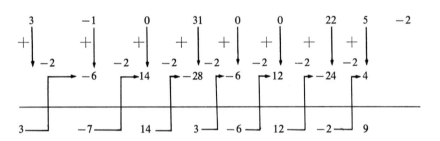

Example 4–25

Calculate $P(-5)$ if

$$P(x) = 3x^5 + 11x^4 + 90x^2 - 19x + 53.$$

Solution.

	3	11	0	90	-19	53	-5
		-15	20	-100	50	-155	
	3	-4	20	-10	31	-102	

Therefore, $P(-5) = -102$.

Example 4–26

If $P(x) = 5x^5 - x^3 + x + 2$, compute $P(3)$.

Solution.

5	0	−1	0	1	2	3
	15	45	132	396	1191	
5	15	44	132	397	1193	

Hence, $P(3) = 1193$.

It may have occurred to the student that synthetic division *seems* to be a more efficient way of calculating $P(c)$ as opposed to the obvious straightforward way. Is this actually the case or does it just seem so? And, if this process is really advantageous, to what is the advantage attributable? To answer this question, let us consider our starting point. We illustrate again with the polynomial

$$P(x) = a_0 + a_1 x + a_3 x^2 + a_3 x^3.$$

The original problem was to calculate $P(c)$, and to do this we wrote

$$P(c) = a_0 + a_1 c + a_2 c^2 + a_3 c^3$$
$$= a_0 + (a_1 c + [a_2 + a_3 c]c)c.$$

This is possible by making repeated use of the distributive law:

$$a(b + d) = a \cdot b + a \cdot d.$$

Notice that on the left side of this equation there are 2 operations (one multiplication and one addition) and on the right-hand side there are 3 operations (two multiplications and one addition). Therefore, in expressing $P(c)$ as

$$a_0 + (a_1 c + [a_2 + a_3 c]c)c$$

we have replaced expressions of the form $a \cdot b + a \cdot d$ by ones of the form $a(b + d)$ and saved one operation each time. There is, then, a real advantage in using the synthetic division process, and the advantage stems from the distributive law.

27. If $f(x) = x^3 - x^2 + x - 1$, compute $f(1)$, $f(2)$, $f(3)$, $f(-1)$ and $f(-2)$.

28. If $f(x) = 4x^2 + 5x + 7$, compute $f(1), f(-1), f(2), f(-2), f(-3), f(\frac{1}{2})$.

29. If $f(x) = 8x^4 - 3x^2 + x - 1$, compute $f(1)$, $f(-1)$, $f(\frac{1}{2})$, $f(-\frac{1}{2})$, $f(-2)$.

30. If $f(x) = 2x^5 - x^4 + 2x^3 - x + 2$, compute $f(1)$, $f(-1)$, $f(\frac{1}{2})$, $f(-2)$ and $f(2)$.

31. If $f(x) = 2x^3 + x^2 - 5x + 3$, compute $f(1)$, $f(2)$, $f(5)$, $f(-1)$, $f(-3), f(-6)$.

32. Compute $f(3)$ if
$$f(x) = 5x^5 - 6x^4 - 8x^3 + 7x^2 + 6x + 2.$$

33. Compute $f(a)$ if
$$f(x) = x^3 - (a + b + c)x^2 + (ab + ac + bc)x - abc.$$

34. If $P(x) = x^3 + mx^2 - 20x + 6$, determine m so that 3 is a zero of P. [*Hint:* Use synthetic division.]

35. If $C(x) = 2x^3 - x^2 + kx + n$, choose k and n so that $C(-2) = 0$ and $C(4) = 0$.

4.6

The Factor and Remainder Theorem

The reader is no doubt familiar with the usual process of dividing one polynomial function $P(x)$ by another, $D(x)$, to obtain a quotient and a remainder. This division is justified by the following theorem; the proof of this will not be given.

Theorem 13

The Division Algorithm. Let $P(x)$ be a polynomial of degree n and let $D(x)$ be a polynomial of degree m. Then, there are unique polynomials $R(x)$ and $Q(x)$ such that

$$P(x) = D(x) \cdot Q(x) + R(x), \text{ for all } x \in \mathfrak{D}(P)$$

and $R(x)$ is either the zero polynomial or has a degree less than m. (Q is called the *quotient* and R the *remainder*.)

It is convenient to say that $D(x)$ *divides* $P(x)$ or is a factor of $P(x)$ in case

$$P(x) = D(x) \cdot Q(x)$$

for some polynomial $Q(x)$.

Let us consider the above theorem applied to polynomials $P(x)$ and $D(x) = x - c$, for some $c \in \mathfrak{D}(P)$. Then, by the theorem, there is $Q(x)$, $R(x)$ such that

$$P(x) = (x - c) \cdot Q(x) + R(x).$$

And in addition $R(x)$ is either the zero polynomial or has degree less than the degree of $(x - c)$; i.e., $R(x)$ is either the zero polynomial or a constant polynomial: $R(x) = a$ for all $x \in \mathfrak{D}(R)$. Since the above is true for all $x \in \mathfrak{D}(P)$, it is in particular true for c:

$$
\begin{aligned}
P(c) &= (c - c) \cdot Q(c) + R(c) \\
&= 0 \cdot Q(c) + R(c) \\
&= R(c).
\end{aligned}
$$

Therefore, $R(x)$ is the constant $P(c)$: $R(x) = P(c)$ for all $x \in \mathfrak{D}(R)$. Hence, for any polynomial function $P(x)$ and any $c \in \mathfrak{D}(P)$ we may write

$$P(x) = (x - c) \cdot Q(x) + P(c) \qquad \textbf{(4-2)}$$

for some Q.

The number $c \in \mathfrak{D}(P)$ is a zero of $P(x)$ if and only if $x - c$ divides $P(x)$.

Theorem 14

Proof: If $c \in \mathfrak{D}(P)$ is a zero of $P(x)$, then $P(c) = 0$ and from **(4-2)** above

$$P(x) = (x - c) \cdot Q(x)$$

so that $(x - c)$ divides $P(x)$.

Now suppose that $(x - c)$ divides $P(x)$. Then by definition

$$P(x) = (x - c) \cdot L(x)$$

for some $L(x)$. Then we have an instance of the Division Algorithm

$$P(x) = (x - c) \cdot L(x) + 0$$

with quotient $L(x)$ and remainder 0. But since the quotient and remainder are unique, and **(4-2)** states that the remainder is always $P(c)$, we must have $P(c) = 0$ so that c is a zero of P.

The synthetic division process is extremely useful in writing a polynomial $P(x)$ in the form given in **(4-2)**:

$$P(x) = (x - c) \cdot Q(x) + P(c)$$

where $c \in \mathfrak{D}(P)$. We illustrate with

$$P(x) = a_0 + a_1 x + a_2 x^2 + a_3 x^3.$$

First, we compute $P(c)$ by synthetic division:

a_3	a_2	a_1	a_0	c
	a_3c	$(a_2 + a_3c)c$	$[a_1 + (a_2 + a_3c)c]c$	
a_3	$a_2 + a_3c$	$a_1 + (a_2 + a_3c)c$	$a_0 + [a_1 + (a_2 + a_3c)c]c = P(c)$	

Now take the first three numbers in the last row as the coefficients of a new polynomial:

$$a_3x^2 + (a_2 + a_3c)x + [a_1 + (a_2 + a_3c)c].$$

It turns out that this new polynomial is the $Q(x)$ that we desire in **(4-2)**:

$$
\begin{aligned}
(x &- c) \cdot \{a_3x^2 + (a_2 + a_3c)x + [a_1 + (a_2 + a_3c)c]\} \\
&= \{a_3x^2 + (a_2 + a_3c)x + [a_1 + (a_2 + a_3c)c]\} \cdot x \\
&\quad - \{a_3x^2 + (a_2 + a_3c)x + [a_1 + (a_2 + a_3c)c]\} \cdot c \\
&= a_3x^3 + (a_2 + a_3c)x^2 + [a_1 + (a_2 + a_3c)c]x \\
&\quad - a_3cx^2 - (a_2 + a_3c)cx - [a_1 + (a_2 + a_3c)c]c \\
&= a_3x^3 + a_2x^2 + a_1x - [a_1 + (a_2 + a_3c)c]c \\
&= a_3x^3 + a_2x^2 + a_1x + a_0 - a_0 - [a_1 + (a_2 + a_3c)c]c \\
&= P(x) - P(c).
\end{aligned}
$$

Therefore, we have

$$
\begin{aligned}
P(x) &= (x - c) \cdot \{a_3x^2 + (a_2 + a_3c)x + [a_1 + (a_2 + a_3c)c]\} + P(c) \\
&= (x - c) \cdot Q(x) + P(c).
\end{aligned}
$$

The general situation for polynomials of degree different from 3 presents no new difficulties.

Example 4–27

To divide $P(x) = 5x^5 - x^3 + x + 2$ by $x - 3$ we first compute $P(3)$ by synthetic division thus obtaining the remainder $P(3)$. And the numbers (except the last) in the third row will be the coefficients of the quotient:

5	0	-1	0	1	2	3
	15	45	132	396	1191	
5	15	44	132	397	$1193 = P(3)$	

$$5x^4 + 15x^3 + 44x^2 + 132x + 397 = Q(x)$$

Thus,

$$5x^5 - x^3 + x + 2$$
$$= (x - 3) \cdot (5x^4 + 15x^3 + 4x^2 + 132x + 397) + 1193.$$

Divide $3x^7 - x^6 + 31x^4 + 22x + 5$ by $x + 2$ and obtain the quotient and remainder.

Example
4–28

Solution. Notice that we are dividing by $x + 2 = x - (-2)$ so that we must use -2 in the synthetic division:

3	-1	0	31	0	0	22	5	-2
	-6	14	-28	-6	12	-24	4	
3	-7	14	3	-6	12	-2	$9 = P(-2)$	

The remainder is therefore $P(-2) = 9$ and the quotient is

$$3x^6 - 7x^5 + 14x^4 + 3x^3 - 6x^2 + 12x - 2.$$

Use synthetic division to find the quotient and remainder when $4x^3 + x + 1$ is divided by $3x - 1$.

Example
4–29

Solution. In order to solve this problem we first divide $4x^3 + x + 1$ by $x - 1/3$:

4	0	1	1	$\frac{1}{3}$
	$\frac{4}{3}$	$\frac{4}{9}$	$\frac{13}{27}$	
4	$\frac{4}{3}$	$\frac{13}{9}$	$\frac{40}{27}$	

Then

$$4x^3 + x + 1 = \left(x - \frac{1}{3}\right)\left(4x^2 + \frac{4}{3}x + \frac{13}{9}\right) + \frac{40}{27}$$

Next, we may factor a 3 from $4x^2 + \frac{4}{3}x + \frac{13}{9}$:

$$4x^2 + \frac{4}{3}x + \frac{13}{9} = 3\left(\frac{4}{3}x^2 + \frac{4}{9}x + \frac{13}{27}\right)$$

Then

$$\left(x - \frac{1}{3}\right)\left(4x^2 + \frac{4}{3}x + \frac{13}{9}\right) = \left(x - \frac{1}{3}\right)\left[3\left(\frac{4}{3}x^2 + \frac{4}{9}x + \frac{13}{27}\right)\right]$$

$$= (3x - 1)\left(\frac{4}{3}x^2 + \frac{4}{9}x + \frac{13}{27}\right)$$

so that

$$4x^2 + x + 1 = (3x - 1)\left(\frac{4}{3}x^2 + \frac{4}{9}x + \frac{13}{27}\right) + \frac{40}{27}.$$

Our quotient is therefore $\frac{4}{3}x^2 + \frac{4}{9}x + \frac{13}{27}$ and the remainder is $\frac{40}{27}$.

Example 4–30

Show that $x + 4$ is a factor of $P(x) = x^3 + 5x^2 + 5x + 4$.

Solution. By Theorem 14, $x + 4$ is a factor of $P(x)$ if and only if $P(-4) = 0$. Hence, we compute $P(-4)$ by synthetic division:

1	5	5	4	-4
	-4	-4	-4	
1	1	1	0	

Therefore, $P(-4) = 0$ and $x + 4$ is a factor of $x^3 + 5x^2 + 5x + 4$.

Exercises

36. Find the quotient and remainder if:
    (a) $f(x) = x^2 - x + 1$ is divided by $x + 1$.
    (b) $g(x) = 3x^3 - x^2 + 10x - 1$ is divided by $x - 1$.
    (c) $m(x) = x^5 - 4x^3 - 3x^2 - 8$ is divided by $x + 3$.
    (d) $n(x) = x^4 - 8x^2 - x + \frac{3}{2}$ is divided by $x + 1$
    (e) $r(x) = 5x^3 - 3x^2 + 7x - 1$ is divided by $x - 2$.
    (f) $g(x) - 3x^4 - 2x^3 + x$ is divided by $3x + 7$.

37. (a) Show that $x - 1$ is not a factor of $P(x) = 9x^3 + x^2 - 7x + 2$.
    (b) Show that $x - 1$ is not a factor of $P(x) = 4x^3 - 2x^2 + x - 1$.
    (c) Show that $x + 2$ is not a factor of $P(x) = -x^5 + 16x^4 + x - 1$.
    (d) Show that $x - 3$ is not a factor of $P(x) = 3x^5 - 2x^4 - x^3 - 3x^2 - x + 3$.
    (e) Show that $2x - 4$ is not a factor of $P(x) = 3x^5 - x^4 + 3x^3 + x^2 - x - 4$.

(f) Show that $x + 1$ is a factor of $P(x) = 3x^3 + 4x^2 + 5x + 4$.

(g) Show that $x - 7$ is a factor of $P(x) = x^4 - 8x^3 + 8x^2 - 8x + 7$.

(h) Show that $x + 3$ is a factor of $P(x) = -2x^4 - x^3 + 3x^2 - 37x - 3$.

38. Use synthetic division to show that
    (a) -1 is a zero of $f(x) = x^3 + 2x^2 + 2x + 1$.
    (b) 1 is a zero of $f(x) = 2x^4 - x^3 - 4x^2 + 3x$.
    (c) 4 is a zero of $f(x) = 2x^4 - 7x^3 - 7x^2 + 14x - 8$.
    (d) 2 is not a zero of $f(x) = 3x^5 - x^4 + 3x^3 + x^2 - x - 4$.
    (e) $1/3$ is not a zero of $f(x) = 3x^5 - x^4 + 3x^3 - x^2 - x - 4$.

39. Let b be a number, not zero. Show that if n is an odd integer $x + b$ is a factor of $x^n + b^n$ and is not a factor if n is even.

We have defined the function f/g of two given functions f, g with $g(x) \neq 0$ for all $x \in \mathfrak{D}(f) \cap \mathfrak{D}(g)$. If f and g are polynomial functions, the function f/g is called a *rational* function. It is possible to show that a *rational function can always be obtained as a sum of rational functions of the form*

$$P(x), \quad \frac{A}{(x - a)^k} \quad \text{and} \quad \frac{Ax + B}{(x^2 + cx + d)^t} \qquad \textbf{(4-3)}$$

where $P(x)$ is a polynomial, a, A, B, c, d are numbers, and k, t are positive integers. Writing a rational function in this fashion affects a simplification relative to certain problems which arise in calculus. We shall not attempt to prove this result here, but rather describe a technique for writing a rational function as a sum of functions of the form given in **(4-3)**.

It first may be observed that if in the rational function P/Q, P has a degree greater than or equal to the degree of Q, we may find polynomials F and R such that

$$P/Q = F + R/Q, \qquad \textbf{(4-4)}$$

and the degree of R is *less* than the degree of Q (if $R \neq 0$). This follows from the Division Algorithm. For the polynomials P, Q there are polynomials F and R such that

$$P(x) = Q(x) \cdot F(x) + R(x),$$

and $R(x)$ is either zero or has degree less than the degree of $Q(x)$. Then dividing by $Q(x)$ gives **(4-4)**.

Every rational function R/Q, where R has degree less than the degree of Q is a sum of terms of the form

$$\frac{A}{(x-a)^k} \quad \text{and} \quad \frac{Ax+B}{(x^2+cx+d)^i}.$$

These terms are usually referred to as *partial fractions*. The problem then is to determine the partial fractions

$$\frac{A}{(x-a)^k} \quad \text{and} \quad \frac{Ax+B}{(x^2+cx+d)^i}$$

that we need to express a given function R/Q. This is done by factoring the polynomial Q. It can be shown that every polynomial with real number coefficients can be factored as a constant multiplied by a product of terms of the form

$$(x-a)^k \quad \text{and} \quad (x^2+cx+d)^i$$

where $c^2 - 4d < 0$. Once this factorization has been determined, the terms

$$\frac{A}{(x-a)^k} \quad \text{and} \quad \frac{Bx+C}{(x^2+cx+d)^i}$$

needed to obtain R/Q as a sum may be determined. This determination is made as follows:

(a) If $(x-a)^k$ is a factor of Q and this is the highest power of $(x-a)$ that occurs as a factor of Q, then to obtain R/Q as a sum we need one each of the following summands:

$$\frac{A_1}{x-a}, \frac{A_2}{(x-a)^2}, \frac{A_3}{(x-a)^3}, \ldots, \frac{A_k}{(x-a)^k}$$

where $A_1, A_2, \ldots, A_k$ are numbers to be determined.

(b) If $(x^2+cx+d)^i$ is a factor of Q with $c^2 - 4d < 0$ and this is the highest power of (x^2+cx+d) that occurs as a factor of Q, then to obtain R/Q as a sum we need one each of the following summands:

$$\frac{B_1x+C_1}{x^2+cx+d}, \frac{B_2x+C_2}{(x^2+cx+d)^2}, \ldots, \frac{B_ix+C_i}{(x^2+cx+d)^i}$$

where $B_1, C_1, B_2, C_2, \ldots, B_i, C_i$ are numbers to be determined.

Example 4–31

Express $\dfrac{x^4}{x^3+2x^2+2x+1}$ as a sum of a polynomial and partial fractions.

Solution. By dividing x^4 by $x^3 + 2x^2 + 2x + 1$,

$$x^4 = (x-2)(x^3-2x^2+2x+1) + (2x^2+3x+2)$$

so that

$$\frac{x^4}{x^3 + 2x^2 + 2x + 1} = (x - 2) + \frac{2x^2 + 3x + 2}{x^3 + 2x^2 + 2x + 1}.$$

Next, $x^3 + 2x^2 + 2x + 1 = (x + 1) \cdot (x^2 + x + 1)$.

(Note that we cannot take the factorization any further since the discriminant of $x^2 + x + 1$ is negative.) Then

$$\frac{2x^2 + 3x + 2}{x^3 + 2x^2 + 2x + 1} = \frac{2x^2 + 3x + 2}{(x + 1) \cdot (x^2 + x + 1)}$$

$$= \frac{A}{x + 1} + \frac{Bx + C}{x^2 + x + 1}.$$

The task now is to determine the numbers A, B, C.

$$\frac{A}{x + 1} + \frac{Bx + C}{x^2 + x + 1} = \frac{A(x^2 + x + 1) + (Bx + C)(x + 1)}{(x + 1)(x^2 + x + 1)}$$

$$= \frac{(A + B)x^2 + (A + B + C)x + A + C}{(x + 1)(x^2 + x + 1)}$$

so that

$$\frac{2x^2 + 3x + 2}{(x + 1)(x^2 + x + 1)} = \frac{(A + B)x^2 + (A + B + C)x + A + C}{(x + 1)(x^2 + x + 1)}.$$

and it follows that

$$2x^2 + 3x + 2 = (A + B)x^2 + (A + B + C)x + A + C.$$

Since this is true for *all* x we get

$$2 = A + C \text{ if } x = 0 \qquad\qquad \textbf{(4-5)}$$
$$7 = 3A + 2B + 2C \text{ if } x = 1 \qquad \textbf{(4-6)}$$
$$1 = A \text{ if } x = -1. \qquad\qquad \textbf{(4-7)}$$

Therefore, from **(4-5)** and **(4-7)** $C = 1$ and from **(4-6)**

$$7 = 3 + 2B + 2$$
$$7 = 5 + 2B$$
$$2 = 2B$$

or

$$B = 1.$$

Thus, we conclude that $A = B = C = 1$.

As a check we add $\dfrac{1}{x+1}$ and $\dfrac{x+1}{x^2+x+1}$

$$\frac{1}{x+1} + \frac{x+1}{x^2+x+1} = \frac{(x^2+x+1)+(x+1)^2}{(x+1)(x^2+x+1)}$$

$$= \frac{2x^2+3x+2}{(x+1)(x^2+x+1)}.$$

Hence,

$$\frac{x^4}{x^3+2x^2+2x+1} = (x-2) + \frac{1}{x+1} + \frac{x+1}{x^2+x+1}.$$

Example 4–32

Express $\dfrac{x+1}{x^2(x-1)(x^2-x+1)}$ as a sum of partial fractions.

Solution. We need to find numbers A, B, C, D, E such that

$$\frac{x+1}{x^2(x-1)(x^2-x+1)} = \frac{A}{x} + \frac{B}{x^2} + \frac{C}{x-1} + \frac{Dx+E}{x^2-x+1}.$$

From

$$\frac{A}{x} + \frac{B}{x^2} + \frac{C}{x-1} + \frac{Dx+E}{x^2-x+1}$$

$$= \frac{Ax+B}{x^2} + \frac{C}{x-1} + \frac{Dx+E}{x^2-x+1}$$

$$= \frac{(A+C)x^2 + (B-A)x - B}{x^2(x-1)} + \frac{Dx+E}{x^2-x+1}$$

$$= \frac{(A+C+D)x^4 + (B-2A-C+E-D)x^3 + (2A-2B+C-E)x^2 + (2B-A)x - B}{x^2(x-1)(x^2-x+1)},$$

we conclude that

$$(A+C+D)x^4 + (B-2A-C+E-D)x^3$$
$$+ (2A-2B+C-E)x^2 + (2B-A)x - B = x+1.$$

Since this must be true for all x, we choose $x = 0$, $x = 1$, $x = -1$, $x = 2$, $x = -2$ and obtain the following equations:

For $x = 0$, $-B = 1$ or $B = -1$.

For $x = 1$, $C = 2$.

For $x = -1$, $6A + 3C + 2D - 6B - 2E = 0$.

For $x = 2$, $6A + 3B + 12C + 8D + 4E = 3$.

For $x = -2$, $42A - 21B + 28C + 24D - 12E = -1$.

Using the two values for B and C ($B = -1$, $C = 2$) in the last three equations gives:

$$6A + 2D - 2E + 12 = 0. \tag{4-8}$$
$$6A + 8D + 4E + 21 = 3. \tag{4-9}$$
$$42A + 24D - 12E + 77 = -1. \tag{4-10}$$

Subtracting Eq. **(4-8)** from **(4-9)**:

$$6D + 6E + 9 = 3 \quad \text{or} \quad D + E = -1.$$

Now, multiplying Eq. **(4-9)** by 7:

$$42A + 56D + 28E + 147 = 21,$$

and subtracting this from Eq. **(4-10)**:

$$-32D - 40E - 70 = -22 \quad \text{or} \quad -32D - 40E = 48.$$

Dividing this last equation by 8 gives:

$$-4D - 5E = 6.$$

Since we already have that $D + E = -1$, we may substitute $D = -1 - E$ in the last equation to obtain

$$-4(-1 - E) - 5E = 6$$
$$4 + 4E - 5E = 6$$
$$-E = 2 \quad \text{or} \quad E = -2.$$

Then

$$D = -1 - (-2) = -1 + 2 = 1.$$

We have thus obtained

$$B = -1, C = 2, D = 1, E = -2.$$

It only remains to determine A, and this can be done from Eq. **(4-8)** now that we have $D = 1$ and $E = -2$:

$$6A + 2(1) - 2(-2) + 12 = 0$$
$$6A + 18 = 0$$
$$A + 3 = 0$$
$$A = -3.$$

It is easy to verify that these values for A, B, C, D, E work. Hence,

$$\frac{-3}{x} + \frac{-1}{x^2} + \frac{2}{x - 1} + \frac{x - 2}{x^2 - x + 1} = \frac{x + 1}{x^2(x - 1)(x^2 - x + 1)}.$$

Partial Fractions

Example
4–33

Express $\dfrac{x + 1}{2x^2((\frac{1}{2})x - \frac{1}{2})(3x^2 - 3x + 3)}$ as a sum of partial fractions.

Solution.

$$\frac{x + 1}{2x^2((\frac{1}{2})x - \frac{1}{2})(3x^2 - 3x + 3)} = \frac{x + 1}{x^2(x - 1)(3x^2 - 3x + 3)}$$

$$= \frac{x + 1}{3x^2(x - 1)(x^2 - x + 1)}$$

$$= \frac{1}{3} \cdot \frac{x + 1}{x^2(x - 1)(x^2 - x + 1)}.$$

Thus, the problem is reduced to Example 4–32 above:

$$\frac{1}{3} \cdot \frac{x + 1}{x^2(x - 1)(x^2 - x + 1)} = \frac{1}{3}\left[\frac{-3}{x} + \frac{-1}{x^2} + \frac{2}{x - 1} + \frac{x - 2}{x^2 - x + 1} \right]$$

$$= \frac{-1}{x} + \frac{-1}{3x^2} + \frac{2}{3x - 3} + \frac{x - 2}{3x^2 - 3x + 3}.$$

Example
4–34

Express $\dfrac{x + 1}{x^2(x - 1)^3(x^2 - x + 1)^2}$ as a sum of partial fractions.

Solution. Numbers $A_1, A_2, A_3, A_4, A_5, A_6, A_7, A_8, A_9$ must be determined so that

$$\frac{A_1}{x} + \frac{A_2}{x^2} + \frac{A_3}{x - 1} + \frac{A_4}{(x - 1)^2} + \frac{A_5}{(x - 1)^3} + \frac{A_6x + A_7}{x^2 - x + 1}$$

$$+ \frac{A_8x + A_9}{(x^2 - x + 1)^2} = \frac{x + 1}{x^2(x - 1)^3(x^2 - x + 1)^2}.$$

The procedure is similar to that in Example 4–32 except that the calculations are much more involved and tedious. We leave the details to the reader.

Express each of the following rational functions as the sum of a polynomial and/or partial fractions.

40. $\dfrac{1}{x(x + 1)}$

41. $\dfrac{x}{(x + 1)(x - 1)}$

42. $\dfrac{4x + 1}{2x^2 + 4x - 6}$

43. $\dfrac{x^2}{(x^2 - 1)}$

44. $\dfrac{2x + 1}{x^3 + 1}$

45. $\dfrac{x^3 + x^2 + x - 1}{(2x + 1)(x^2 + 1)^2}$

46. $\dfrac{x}{(x + 1)^2}$

47. $\dfrac{x}{(x + 1)^3}$

48. $\dfrac{x^2}{(x + 1)^3}$

49. $\dfrac{2x^2 + 1}{x^2(x + 1)^2}$

It is convenient at this point to consider inequalities again in the light of our considerations concerning polynomial functions. In Chapter 2 inequalities of the type

$$x^2 - x - 6 \geq 0$$

were studied. This is obviously an instance of an inequality of the kind

$$P(x) \geq 0$$

where $P(x)$ is a quadratic function. Here the domains of our polynomial functions will always be the set of all real numbers. The problem is to describe all those numbers x for which it is true that

$$P(x) \geq 0.$$

We were able to solve the inequality

$$x^2 - x - 6 \geq 0,$$

by factoring

$$(x - 3)(x + 2) \geq 0$$

and then reasoning that this product is non-negative only if

(a) $x - 3 \geq 0$ and $x + 2 \geq 0$ **(4-11)**

or

(b) $x - 3 \leq 0$ and $x + 2 \leq 0.$ **(4-12)**

From **(4-11)**

$$x \geq 3 \quad \text{and} \quad x \geq -2$$

and from **(4-12)**

$$x \leq 3 \quad \text{and} \quad x \leq -2.$$

4.8

Inequalities Involving Quadratics

Condition **(4-11)** is met if $x \geq 3$ and condition **(4-12)** is met if $x \leq -2$. These two conditions are pictured in Fig. 4–13.

Figure
4–13

$\{x \mid x \leq -2\} \longleftarrow \qquad \longrightarrow \{x \mid x \geq 3\}$

$-2 -1 \ \ 0 \ \ 1 \ \ 2 \ \ 3$

Consider now the graph of

$$Q(x) = x^2 - x - 6.$$

This is given in Fig. 4–14. It is quite clear from Fig. 4–14 why we reached the solution indicated in Fig. 4–13. In fact, we can also immediately see that the inequality

$$x^2 - x - 6 \leq 0$$

has the solution set $[-2, 3]$.

Figure
4–14

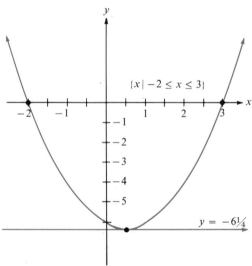

The solution sets for both inequalities

$$Q(x) \geq 0 \qquad \text{and} \qquad Q(x) \leq 0$$

are described in terms of the two numbers $-2, 3$; and these are the zeros of the function $Q(x)$.

Because of the knowledge we now have of the graphs of quadratic functions, inequalities of the form $Q(x) \geq 0$ and $Q(x) \leq 0$ can be solved simply by investigating the discriminant and zeros. In the case above,

$$Q(x) = x^2 - x - 6$$

has a discriminant of $(-1)^2 - 4 \cdot 1(-6) = 25$ and hence two zeros: -2 and 3. Also, the graph of Q has its low point (the coefficient of x^2 is positive) on the line $y = -\frac{25}{4} = -6\frac{1}{4}$ and rises above this line to cross the x-axis at $(-2, 0)$ and $(3, 0)$. Thus, for $x \in [-2, 3]$ the graph of Q lies between the lines $y = -6\frac{1}{4}$ and $y = 0$; i.e.,

$$-6\frac{1}{4} \leq Q(x) \leq 0 \text{ for all } x \in [-2, 3].$$

(See Fig. 4–14.)

Theorem
15

Let $Q(x) = c + bx + ax^2$, $a \neq 0$, and let $d = b^2 - 4ac$ be its discriminant.

(a) If $a > 0$ and $d < 0$, then the solution set for the inequality $Q(x) \leq 0$ is $\varnothing$ and the solution set for the inequality $Q(x) > 0$ is $\mathbf{R}$ (all real numbers).

(b) If $a > 0$ and $d > 0$, then the solution set for the inequality $Q(x) \leq 0$ is $[z_1, z_2]$, where $z_1 < z_2$ are the zeros of Q, and the solution set for the inequality $Q(x) \geq 0$ is $\{x \mid x \leq z_1\} \cup \{x \mid x \geq z_2\}$.

(c) If $a < 0$ and $d < 0$, then the solution set for $Q(x) < 0$ is $\mathbf{R}$ and the solution set for $Q(x) \geq 0$ is $\varnothing$.

(d) If $a < 0$ and $d > 0$, then the solution set for $Q(x) \leq 0$ is $\{x \mid x \leq z_1\} \cup \{x \mid x \geq z_2\}$ where $z_1 < z_2$ and these are the zeros of Q. The solution set for $Q(x) \geq 0$ is $[z_1, z_2]$.

Proof: We prove statement (d) and leave the others for the student. With $d > 0$, we know that Q has two distinct zeros, say, $z_1 < z_2$. Since $a < 0$, the graph of Q falls from its high point and crosses the x-axis at $(z_1, 0)$ and $(z_2, 0)$. (See Fig. 4–15.) The solution sets of the two inequalities are then clearly the ones given.

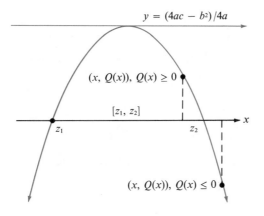

$$y = (4ac - b^2)/4a$$

$(x, Q(x))$, $Q(x) \geq 0$

$[z_1, z_2]$

z_1 z_2 x

$(x, Q(x))$, $Q(x) \leq 0$

Figure
4–15

Inequalities Involving Quadratics

**Example
4–35**

Find the solution set for $x^2 - x + 1 > 0$.

Solution. $d = (-1)^2 - 4(1)(1) = 1 - 4 = -3 < 0$ and $a = 1 > 0$.
Hence, $x^2 - x + 1 > 0$ for all real numbers.

**Example
4–36**

Find the solution set for $-x^2 - x + 1 \geq 0$.

Solution. $d = (-1)^2 - 4(-1)(1) = 1 + 4 = 5$ and $a = -1 < 0$. The
zeros of the function are

$$\frac{1 + \sqrt{5}}{-2} = -\frac{1}{2} - \frac{1}{2}\sqrt{5}$$

and

$$\frac{1 - \sqrt{5}}{-2} = -\frac{1}{2} + \frac{1}{2}\sqrt{5}.$$

The solution set is then

$$\left[-\frac{1}{2} - \frac{1}{2}\sqrt{5}, -\frac{1}{2} + \frac{1}{2}\sqrt{5} \right].$$

**Example
4–37**

Obtain the solution set of $x < x^2$.

Solution. This is equivalent to

$$x^2 - x > 0,$$

and the quadratic $x^2 - x$ has a discriminant $d = (-1)^2 - 4(1)\cdot 0 = 1 > 0$
and $a = 1$. The zeros are 0 and 1 so that the solution set is

$$\{x \mid x > 1\} \cup \{x \mid x < 0\}.$$

**Example
4–38**

Find the solution set for $x^2 \leq 1$.

Solution. This is equivalent to $x^2 - 1 \leq 0$; the quadratic $x^2 - 1$ has the
two zeros $1, -1$ so $[-1, 1]$ is the solution set.

Exercises

50. Answer the questions below for each of the quadratics, $Q(x)$, given in
parts (a)–(k).

(i) What is the discriminant of Q?
(ii) How many zeros does Q have? If Q has zeros, what are they?
(iii) What is the solution set for $Q(x) \geq 0$?
(iv) What is the solution set for $Q(x) \leq 0$?

(a) $Q(x) = -x^2 + 3x$ (b) $Q(x) = 3x^2 + 1$
(c) $Q(x) = 2x^2 - 5$ (d) $Q(x) = -2x^2 + x$
(e) $Q(x) = 2x^2 - x - 1$ (f) $Q(x) = x^2 - 10x + 2$
(g) $Q(x) = x^2 - 5x + 7$ (h) $Q(x) = -\frac{1}{4}x^2 + 4x - 17$
(i) $Q(x) = -3x^2 - x + 2$ (j) $Q(x) = \frac{3}{8}x^2 - 7x + 2$
(k) $Q(x) = (4\sqrt{2})^{-1}x^2 + \sqrt{3}x + \sqrt{2}$

51. Obtain the solution set of each of the following inequalities:
(a) $2x^2 - 8x + 1 \geq 0$
(b) $(\frac{2}{3})x^2 + x - 1 \leq 0$
(c) $x^2 + \sqrt{2}x + \sqrt{3} < 0$
(d) $-x^2 + \sqrt{2}x + \sqrt{3} < 0$
(e) $x^2 - x \leq 0$
(f) $x + 2 > 5x^2$
(g) $(1 + \sqrt{2})x^2 \geq (10 + 11\sqrt{2})x + 11$

52. Prove the (a), (b), (c) parts of Theorem 15.

We consider now a type of inequality which involves rational functions. Suppose, for example, that we wish to find the solution set of

$$\frac{3x + 1}{(x - 1)(x + 7)} > 0 \quad \text{or} \quad \frac{3x + 1}{(2 - 1)(x + 7)} < 0.$$

We observe the following:

(i) $3x + 1 = 3(x + \frac{1}{3}) > 0$ for $x > -\frac{1}{3}$
      $3x + 1 < 0$ for $x < -\frac{1}{3}$.
(ii) $x - 1 > 0$ for $x > 1$
      $x - 1 < 0$ for $x < 1$.
(iii) $x + 7 > 0$ for $x > -7$
       $x + 7 < 0$ for $x < -7$.

Taking now the numbers $-\frac{1}{3}$, 1, and -7 in order, we make the following table:

Inequalities Involving Rational Functions

If ... ,	then ...
$x < -7$	$x + 7 < 0 \quad x - 1 < 0 \quad 3x + 1 < 0 \quad \dfrac{3x + 1}{(x - 1)(x + 7)} < 0$
$-7 < x < -\frac{1}{3}$	$x + 7 > 0 \quad x - 1 < 0 \quad 3x + 1 < 0 \quad \dfrac{3x + 1}{(x - 1)(x + 7)} > 0$
$-\frac{1}{3} < x < 1$	$x + 7 > 0 \quad x - 1 < 0 \quad 3x + 1 > 0 \quad \dfrac{3x + 1}{(x - 1)(x + 7)} < 0$
$1 < x$	$x + 7 > 0 \quad x - 1 > 0 \quad 3x + 1 > 0 \quad \dfrac{3x + 1}{(x - 1)(x + 7)} > 0$

The information in the above table is pictured graphically in Fig. 4–16, where $P(x)$ is equal to $\dfrac{3x + 1}{(x - 1)(x + 7)}$.

Figure
4–16

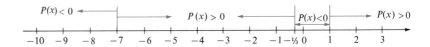

The information obtained for the rational function

$$\frac{3x + 1}{(x - 1)(x + 7)}$$

made use of the fact that it is a quotient of linear factors, and that linear functions are positive on one side of their zero and negative on the other side.

Example 4–39

Obtain the solution set of

$$\frac{x^2 + x + 6}{x^2 - 4} > 0.$$

Solution. $x^2 + x + 6 > 0$ for all x since $a = 1 > 0$ and $d = 1^2 - 4(1)(6) = -23 < 0$. Therefore,

$$\frac{x^2 + x + 6}{x^2 - 4} > 0 \text{ if and only if } x^2 - 4 > 0.$$

But the solution set for $x^2 - 4 > 0$ is $\{x \mid x < -2\} \cup \{x \mid x > 2\}$. Hence, the solution set for

$$\frac{x^2 + x + 6}{x^2 - 4} > 0 \text{ is } \{x \mid x < -2\} \cup \{x \mid x > 2\}.$$

Find the solution set for

Example
4–40

$$\frac{P(x)}{Q(x)} = \frac{(7x - 1)((\frac{2}{3})x - 1)}{(5x + 2)(x - 8)} < 0.$$

Solution. Since the zeros of the linear factors involved are $\frac{1}{7}$, $\frac{3}{2}$, $-\frac{2}{5}$, and 8 we have

If . . . ,   then . . .

$x < -\frac{2}{5}$	$7x - 1 < 0$	$(\frac{2}{3})x - 1 < 0$	$5x + 2 < 0$	$x - 8 < 0$	$P(x)/Q(x) > 0$
$-\frac{2}{5} < x < \frac{1}{7}$	$7x - 1 < 0$	$(\frac{2}{3})x - 1 < 0$	$5x + 2 > 0$	$x - 8 < 0$	$P(x)/Q(x) < 0$
$\frac{1}{7} < x < \frac{3}{2}$	$7x - 1 > 0$	$(\frac{2}{3})x - 1 < 0$	$5x + 2 > 0$	$x - 8 < 0$	$P(x)/Q(x) > 0$
$\frac{3}{2} < x < 8$	$7x - 1 > 0$	$(\frac{2}{3})x - 1 > 0$	$5x + 2 > 0$	$x - 8 < 0$	$P(x)/Q(x) < 0$
$8 < x$	$7x - 1 > 0$	$(\frac{2}{3})x - 1 > 0$	$5x + 2 > 0$	$x - 8 > 0$	$P(x)/Q(x) > 0$

We conclude that the solution set of

$$P(x)/Q(x) < 0 \quad \text{is} \quad (-\tfrac{2}{5}, \tfrac{1}{7}) \cup (\tfrac{3}{2}, 8).$$

Show that

Example
4–41

$$\frac{1}{x - 1} + \frac{1}{x + 1} < 0$$

for $x \in (0, 1)$.

Solution. Since

$$\frac{1}{x - 1} + \frac{1}{x + 1} = \frac{2x}{(x - 1)(x + 1)} = \frac{P(x)}{Q(x)},$$

we have

$$\frac{P(x)}{Q(x)} < 0 \quad \text{for} \quad x < -1,$$

$$\frac{P(x)}{Q(x)} > 0 \quad \text{for} \quad -1 < x < 0,$$

$$\frac{P(x)}{Q(x)} < 0 \quad \text{for} \quad 0 < x < 1,$$

$$\frac{P(x)}{Q(x)} > 0 \quad \text{for} \quad 1 < x.$$

53. Find the solution sets for the following:

(a) $\dfrac{x+1}{x-4} < 0$

(b) $\dfrac{x+1}{(x-4)(x+5)} > 0$

(c) $\dfrac{2}{3x+1} < \dfrac{1}{5x-8}$

(d) $\dfrac{1}{x-2} < \dfrac{1}{x^2-x-6}$

(e) $\dfrac{2}{-2x^2+5} \geq \dfrac{1}{-x^2+\frac{1}{2}}$

(f) $\dfrac{-x^2+2x+3}{x^2-x-6} < 0$

54. For each of the functions $g(x)$ given below find the solution set for $g(x) > 0$, $g(x) < 0$, $g(x) \geq 0$ and $g(x) \leq 0$.

(a) $g(x) = \dfrac{x-1}{x+2}$

(b) $g(x) = \dfrac{(x-1)(x-3)}{(x+2)}$

(c) $g(x) = \dfrac{(2x-1)}{(x+3)(3x+1)}$

(d) $g(x) = \dfrac{x^2+x+1}{(2x-7)(3x+2)}$

4.10

Zeros of Polynomial Functions

This section contains some results about polynomial functions and their zeros. Here we consider only polynomial functions whose domains are the set of all real numbers. Given a polynomial function P we wish to know if there is a real number c such that $P(c) = 0$ and if there is such a number, how many are there and how may these be determined?

Some functions have infinitely many zeros (*e.g.*, the greatest integer function), but polynomial functions have only a limited number and this number depends upon the degree of the polynomial.

Theorem 16

A polynomial function of degree n has at most n zeros.

The proof of the above theorem is too difficult for this text; the interested reader will find a discussion of this and related results in *What is Mathe-*

matics by Courant and Robbins, Oxford University Press, N.Y., 1951, pp. 101–103, pp. 269–271.

For examples relating to the above theorem consider the polynomials $P(x) = x^2 + 20$ and $Q(x) = x^3 - 6x^2 + 11x - 6$. If c is a zero of P, then $P(c) = c^2 + 20 = 0$ and $c^2 = -20$. But there are no real numbers c such that $c^2 < 0$. Hence, P has no real zeros. On the other hand,

$$Q(x) = x^3 - 6x^2 + 11x - 6 = (x - 1)(x - 2)(x - 3)$$

so that Q has the zeros 1, 2, 3; the number of zeros is thus exactly the degree of Q.

Let P be a polynomial function and let a, b be numbers such that $P(a) < P(b)$. If k is a number such that $P(a) < k < P(b)$, then there is a number c which lies between a and b such that $P(c) = k$.

Theorem
17

This theorem is illustrated in Fig. 4–17; its proof may be found in most standard calculus texts. We wish to use this result to isolate the zeros of polynomials. For a polynomial P it may be possible to find numbers a, b such that $P(a)$ is negative and $P(b)$ is positive. Then $P(a) < 0 < P(b)$ and the theorem implies that there is a number c which lies between a and b such that $P(c) = 0$.

Show that the polynomial $P(x) = x^4 + 3x^3 - x^2 + 12x - 5$ has a zero between 0 and 1.

Example
4–42

Solution. We compute $P(0)$ and $P(1)$. It is easy to see that $P(0) = -5$; $P(1)$ may be computed by synthetic division:

$$
\begin{array}{rrrrr|r}
1 & 3 & -1 & 12 & -5 & \quad 1 \\
 & 1 & 4 & 3 & 15 & \\
\hline
1 & 4 & 3 & 15 & 10 &
\end{array}
$$

Therefore, $P(1) = 10$ and it may be concluded by Theorem 16 that P has a zero between 0 and 1.

Theorem 17 supplies us with no constructive method for determining the zeros of a polynomial; it tells us only that it is possible to "trap" the zeros between certain other numbers. The following theorem gives information about upper bounds for the zeros of a polynomial. Recall that by Theorem 13 (The Division Algorithm) if P is a polynomial and c is a number we may write

$$P(x) = (x - c)Q(x) + P(c).$$

Theorem 18

Let P be a polynomial and let c be a positive real number. If $P(x) = (x - c)Q(x) + P(c)$, where $P(c) > 0$ and the coefficients of Q are non-negative, then P has no zero as great as c.

Proof. Suppose d is a number such that $d > c$. Then $d - c > 0$ and $Q(d) \geq 0$. Therefore,

$$(d - c)Q(d) + P(c) = P(d) > 0$$

and this implies that any number greater than c is not a zero of P.

In order to apply Theorem 18 we only need to look at the numbers in the last row of the synthetic division scheme. For example, consider $P(x) = 3x^4 + 14x^3 + 14x^2 - 20x - 8$. The following synthetic division shows (since there are no negative numbers in the third row) that P has no zero as large as 1.

$$
\begin{array}{rrrrr|r}
3 & 14 & 14 & -20 & -8 & 1 \\
 & 3 & 17 & 31 & 11 & \\
\hline
3 & 17 & 31 & 11 & 3 &
\end{array}
$$

We say in this case that 1 is an *upper bound* for the zeros of P.

To determine a lower bound for the zeros of a polynomial it is convenient to use the polynomial $R(x) = P(-x)$.

Theorem 19

If the polynomial R has no zero greater than the positive number c, then P has no zero less than $-c$.

Proof. Suppose that d is a positive number such that $d > c$. Then $R(d) \neq 0$ by hypothesis. But $R(d) = P(-d)$ so that $P(-d) \neq 0$. Since $-d < -c$ this implies that P has no zero less than $-c$.

Find an upper and a lower bound for the zeros of the polynomial
$P(x) = x^5 + 3x^4 - 2x^3 - 3x^2 + 4x + 1$.

Example
4–43

Solution. We first use Theorem 17 to find an upper bound for the zeros of
P. We start by trying 1:

$$
\begin{array}{rrrrrr|r}
1 & 3 & -2 & -3 & 4 & 1 & 1 \\
  & 1 & 4 & 2 & -1 & 3 & \\
\hline
1 & 4 & 2 & -1 & 3 & 4 &
\end{array}
$$

From this we cannot conclude that 1 is an upper bound for the zeros of P
since one of the numbers in row 3 is negative. Now we try 2:

$$
\begin{array}{rrrrrr|r}
1 & 3 & -2 & -3 & 4 & 1 & 2 \\
  & 2 & 10 & 16 & 26 & 60 & \\
\hline
1 & 5 & 8 & 13 & 30 & 61 &
\end{array}
$$

Since all numbers in the third row are positive, we may conclude that P
has no zero as great as 2. To obtain a lower bound we construct
$R(x) = P(-x)$. In this case

$$P(-x) = (-x)^5 + 3(-x)^4 - 2(-x)^3 - 3(-x)^2 + 4(-x) + 1$$
$$= -x^5 + 3x^4 + 2x^3 - 3x^2 - 4x + 1.$$

Here the leading coefficient is -1 and we can see that it is not possible to
obtain all non-negative numbers in the third row of the synthetic division
scheme. But $P(-x)$ has the same zeros as $-P(-x)$. Hence, we consider

$$-P(-x) = x^5 - 3x^4 - 2x^3 + 3x^2 + 4x - 1.$$

Using Theorem 18 we look for an upper bound for the zeros of $-P(-x)$.
Again we start with 1:

$$
\begin{array}{rrrrrr|r}
1 & -3 & -2 & 3 & 4 & -1 & 1 \\
  & 1 & -2 & -4 & -1 & 3 & \\
\hline
1 & -2 & -4 & -1 & 3 & 2 &
\end{array}
$$

Since some of the numbers are negative in the third row, no conclusion can be drawn. Now we continue with 2, 3, 4, . . . until a third row is obtained with no negative entries.

$$
\begin{array}{rrrrrr|r}
1 & -3 & -2 & 3 & 4 & -1 & 2 \\
  & 2 & -2 & -8 & -10 & -12 & \\
\hline
1 & -1 & -4 & -5 & -6 & -13 & \\
1 & -3 & -2 & 3 & 4 & -1 & 3 \\
  & 3 & 0 & -6 & -9 & -15 & \\
\hline
1 & 0 & -2 & -3 & -5 & -16 & \\
1 & -3 & -2 & 3 & 4 & -1 & 4 \\
  & 4 & 4 & 8 & 44 & 192 & \\
\hline
1 & 1 & 2 & 11 & 48 & 191 & \\
\end{array}
$$

Therefore, 4 is an upper bound for the zeros of $-P(-x)$ and, by Theorem 18, -4 is a lower bound for the zeros of P. We then conclude that all the zeros of P lie in the interval $(-4, 2)$. To be more precise we should say that the zeros of P lie in the interval $(-4, 2)$ *if* P *has any zeros*. Nothing we have done so far shows that P actually has real zeros in the interval $(-4, 2)$. However, a synthetic division computation shows that $P(-4) = -191$ and $P(2) = 61$ and this implies by Theorem 16 that P does have real zeros in the interval $(-4, 2)$.

Exercises

55. In each of the following you are to show that the given polynomial has a zero in the given intervals.
(a) $P(x) = x^3 - x - 3$; $(-1, 3)$
(b) $P(x) = -4x^3 + x^2 - 5x - 4$; $(-5, 0)$
(c) $P(x) = -2x^4 - x^3 + 5x^2 + x - 2$; $(0, 1)$
(d) $P(x) = \frac{1}{2}x^4 - 4x^3 + 5x^2 - 4x + 11$; $(-4, 2)$
(e) $P(x) = x^5 - x^4 + x^3 - x^2 + x + 1$; $(-1, 1)$
(f) $P(x) = -7x^5 - x^4 + 11x^3 + 3x^2 - x - 1$; $(0, 1)$, $(1, 2)$
(g) $P(x) = -5x^5 - 51x^4 - 9x^3 + 11x^2 - 101$; $(-11, -10)$
(h) $P(x) = x^6 + 20x^5 - 22x^4 - 20x^3 + 23x^2 + 43x + 20$; $(-21, 0)$
(i) $P(x) = (1/3)x^7 + (1/3)x^6 - 3x^5 - 3x^4 + (1/9)x^3 - (1/3)x^2 + x - 4$; $(3, 6)$

56. For each polynomial in 54 above find upper and lower bounds for their zeros.

57. Answer the following questions about the polynomial $P(x) = 6x^2 - 7x + 2$.
   (a) Show that 2 is an upper bound for the zeros of P by using Theorem 17.
   (b) Show that -1 is a lower bound for the zeros of P by using Theorem 18.
   (c) Compute $P(-1)$ and $P(0)$. From this can you draw any conclusion about zeros of P in $(-1, 0)$?
   (d) Compute $P(0)$ and $P(1)$. From this can you draw any conclusion about zeros of P in $(0, 1)$?
   (e) Compute $P(1)$ and $P(2)$. From this can you draw any conclusion about zeros of P in $(1, 2)$?
   (f) Compute the discriminant of P. What can you conclude about zeros of P from the discriminant?
   (g) Find the zeros of P. Why did your computations above not indicate that P had zeros in the interval $(0, 1)$?
   (h) Compute $P(0)$ and $P(7/12)$.
   (i) Compute $P(7/12)$ and $P(1)$.

58. For each given polynomial P and number k below use Theorem 16 and show that there is a number c such that $P(c) = k$.
   (a) $P(x) = x^3 - 5x^2 + 8x - 9$; $k = 6$
   (b) $P(x) = -3x^4 + 8x^3 + 5x^2 - 7x - 1$; $k = -75$
   (c) $P(x) = -3x^4 + 8x^3 + 5x^2 - 7x - 1$; $k = 16\frac{1}{2}$
   (d) $P(x) = -x^5 - 2x^3 + x^2 + 74x - 10$; $k = -1$
   (e) $P(x) = -x^5 - 2x^3 + x^2 + 74x - 10$; $k = 40/13$
   (f) $P(x) = -x^5 - 2x^3 + x^2 + 74x - 10$; $k = \sqrt{2}$

59. Show that there is a number c in the interval $(-2, 0)$ such that $P(c) = 1$, where

$$P(x) = x^6 - x^5 - 5x^4 + (5/2)x^3 + x^2 + \tfrac{1}{2}x + 3/2.$$

60. Show that a non-zero polynomial with non-negative coefficients cannot have a positive zero.

Before the next theorem is considered we wish to review for the reader some properties of the integers which we will need.

4.11
Rational Zeros of
Polynomials

If a, b are integers, we say that a is a *factor* of b (or a *divides b*) if and only if there is an integer k such that $a \cdot k = b$.

Definition

Example
4–44

The integer 3 is a factor of 12 since $3 \cdot 4 = 12$; 4 is a factor of 2^{12} since $4 \cdot 2^{10} = 2^2 \cdot 2^{10} = 2^{12}$; 1 is a factor of every integer since $1 \cdot x = x$ for all x; also, each integer x is a factor of itself since $x = 1 \cdot x$.

Definition

(*a*) If the integer m is a factor of two integers a and b, then m is called a *common* factor of a, b. (*b*) Integers a, b are *relatively prime* if and only if the only positive common factor of a, b is 1.

Example
4–45

Find the common factors of the integers 36 and 48.

Solution. We will first find all the factors of each of these integers separately and then find the common ones. To find the factors of 36 we write

$$36 = 4 \cdot 9 = 2^2 \cdot 3^2.$$

Then the positive factors of 36 are 1, 2, 3, 2^2, 6, 3^2, $2^2 \cdot 3$, $2 \cdot 3^2$, $2^2 \cdot 3^2$, *i.e.*, 1, 2, 3, 4, 6, 9, 12, 18, 36. In a similar way we write $48 = 16 \cdot 3 = 2^4 \cdot 3$ and see that the positive factors are 1, 2, 2^2, 2^3, 2^4, 3, $2 \cdot 3$, $2^2 \cdot 3$, $2^3 \cdot 3$, $2^4 \cdot 3$ or 1, 2, 3, 4, 6, 8, 12, 16, 24, 48. Comparing the set of factors of 36 with the factors of 48, the common ones are 1, 2, 3, 4, 6, 12. Of course the negative integers -1, -2, -3, -4, -6, -12 are also common factors of these two integers so that all the common factors are: ± 1, ± 2, ± 3, ± 4, ± 6, ± 12.

Example
4–46

Show that the integers 24 and 25 are relatively prime.

Solution. From $24 = 2^3 \cdot 3$ and $25 = 5^2$ we see that the positive factors of 24 are 1, 2, 3, 4, 6, 8, 12, 24 and the positive factors of 25 are 1, 5, 25. Hence, the only positive common factor of 24 and 25 is 1. Therefore, 24 and 25 are relatively prime by definition.

Theorem
20

If a, b are integers such that (i) a is a factor of b, (ii) $b = m \cdot n$ for some integers m, n and (iii) a, m are relatively prime, then a is a factor of n.

The proof of Theorem 20 will not be given here; for a proof of this theorem and a further discussion of these topics the reader may consult the author's *Elementary Exercises in Group Theory* (Chapter 3).

If a, b are relatively prime integers, then a, b^n are relatively prime for all positive integers n.

Proof. Let k be a positive integer which is a common factor of a, b^n. Then $a = k \cdot t$ and $b^n = k \cdot s$ for some integers t, s. From the equation $b^n = k \cdot s$ we obtain

$$t \cdot b^n = (k \cdot t) \cdot s$$

by multiplying by t. Into this last equation we may substitute a for $k \cdot t$ to get

$$t \cdot b^n = a \cdot s.$$

This shows that a is a factor of $(t \cdot b^{n-1}) \cdot b$. Hence, by Theorem 20, since a, b are relatively prime, a is a factor of $t \cdot b^{n-1}$; applying the same theorem to a and $(t \cdot b^{n-2}) \cdot b$ shows that a is a factor of $t \cdot b^{n-2}$. Continuing in this way we finally conclude that a is a factor of t. Then $t = a \cdot u$ for some integer u. Now substitute $a \cdot u$ for t in the equation $a = k \cdot t$:

$$a = k \cdot (a \cdot u).$$

Then $k \cdot u = 1$. The only product of two integers which is 1 is $1 \cdot 1$ and $(-1) \cdot (-1)$. Since we assumed that k was positive, we conclude that $k = 1$ and this shows that a, b^n are relatively prime.

Let $P(x)$ be a polynomial function:

$$P(x) = a_0 + a_1 x + \cdots + a_n x^n$$

where a_i is an integer for each $i = 0, 1, 2, \ldots, n$. Then if a/b is a rational zero of P and a, b are relatively prime integers, a is a factor of a_0 and b is a factor of a_n.

Proof. Since a/b is a zero of P, $P(a/b) = 0$ or

$$a_0 + a_1\left(\frac{a}{b}\right) + a_2\left(\frac{a}{b}\right)^2 + \cdots + a_{n-1}\left(\frac{a}{b}\right)^{n-1} + a_n\left(\frac{a}{b}\right)^n = 0.$$

This equation may be multiplied by b^n to give

$$b^n a_0 + b^{n-1} a_1 a + b^{n-2} a_2 a^2 + \cdots + b a_{n-1} a^{n-1} + a_n a^n = 0. \quad \textbf{(4-13)}$$

Then from this we have

$$b^n a_0 = (-b^{n-1} a_1 - b^{n-2} a_2 a - \cdots - a_n a^{n-1}) a.$$

But this shows that a is a factor of $b^n a_0$ and since a, b are relatively prime, so are a, b^n by Theorem 21. Then by Theorem 20 a is a factor of a_0. To show that b is a factor of a_n we write equation **(4-13)** as $a_n a^n = (-b^{n-1} a_0 - b^{n-2} a_1 a - \cdots - a_{n-1} a^{n-1}) b$. Then b is a factor of $a_n a^n$ and because b, a^n are relatively prime b is a factor of a_n.

Example 4-47

Discuss the rational zeros of the polynomial $P(x) = 2 - 3x + x^3$.

Solution. If a/b is a rational zero of P, then by Theorem 21 a is a factor of 2 and b is a factor of 1. Therefore, the only possibilities for a, b are:

$$a: 1, -1, 2, -2 \text{ (the only factors of 2)}$$
$$b: 1, -1 \text{ (the only factors of 1)}$$

Then the possibilities for a/b are:

$$\frac{a}{b}: \quad \frac{1}{-1}, \frac{1}{1}, \frac{-1}{1}, \frac{-1}{-1}, \frac{2}{1}, \frac{2}{-1}, \frac{-2}{1}, \frac{-2}{-1}.$$

Some of the numbers above are equal to each other so that the only choices for a/b are $1, -1, 2, -2$. These numbers may be tested by synthetic division to determine if they are zeros:

$$
\begin{array}{rrrr|l}
1 & 0 & -3 & 2 & \; 1 \\
  & 1 & 1 & -2 & \\
\hline
1 & 1 & -2 & 0 & = P(1)
\end{array}
$$

$$
\begin{array}{rrrr|l}
1 & 0 & -3 & 2 & \; -1 \\
  & -1 & 1 & 2 & \\
\hline
1 & -1 & -2 & 4 & = P(-1) \neq 0
\end{array}
$$

$$
\begin{array}{rrrr|l}
1 & 0 & -3 & 2 & \; 2 \\
  & 2 & 4 & 2 & \\
\hline
1 & 2 & 1 & 4 & = P(2) \neq 0
\end{array}
$$

$$
\begin{array}{rrrr|l}
1 & 0 & -3 & 2 & \; -2 \\
  & -2 & 4 & -2 & \\
\hline
1 & -2 & 1 & 0 & = P(-2)
\end{array}
$$

We then conclude that 1 and -2 are the only rational zeros of the function $P(x) = 2 - 3x + x^3$.

In Example 4–47 the only possible rational zeros are integers; this is true because the a_n in this case is 1 and for a/b to be a zero, b must be a factor of a_n. But if b is a factor of $a_n = 1$, then $b = \pm 1$ and $a/b = \pm a$ is an integer.

Although Theorem 22 applies only to polynomials with integeral co-efficients, it is possible to gain information about polynomials with ra-tional coefficients by a process which we now describe. If $a \neq 0$ is a number and P is a polynomial function, observe that the two polynomials P and $a \cdot P$ have the same zeros: $P(c) = 0$ if and only if $a \cdot P(c) = 0$. If a polynomial P has rational coefficients which are not integral, we may choose a suitable integer a such that $a \cdot P$ has integral coefficients. We may then investigate the rational zeros of $a \cdot P$ by using Theorem 21.

Find the rational zeros of

Example 4–48

$$P(x) = \frac{x^4}{6} + x^3 + \frac{13}{6}x^2 + 2x + \frac{2}{3}.$$

Solution. First, multiply $P(x)$ by 6:

$$6P(x) = x^4 + 6x^3 + 13x^2 + 12x + 4.$$

Now apply Theorem 21 to $6P(x)$. Since $a_4 = 1$, the only possible zeros are $-1, 1, 2, -2, -4, 4$.

$$
\begin{array}{rrrrr|r}
1 & 6 & 13 & 12 & 4 & -1 \\
 & -1 & -5 & -8 & -4 & \\
\hline
1 & 5 & 8 & 4 & 0 &
\end{array}
$$

Thus, $6P(-1) = 0$; $P(-1) = 0$. Also $6P(x) = (x + 1)(x^3 + 5x^2 + 8x + 4)$. To test for the remaining zeros we only have to use $x^3 + 5x^2 + 8x + 4 = Q(x)$. The numbers $-1, 1, 2, -2, -4, 4$ are also seen to be the pos-sible rational zeros of $Q(x)$.

$$
\begin{array}{rrrr|r}
1 & 5 & 8 & 4 & -1 \\
 & -1 & -4 & -4 & \\
\hline
1 & 4 & 4 & 0 &
\end{array}
$$

Then

$$Q(x) = x^3 + 5x^2 + 8x + 4 = (x + 1)(x^2 + 4x + 4)$$

and

$$6P(x) = (x + 1)(x + 1)(x^2 + 4x + 4)$$
$$= (x + 1)(x + 1)(x + 2)(x + 2).$$

The only rational zeros of $6P(x)$ are -1 and -2; hence, the rational zeros of P are -1 and -2.

Theorem 23

If P is a polynomial with integral coefficients and a is an integral zero of P, then $m - a$ is a factor of $P(m)$ for all integers m.

Proof. If a is a zero of P, then $P(a) = 0$ and we may write

$$P(x) = (x - a)Q(x)$$

for all x by Theorem 14. Then for any integer m

$$P(m) = (m - a)Q(m)$$

This implies that $m - a$ is a factor of $P(m)$ since Q has integral coefficients and $Q(m)$ is an integer.

Theorem 23 may be used in some cases to shorten the work of deciding if integers are zeros of a polynomial. Suppose that we wish to decide if an integer a is a zero of a polynomial P with integral coefficients. If we can find an integer m such that $m - a$ is not a factor of $P(m)$, then we know that a is not a zero of P. The fewer factors the number $P(m)$ has, the more likely we are to rule out several possible integral zeros. Consider, for example, the polynomial $P(x) = 3x^4 + 14x^3 + 14x^2 - 20x - 8$. Here $a_0 = -8$ and $a_4 = 3$. Then if a/b is a rational zero of P, a must be a factor of -8 and b a factor of 3. The only integral zeros occur when $b = \pm 1$. Therefore, the only integral zeros which are possible are the factors of -8; these are $\pm 1, \pm 2, \pm 4, \pm 8$. These will be tested by a use of Theorem 23. Choose a convenient value for m, say $m = 1$. Now compute $P(m) = P(1)$.

$$
\begin{array}{rrrrr|l}
3 & 14 & 14 & -20 & -8 & \quad 1 = m \\
  & 3 & 17 & 31 & 11 & \\
\hline
3 & 17 & 31 & 11 & 3 & = P(m)
\end{array}
$$

Thus $P(1) = 3$. The table below may now be constructed:

a	$m - a$	$m - a$ is a factor of $P(m) = 3$	a is a possible zero of P
1	0	No	No
-1	2	No	No
2	-1	Yes	Yes
-2	3	Yes	Yes
4	-3	Yes	Yes
-4	5	No	No
8	-7	No	No
-8	9	No	No

From the table we see that the only possible integral zeros are 2, -2, 4. We check these by synthetic division:

$$
\begin{array}{rrrrr|r}
3 & 14 & 14 & -20 & -8 & 2 \\
& 6 & 40 & 108 & 176 & \\
\hline
3 & 20 & 54 & 88 & 168 & \\
\end{array}
$$

$$
\begin{array}{rrrrr|r}
3 & 14 & 14 & -20 & -8 & -2 \\
& -6 & -16 & 4 & 32 & \\
\hline
3 & 8 & -2 & -16 & 24 & \\
\end{array}
$$

$$
\begin{array}{rrrrr|r}
3 & 14 & 14 & -20 & -8 & 4 \\
& 12 & 104 & 472 & 1808 & \\
\hline
3 & 26 & 118 & 452 & 1800 & \\
\end{array}
$$

Therefore, P has no integral zeros. In this case we chose m so that $P(m) = 3$. It is not always possible to choose such a convenient value for m and usually some experimentation is required to find a suitable value (if one exists).

61. In each of the following draw a conclusion from the given information.

    (a) Given: $x = 2y$, $x, y \in Z$. Draw a conclusion about 2 and y.
       Solution. 2 is a factor of x and y is a factor of x.

Exercises

(b) Given: $3a = x + y$, $x, y \in Z$. Draw a conclusion about 3 and a.

(c) Given: x is a common factor of 6 and 8 and $x > 0$. Draw a conclusion about x.

(d) Given: x is a common factor of 4 and 9 and $x > 0$. Draw a conclusion about x.

(e) Given: x is a common factor of 2^n and 3^m, where m, n are positive integers. Draw a conclusion about x.

(f) Given: a, b are integers and 1 is the only positive common factor of a, b. Draw a conclusion about a, b.

(g) Given: $2w = 3u$, w, u are integers. Draw a conclusion about 2, u and 3, w.

(h) Given: $P(x) = 3 + ax + bx^2 + x^3$, a, b are integers and w/z is a rational zero of P. Draw a conclusion about w, z.

62. Write down all the positive factors of each number given below.

(a) 27   (b) 16   (c) 125   (d) 64   (e) 12
(f) 18   (g) 36   (h) 30   (i) 60   (j) 180

63. Based upon your experience with problem 62, can you describe a routine method for writing down all the positive factors of any given positive integer?

64. Find the positive common factors of each pair of integers below.

(a) 6, 8   (b) 10, 12   (c) 24, 36
(d) 20, 50   (e) 28, 42   (f) 100, 120

65. Let $P(x) = a_0 + a_1x + a_2x^2 + \cdots + a_nx^n$, where each a_i is an integer for $i = 0, 1, 2, \ldots, n$. Suppose that a, b are integers, $b \neq 0$, and that a/b is a zero of P. For each value of a_0 and a_n given below determine the possible values of a/b.

(a) $a_0 = 1$, $a_n = 1$ (b) $a_0 = -1$, $a_n = 1$
(c) $a_0 = 1$, $a_n = -1$ (d) $a_0 = 2$, $a_n = 1$
(e) $a_0 = 1$, $a_n = 2$ (f) $a_0 = -2$, $a_n = -1$
(g) $a_0 = 2$, $a_n = -3$ (h) $a_0 = 6$, $a_n = -8$
(i) $a_0 = -8$, $a_n = 10$ (j) $a_0 = 16$, $a_n = 24$

66. Suppose that a is a zero of a polynomial function P. Then by Theorem 14

$$P(x) = (x - a)Q(x)$$

for some polynomial Q. Show that if b is a zero of P and $b \neq a$, then b is a zero of Q.

67. Find all rational zeros of each of the following.

(a) $P(x) = x^4 + x^3 + 3x^2 - x - 4$. *Solution.* The possible rational zeros are $\pm 1, \pm 2, \pm 4$. Using synthetic division we have

$$
\begin{array}{rrrrr|r}
1 & 1 & 3 & -1 & -4 & 1 \\
  & 1 & 2 & 5 & 4 & \\
\hline
1 & 2 & 5 & 4 & 0 = P(1) &
\end{array}
$$

Then $P(x) = (x - 1)(x^3 + 2x^2 + 5x + 4)$ and to continue, we need to check for the rational zeros of $x^3 + 2x^2 + 5x + 4$ by exercise 66 above and these must also be found among ± 1, $\pm 2, \pm 4$.

$$
\begin{array}{rrrr|r}
1 & 2 & 5 & 4 & 1 \\
  & 1 & 3 & 8 & \\
\hline
1 & 3 & 8 & 12 = P(1) \neq 0 &
\end{array}
$$

This shows that 1 is an upper bound for the zeros of $x^3 + 2x^2 + 5x + 4$ so that we do not need to try 2, 4.

$$
\begin{array}{rrrr|r}
1 & 2 & 5 & 4 & -1 \\
  & -1 & -1 & -4 & \\
\hline
1 & 1 & 4 & 0 = P(-1) &
\end{array}
$$

Therefore, $x^3 + 2x^2 + 5x + 4 = (x + 1)(x^2 + x + 4)$ and $P(x) = (x - 1)(x + 1)(x^2 + x + 4)$. The only other zeros that P can have are those which are zeros of $x^2 + x + 4$. But for this quadratic we have that the discriminant, $d = b^2 - 4ac = 1^2 - 4 \cdot 1 \cdot 4 < 0$. Hence, $x^2 + x + 4$ has no real zeros and we conclude that ± 1 are the only rational zeros of P.

(b) $P(x) = x^3 + 3x^2 + 3x + 2$
(c) $P(x) = 2x^3 + 5x^2 + 3x + 2$
(d) $P(x) = 2x^3 + x^2 + x - 1$
(e) $P(x) = 12x^3 + 16x^2 - 5x - 3$
(f) $P(x) = 2x^3 - 2x^2 + (5/2)x - 1$
(g) $P(x) = 4x^4 - 2x^3 + 3x^2 + \frac{1}{2}x - 1$

68. In each of the following a polynomial, P, and an integer, m, are given. In each case compute $P(m)$ and use Theorem 22 to rule out some of the possible integral zeros of P. Then check the remaining possibilities to see if they are zeros by synthetic division.

    (a) $P(x) = x^4 - x^3 - x^2 - x - 3$, $m = 2$
    (b) $P(x) = x^5 + 2x^4 - 5x^3 + 7x^2 - 20x - 14$, $m = 2$
    (c) $P(x) = x^5 + 15x^4 + 21x^3 - 3x^2 - 7x + 24$, $m = -1$
    (d) $P(x) = x^6 - 22x^5 - x^4 - 3x^3 + x^2 - 27x + 18$, $m = 1$
    (e) $P(x) = x^8 - 4x^7 + 5x^6 - 10x^5 + 11x^4 + 13x^3 - 35x^2$
              $+ 14x + 70$, $m = 3$

69. Find the rational zeros of each of the following:
    (a) $P(x) = x^2 - 5$
    (b) $P(x) = x^3 - 1$
    (c) $P(x) = 2x^3 + 4x - \frac{1}{2}$
    (d) $P(x) = 12x^3 - 20x^2 - x + 3$
    (e) $P(x) = x^3 + (8/3)x^2 + (1/12)x - 5/6$
    (f) $P(x) = 12x^4 + 92x^3 + 51x^2 - 36x - 7$

70. (a) Show that $P(x) = x^2 - 2$ has no rational zeros.
    (b) Show that $\sqrt{2}$ is a zero of $P(x) = x^2 - 2$.
    (c) Draw a conclusion about $\sqrt{2}$ from (a) and (b).

71. (a) Show that $P(x) = x^2 - 3$ has no rational zeros.
    (b) Show that $\sqrt{3}$ is a zero of $P(x) = x^2 - 3$.
    (c) Draw a conclusion about $\sqrt{3}$ from (a) and (b).

72. Show that each of the following numbers is irrational by considering appropriate polynomial functions as in 70 and 71 above.
    (a) $\sqrt{5}$ (b) $\sqrt{7}$ (c) $\sqrt{6}$
    (d) $\sqrt{2} + \sqrt{3}$ [Hint: $x^4 - 10x^2 + 1$]
    (e) $\sqrt{2} - \sqrt{3}$ [Hint: $x^4 - 10x^2 + 1$]
    (f) $\sqrt{2} + \sqrt{5}$ [Hint: $x^4 - 14x^2 + 9$]
    (g) $\sqrt{3} - 2\sqrt{2}$ [Hint: $x^4 - 22x^2 + 25$]
    (h) $\sqrt{5} - 2\sqrt{3}$ [Hint: $x^4 - 34x^2 + 49$]
    (i) $2\sqrt{5} - \sqrt{7}$ [Hint: $x^4 - 54x^2 + 169$]

THE TRIGONOMETRIC FUNCTIONS

It will be assumed here that the reader is familiar with the measurement of angles in degrees and also that he knows the elementary facts concerning similar triangles. With these few essentials it is possible to define the elementary functions of trigonometry associated with right triangles.

Let a right triangle ABC be given as in Fig. 5–1. The length of the side opposite $\angle A$ (*angle A*) has been labeled a, the length of the side opposite $\angle B$ is denoted by b, and c denotes the length of the side opposite $\angle C$. Referring to Fig. 5–1, we may now define functions s, c, t as follows:

$$s(\angle A) = \frac{a}{c} = \frac{\text{length of side opposite } \angle A}{\text{length of hypotenuse}}$$

$$c(\angle A) = \frac{b}{c} = \frac{\text{length of side adjacent to } \angle A}{\text{length of hypotenuse}}$$

$$t(\angle A) = \frac{a}{b} = \frac{\text{length of side opposite to } \angle A}{\text{length of side adjacent to } \angle A}$$

Figure
5–1

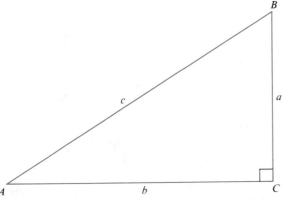

The definitions of s, c, t appear to depend on the right triangle ABC; actually, this is not the case. If $\triangle A'B'C'$ is any other right triangle and $\angle A$ is congruent to $\angle A'$ (*i.e.*, they have the same measure), then $\angle B$ is congruent to $\angle B'$ and the right triangles ABC and $A'B'C'$ are similar (see Fig. 5–2). Since the two triangles are similar, it may be concluded that

$$\frac{a}{a'} = \frac{c}{c'}, \quad \frac{b}{b'} = \frac{c}{c'}, \quad \frac{a}{a'} = \frac{b}{b'}$$

and from these equations it follows that

$$\frac{a}{c} = \frac{a'}{c'}, \quad \frac{b}{c} = \frac{b'}{c'}, \quad \frac{a}{b} = \frac{a'}{b'}.$$

Then from the definition of s, c, t

$$s(\angle A) = \frac{a}{c} = \frac{a'}{c'} = s(\angle A')$$

$$c(\angle A) = \frac{b}{c} = \frac{b'}{c'} = c(\angle A')$$

$$t(\angle A) = \frac{a}{b} = \frac{a'}{b'} = t(\angle A')$$

**Figure
5–2**

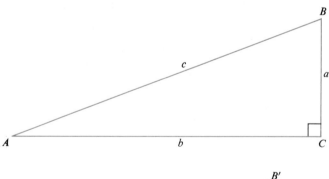

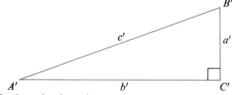

We see from this that the functions s, c, t have equal values at congruent angles. This amounts to saying that the value of the functions s, c, t at a given angle depends only on the "size" of the angle.

The functions s, c, t are called *sine*, *cosine* and *tangent* respectively and it is customary to write sin $(\angle A)$, cos $(\angle A)$, tan $(\angle A)$ for $s(\angle A)$, $c(\angle A)$, $t(\angle A)$.

If the measure of $\angle A$ is $d°$, sin $(d°)$, cos $(d°)$, tan $(d°)$ are defined by the following equations: $\sin (d°) = \sin (\angle A)$, $\cos (d°) = \cos (\angle A)$, tan $(d°) = \tan (\angle A)$.

Compute sin (45°), cos (45°) and tan (45°).

Example
5–1

Solution. For these computations we may choose any right triangle having an angle whose measure is 45°. Then the other angle also has measure 45° so that the triangle is isosceles. The lengths of the two short sides may conveniently be taken to have length 1 each as in Fig. 5–3. If c is the length of the hypotenuse, the Pythagorean Theorem implies that $c^2 = 1^2 + 1^2 = 2$ or $c = \sqrt{2}$. Then we have

$$\sin (45°) = \frac{a}{c} = \frac{1}{\sqrt{2}}$$

$$\cos (45°) = \frac{b}{c} = \frac{1}{\sqrt{2}}$$

$$\tan (45°) = \frac{a}{b} = \frac{1}{1} = 1.$$

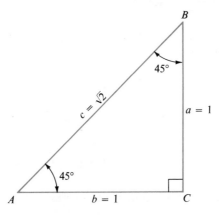

Figure
5–3

If $\angle A$, $\angle B$ are complementary, show that $\sin (\angle A) = \cos (\angle B)$, cos $(\angle A) = \sin (\angle B)$.

Example
5–2

Solution. Since $\angle A$, $\angle B$ are complementary, the sum of their measures is 90°. Hence, any right triangle which contains an angle congruent to $\angle A$ also contains an angle congruent to $\angle B$; in other words, the two acute angles of any right triangle are complementary. Referring to Fig. 5–1 we have

$$\sin(\angle A) = \frac{a}{c}, \quad \sin(\angle B) = \frac{b}{c}$$

$$\cos(\angle A) = \frac{b}{c}, \quad \cos(\angle B) = \frac{a}{c}$$

and from these equations the desired results follow.

Definition

The functions *cotangent, secant, cosecant* are abbreviated cot, sec, csc respectively and are defined as follows: if $\angle A$ is an angle of a right triangle, then

$$\cot(\angle A) = \frac{1}{\tan(\angle A)}, \quad \sec(\angle A) = \frac{1}{\cos(\angle A)}, \quad \csc(\angle A) = \frac{1}{\sin(\angle A)}.$$

Example 5–3

Compute cot (45°), sec (45°), csc (45°).

Solution. By definition of cot, sec, csc

$$\cot(45°) = \frac{1}{\tan(45°)}, \quad \sec(45°) = \frac{1}{\cos(45°)}, \quad \csc(45°) = \frac{1}{\sin(45°)}.$$

Then from Example 5–1 $\sin(45°) = 1/\sqrt{2}$, $\cos(45°) = 1/\sqrt{2}$ and $\tan(45°) = 1$. Therefore,

$$\cot(45°) = \frac{1}{1} = 1$$

$$\sec(45°) = \frac{1}{1/\sqrt{2}} = \sqrt{2}$$

$$\csc(45°) = \frac{1}{1/\sqrt{2}} = \sqrt{2}.$$

By using the definitions it is possible to express cot, sec, csc as ratios of sides of triangles:

$$\cot(\angle A) = \frac{1}{\tan(\angle A)} = \frac{1}{a/b} = \frac{b}{a} = \frac{\text{length of side adjacent to } \angle A}{\text{length of side opposite to } \angle A}$$

$$\sec(\angle A) = \frac{1}{\cos(\angle A)} = \frac{1}{b/c} = \frac{c}{b} = \frac{\text{length of hypotenuse}}{\text{length of side adjacent } \angle A}$$

$$\csc(\angle A) = \frac{1}{\sin(\angle A)} = \frac{1}{a/c} = \frac{c}{a} = \frac{\text{length of hypotenuse}}{\text{length of side opposite } \angle A}$$

The six functions sine, cosine, tangent, cotangent, secant, cosecant are called the *trigonometric* functions.

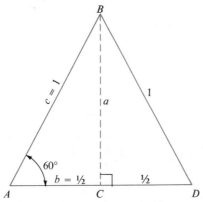

Figure
5–4

Compute the six trigonometric functions of 60°.

Example
5–4

Solutions. We construct an equilateral triangle with each side having a length of 1 (see Fig. 5–4). Then each angle has measure 60°. To compute the trigonometric functions of 60° we drop a perpendicular from B to side AD; this perpendicular bisects AD at point C so that AC has length $\frac{1}{2}$. Using the Pythagorean Theorem on right triangle ABC gives

$$1^2 = a^2 + (\tfrac{1}{2})^2 \quad \text{or} \quad 1 = a^2 + \tfrac{1}{4}.$$

Then $a^2 = 1 - 1/4 = 3/4$ so that $a = \sqrt{3/4} = \frac{1}{2}\sqrt{3}$. Now the trigonometric functions of 60° may be computed:

$$\sin(60°) = \tfrac{1}{2}\sqrt{3}/1 = \tfrac{1}{2}\sqrt{3}$$
$$\cos(60°) = \tfrac{1}{2}/1 = \tfrac{1}{2}$$
$$\tan(60°) = \tfrac{1}{2}\sqrt{3}/\tfrac{1}{2} = \sqrt{3}$$
$$\cot(60°) = \tfrac{1}{2}/\tfrac{1}{2}\sqrt{3} = 1/\sqrt{3}$$
$$\sec(60°) = 1/\tfrac{1}{2} = 2$$
$$\csc(60°) = 1/\tfrac{1}{2}\sqrt{3} = 2/\sqrt{3}$$

On pp. 362–363 will be found Table I which gives the trigonometric functions for numbers from 0° to 90° in increments of 0.5°. The measure of angles is given in the left-hand and right-hand columns with the value of the particular trigonometric function being found at the intersection of the appropriate row and column. A portion of the table is printed below and from it we can see that $\sin(3.5°) = 0.0610$, $\cos(3.5°) = 0.9981$, $\tan(3.5°) = 0.0612$, $\cot(3.5°) = 16.3499$.

Trigonometric Functions $\sin \theta$, $\cos \theta$, $\tan \theta$, and $\cot \theta$

Angle θ		$\sin \theta$	$\cos \theta$	$\tan \theta$	$\cot \theta$			
Radians	Degrees							
0.0000	0.0	0.0000	1.0000	0.0000	—	90.0	1.5708
0.0087	0.5	0.0087	1.0000	0.0087	114.5887	89.5	1.5621
0.0175	1.0	0.0175	0.9998	0.0175	57.2900	89.0	1.5533
0.0262	1.5	0.0262	0.9997	0.0262	38.1885	88.5	1.5446
0.0349	2.0	0.0349	0.9994	0.0349	28.6363	88.0	1.5359
0.0436	2.5	0.0436	0.9990	0.0437	22.9038	87.5	1.5272
0.0524	3.0	0.0523	0.9986	0.0524	19.0811	87.0	1.5184
0.0611	3.5	0.0610	0.9981	0.0612	16.3499	86.5	1.5097
0.0698	4.0	0.0698	0.9976	0.0699	14.3007	86.0	1.5010
0.0785	4.5	0.0785	0.9969	0.0787	12.7062	85.5	1.4923
0.0873	5.0	0.0872	0.9962	0.0875	11.4301	85.0	1.4835
		$\cos \theta$	$\sin \theta$	$\cot \theta$	$\tan \theta$	Degrees	Radians
						Angle θ	

An investigation of the table will indicate that the left-hand column contains readings only between 0° and 45°. This is due to a property of the trigonometric functions which was illustrated in Example 5–2. It was shown there that if $a° + b° = 90°$, then $\sin (a°) = \cos (b°)$. If we wish to find $\sin (64°)$ from the table we could use this result and find $\cos (26°)$ since $64° + 26° = 90°$. From the table $\cos (26°) = 0.8988$ and therefore $\sin (64°) = 0.8988$. For convenience the table is arranged so that we can find 64° in the *right-hand* column and the correct reading (0.8988) as the intersection of the 64°-right-hand row and the column labeled sin at the *bottom* of the page. The same kind of relation holds between the functions tan, cot and between sec, csc. We may summarize by stating the general rule: *to locate trigonometric values of numbers between 0° and 45° one should use the left-hand column and the heading at the top of the page; to locate trigonometric values of numbers between 45° and 90° one should use the right-hand column and the headings at the bottom of the table.*

Example 5–5

Use Table I to find $\sin (38.5°)$, $\cos (81°)$, $\tan (58°)$.

Solution. Since 38.5° is between 0° and 45°, we locate 38.5° in the left-hand column and find $\sin (38.5°)$ in the column labeled sin at the top of the page. We find: $\sin (38.5°) = 0.6225$. For $\cos (81°)$ we must use the right-hand column to locate 81° and the column labeled cos at the bottom: $\cos (81°) = 0.1564$. To find $\tan (58°)$ use the right-hand column and the bottom of the page: $\tan (58°) = 1.6003$.

It should be pointed out that the values given in Table I are only approximations to 4 decimal places of accuracy.

1. Use the right triangles given in Fig. 5–3 and Fig. 5–4 to complete the table below.

a	30	45	60
$\sin (a°)$			
$\cos (a°)$			
$\tan (a°)$			
$\cot (a°)$			
$\sec (a°)$			
$\csc (a°)$			

2. Given $\triangle ABC$ with $\angle C$ a right angle and $a = 6$, $b = 8$, $c = 10$.
   (a) Compute $\sin (\angle A)$, $\tan (\angle A)$, $\sec (\angle A)$.
   (b) Use only the information of part (a) to find $\cos (\angle B)$, $\cot (\angle B)$, $\csc (\angle B)$.
   (c) Use only the information of part (a) to find $\cos (\angle A)$, $\cot (\angle A)$, $\csc (\angle A)$.
   (d) Use only the information of part (b) to find $\sin (\angle B)$, $\tan (\angle B)$, $\sec (\angle B)$.

3. Given $\triangle ABC$ with $\angle C$ a right angle and $a = 1$, $b = 3$.
   (a) Compute c.
   (b) Compute $\sin (\angle A)$, $\tan (\angle A)$, $\sec (\angle A)$.
   (c) Use only the information of part (b) to find $\cos (\angle B)$, $\cot (\angle B)$, $\csc (\angle B)$.
   (d) Use only the information of part (b) to find $\cos (\angle A)$, $\cot (\angle A)$, $\csc (\angle A)$.
   (e) Use only the information of part (c) to find $\sin (\angle B)$, $\tan (\angle B)$, $\sec (\angle B)$.

4. Use Table I to find the following:
   (a) $\cot (28°)$ (b) $\cos (5°)$
   (c) $\tan (46°)$ (d) $\sin (34°)$
   (e) $\cos (15.5°)$ (f) $\cot (61°)$
   (g) $\sin (88.5°)$ (h) $\tan (71.5°)$
   (i) $\cos (0.5°)$ (j) $\sin (60.5°)$

5. Use Table I to find a for each of the following:

Example: sin $(a°)$ = 0.2250. We look in the two columns labeled sin until we find 0.2250. This number is found in the column which is labeled sin at the *top*. Therefore, the a we seek is in the left-hand column. We see that sin $(13.0°)$ = 0.2250. Hence, a = 13.

(a) cot $(a°)$ = 2.0057   (b) cos $(a°)$ = 0.7934
(c) tan $(a°)$ = 0.3346   (d) sin $(a°)$ = 0.7071
(e) cot $(a°)$ = 0.9490   (f) cos $(a°)$ = 0.3256
(g) sin $(a°)$ = 0.8704   (h) tan $(a°)$ = 2.2998
(i) cot $(a°)$ = 57.2900

6. For a right $\triangle ABC$ with $\angle C$ the right angle, a opposite to A, b opposite to B, c the hypotenuse, supply the requested information below. ($m \angle A$ means the measure of $\angle A$).
(a) Given: $m \angle A = 12°$. Find $m \angle B$.
(b) Given: $m \angle A = 12°$, $c = 1$. Find $m \angle B$, b and a.

Solution. Since $m \angle A + m \angle B = 90°$, and $m \angle A = 12°$, $m \angle B = 78°$. By definition sin $(12°)$ = a/c. By Table I sin $(12°)$ = 0.2079 and we are given that $c = 1$. Then

$$0.2079 = a/1 = a.$$

Also by definition cos $(12°)$ = $b/c = b$ (since $c = 1$). Using Table I again gives cos $(12°)$ = 0.9781 so that $b = 0.9781$.
(c) Given: $m \angle B = 38°$, $c = 2$. Find $m \angle A$, a and b.
(d) Given: $m \angle B = 72.5°$, $b = 1$. Find $m \angle A$, a and c.
(e) Given: $a = 0.3256$, $c = 1$. Find $m \angle A$, $m \angle B$ and b.

5.2
Analytical
Trigonometry

In the previous section we defined trigonometric functions which were associated with triangles. These definitions are too restrictive in some respects and for this reason it is desirable to have a generalization of them. While the same names, "sine", "cosine", etc. are used in both Section 5.1 and here, the reader should bear in mind that the functions of this section are different from those of 5.1. It will be shown in 5.4 how the functions of 5.1 are a special case of those defined in this section.

There are several ways of defining the trigonometric functions; the method which will be used here makes use of the unit circle and *assumes* that it is possible to measure arcs along the unit circle. (This assumption is justified in more advanced courses.)

We start with the unit circle

$$\mathcal{C} = \{(x, y) \mid x^2 + y^2 = 1\}$$

whose graph is given in Fig. 5–5. If t is a real number, an arc of the circle $\mathcal{C}$ is measured from the point $I(1, 0)$ having a length $|t|$; the arc is measured in the counter-clockwise direction from I if $t \geq 0$ and is measured in the clockwise direction if $t < 0$. (See Fig. 5–5.) This gives us a method of associating with each real number t a point $P(a, b)$ (the end point of the arc opposite I) of the circle $\mathcal{C}$. In other words, a function F has been defined whose domain is the set of all real numbers and whose range is the set of all points of $\mathcal{C}$.

$$F = \{(t, P(a, b)) \mid t \in \mathbf{R} \text{ and } P \text{ is obtained as described above}\}.$$

**Figure
5–5**

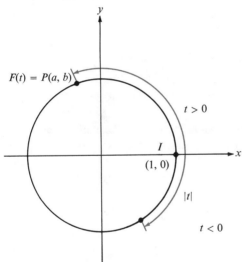

We also may write

$$F(t) = P(a, b).$$

For $t = 0$,

$$F(0) = I(1, 0),$$

and since the circumference of the circle is 2π,

$$F(2\pi) = I(1, 0)$$

and

$$F(-2\pi) = I(1, 0).$$

Suppose that $F(t) = P(a, b)$, that n is an integer and that we wish to compute $F(t + 2n\pi)$. The point on the circle associated with the number $t + 2n\pi$ may be obtained by first finding $F(t) = P(a, b)$ and then measuring from this point an arc having length $2n\pi$. But since n is an integer, it is clear that the end point of the arc associated with $2n\pi$, when we measure from $P(a, b)$, is just $P(a, b)$. Hence,

$$F(t + 2n\pi) = P(a, b)$$

or, since

$$F(t) = P(a, b),$$
$$F(t + 2n\pi) = F(t)$$

for all real numbers t and integers n.

It is now possible to use the function F to define the trigonometric functions *sine* and *cosine*.

Definition

1. Sine $= \{(t, b) \mid t \in \mathbf{R} \text{ and } F(t) = P(a, b)\}$.
2. Cosine $= \{(t, a) \mid t \in \mathbf{R} \text{ and } F(t) = P(a, b)\}$.

The names "sine" and "cosine" are usually abbreviated to "sin" and "cos" so that in the $f(t)$-notation

$$\sin (t) = b$$
$$\cos (t) = a.$$

Observe that $\sin (t)$ and $\cos (t)$ are the second and first coordinates of the point

$$F(t) = P(a, b)$$

on the unit circle. (See Fig. 5–5.)

Example 5–6

Find $\cos (\pi/2)$.

Solution. We start at $I(1, 0)$ and proceed $\pi/2$ units counterclockwise along the circle. Since $\pi/2$ is $\frac{1}{4}$ the total circumference 2π, we stop $\frac{1}{4}$ of the way around the circle. That is,

$$F(\pi/2) = P(0, 1).$$

Therefore,

$$\cos (\pi/2) = 0.$$

Find sin $(\pi/4)$ and cos $(\pi/4)$.

Example
5–7

Solution. To calculate $F(\pi/4)$ our arc from $I(1, 0)$ is in the same direction as the arc for $F(\pi/2)$ and half as long. Hence the point $F(\pi/4)$ is on the line bisecting the first quadrant. (See Fig. 5–6.) Thus, if

$$F(\pi/4) = P(a, b),$$

$a = b$. But $P(a, a)$ is on the circle $\mathcal{C}$.

$$a^2 + a^2 = 1$$

or

$$2a^2 = 1$$
$$a^2 = \tfrac{1}{2}$$

and

$$a = \frac{1}{\sqrt{2}}.$$

(Note that $a = 1/\sqrt{2}$ and not $-1/\sqrt{2}$ since the point is in the first quadrant.) Therefore,

$$\sin (\pi/4) = 1/\sqrt{2}$$
$$\cos (\pi/4) = 1/\sqrt{2}.$$

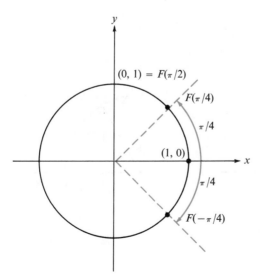

Figure
5–6

**Example
5-8**

Calculate $\sin(-\pi/4)$ and $\cos(-\pi/4)$.

Solution. Referring to Fig. 5–6 it is evident that

$$F(-\pi/4) = P(1/\sqrt{2}, -1/\sqrt{2})$$

and therefore,

$$\sin(-\pi/4) = -1/\sqrt{2} \qquad \text{and} \qquad \cos(-\pi/4) = 1/\sqrt{2}.$$

**Example
5-9**

Calculate $\sin(\pi/3)$ and $\cos(\pi/3)$.

Solution. To calculate $F(\pi/3)$, we observe that $\pi/3 = \frac{2}{3}(\pi/2)$, and hence the arc from $I(1, 0)$ to $F(\pi/3)$ is $\frac{2}{3}$ the length of the arc from $I(1, 0)$ to $F(\pi/2)$. Now using Fig. 5–7 and elementary geometry, the triangle OFI is seen to be equilateral so that the length of the segment OA is $\frac{1}{2}$. Therefore, the first coordinate of $F(\pi/3)$ is $\frac{1}{2}$. If $F(\pi/3) = P(\frac{1}{2}, b)$, then

$$(\tfrac{1}{2})^2 + b^2 = 1$$

and

$$b = \tfrac{1}{2}\sqrt{3}.$$

We then conclude that

$$\sin(\pi/3) = \tfrac{1}{2}\sqrt{3} \qquad \text{and} \qquad \cos(\pi/3) = \tfrac{1}{2}.$$

**Figure
5-7**

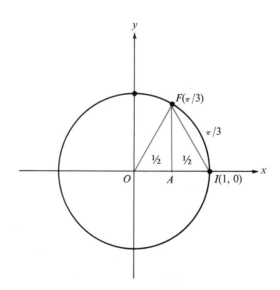

164

Find cos (3π).

Example
5–10

Solution. From the relationship

$$F(t + 2n\pi) = F(t)$$

we have for $n = 1$

$$F(3\pi) = F(\pi + 2\pi) = F(\pi) = P(-1, 0).$$

Thus,

$$\cos(3\pi) = -1.$$

(See Fig. 5–8.)

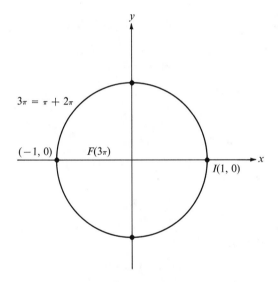

Figure
5–8

In Example 5–10, we used the equation

$$F(t + 2n\pi) = F(t)$$

to find cos (3π). This equation gives rise to the following very important property of the sine and cosine functions:

$$\sin(t + 2n\pi) = \sin(t)$$
$$\cos(t + 2n\pi) = \cos(t)$$

for all real numbers t and integers n.

If $n = 1$ in these equations, we have

$$\sin(t + 2\pi) = \sin(t)$$
$$\cos(t + 2\pi) = \cos(t)$$

for all real numbers t. This shows that the sine and cosine are examples of *periodic* functions.

Definition

A function f is said to be *periodic* if and only if there is some positive number a such that

$$f(t + a) = f(t)$$

for all numbers t and $t + a$ in the domain of f.

It has just been observed that sine and cosine are both *periodic* functions. Now consider

$$f(x) = x - [x].$$

Using the property that

$$[x + 1] = [x] + 1$$

(see Exercise 35(d) in Chapter 3), we have

$$\begin{aligned} f(x + 1) &= (x + 1) - [x + 1] \\ &= (x + 1) - ([x] + 1) \\ &= x - [x] \\ &= f(x). \end{aligned}$$

This then shows that

$$f(x + 1) = f(x)$$

for all x and that $f(x) = x - [x]$ is periodic.

Exercises

7. By using Fig. 5–9, obtain the coordinates of the following points:

(a) $F(0)$ (b) $F(-2\pi)$ (c) $F(\pi/6)$
(d) $F(-\pi/6)$ (e) $F(\pi/4)$ (f) $F(-\pi/3)$
(g) $F(\pi/2)$ (h) $F(-\pi/2)$ (i) $F(3\pi/2)$
(j) $F(2\pi/3)$ (k) $F(5\pi/6)$ (l) $F(\pi)$

(m) $F(5\pi/4)$	(n) $F(7\pi/6)$	(o) $F(4\pi/3)$
(p) $F(5\pi/3)$	(q) $F(11\pi/6)$	(r) $F(-3\pi/2)$
(s) $F(-2\pi/3)$	(t) $F(-5\pi/6)$	(u) $F(-\pi)$
(v) $F(-5\pi/4)$	(w) $F(-7\pi/6)$	(x) $F(-5\pi/3)$

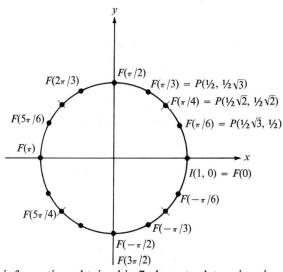

Figure
5–9

8. Use the information obtained in 7 above to determine sin t and cos t for the following values of t.

(a) 0	(b) -2π	(c) $\pi/6$
(d) $-\pi/6$	(e) $\pi/4$	(f) $-\pi/3$
(g) $\pi/2$	(h) $-\pi/2$	(i) $3\pi/2$
(j) $2\pi/3$	(k) $5\pi/6$	(l) π
(m) $5\pi/4$	(n) $7\pi/6$	(o) $4\pi/3$
(p) $5\pi/3$	(q) $-3\pi/2$	(r) $-2\pi/3$
(s) $-5\pi/6$	(t) $-\pi$	(u) $-5\pi/4$
(v) $-7\pi/5$	(w) $-5\pi/3$	(x) $11\pi/6$

9. (a) Given that $F(0.1047) = (0.9945, 0.1045)$, find

$$\cos (0.1047)$$
$$\sin (0.1047)$$
$$F(0.1047 + 2\pi)$$
$$\cos (0.1047 + 2\pi)$$
$$\sin (0.1047 + 2\pi)$$

(b) Given that $F(1.2915) = (0.2756, 0.9613)$, find

$$\cos (1.2915)$$
$$\sin (1.2915)$$
$$F(1.2915 + 6\pi)$$
$$\cos (1.2915 + 6\pi)$$
$$\sin (1.2915 + 6\pi)$$

(c) Given that $F(0.6271) = (0.7790, 0.6271)$, find

$$\cos (0.6271)$$
$$\sin (0.6271)$$
$$F(0.6271 - 22\pi)$$
$$\cos (0.6271 - 12\pi)$$
$$\sin (0.6271 - 42\pi)$$

(d) Given that $F(1.4050) = (0.1650, 0.9863)$, find

$$\cos (1.4050)$$
$$\sin (1.4050)$$
$$F(1.4050 + 98\pi)$$
$$\cos (1.4050 + 98\pi)$$
$$\sin (1.4050 - 32\pi)$$

(e) Given that $F(0.0116) = (0.9999, 0.0116)$, find

$$\cos (0.0116)$$
$$\sin (0.0116)$$
$$F(0.0116 - 1000\pi)$$
$$\cos (0.0116 - 182\pi)$$
$$\sin (0.0116 + 348\pi)$$

10. In each case below a real number t is given. You are to tell which quadrant $F(t)$ is in, and whether cos (t) and sin (t) are positive, negative, or zero.

(a) $t \in (\pi/2, \pi)$ (b) $t \in (0, \pi/2)$
(c) $t \in (-\pi/2, 0)$ (d) $t \in (3\pi, (7/2)\pi)$
(e) $t \in (-\frac{3.9}{2}\pi, -19\pi)$

11. From the given information in each case below you are to determine in which quadrant $F(t)$ is located.
(a) $\cos (t) > 0$ and $\sin (t) < 0$
(b) $\cos (t) < 0$ and $\sin (t) > 0$

(c) $\cos(t) > 0$ and $\sin(t) = 3/4$
(d) $\cos(t) < 0$ and $\sin(t) < 0$
(e) $\cos(t) = -0.1368$ and $\sin(t) > 0$

12. Find:

(a) $\cos(9\pi/2)$ (b) $\sin(25\pi/6)$ (c) $\cos(10\pi/3)$
(d) $\sin(-7\pi/3)$ (e) $\sin(11\pi/4)$ (f) $\cos(-5\pi)$

13. Show that the following are periodic functions and draw their graphs.

(a) $f(x) = 2$
(b) $g(x) = [x] + 1 - x$ [Hint: Compute $g(x + 1)$.]
(c) $N(x) = \begin{cases} x - [x] \text{ if } x \le [x] + \frac{1}{2} \\ [x] + 1 - x \text{ if } x > [x] + \frac{1}{2} \end{cases}$
(d) $T(x) = (x - [x])^2$
(e) $L(x) = \begin{cases} x - [x]^2 \text{ if } [x] \text{ is even} \\ -([x] + 1 - x)^2 \text{ if } [x] \text{ is odd} \end{cases}$

There are four trigonometric functions other than sine and cosine; *tangent, cotangent, secant* and *cosecant*. These are defined in terms of sine and cosine as follows:

(a) tangent $= \left\{ \left(t, \dfrac{\sin t}{\cos t} \right) \middle| t \in \mathbf{R} \text{ and } \cos t \ne 0 \right\}$

(b) cotangent $= \left\{ \left(t, \dfrac{\cos t}{\sin t} \right) \middle| t \in \mathbf{R} \text{ and } \sin t \ne 0 \right\}$

(c) secant $= \left\{ \left(t, \dfrac{1}{\cos t} \right) \middle| t \in \mathbf{R} \text{ and } \cos t \ne 0 \right\}$

(d) cosecant $= \left\{ \left(t, \dfrac{1}{\sin t} \right) \middle| t \in \mathbf{R} \text{ and } \sin t \ne 0 \right\}$

These are usually abbreviated respectively as follows: "tan," "cot," "sec," "csc." Then in the $f(t)$-notation:

(a') $\tan(t) = \dfrac{\sin(t)}{\cos(t)}$ (c') $\sec(t) = \dfrac{1}{\cos(t)}$

(b') $\cot(t) = \dfrac{\cos(t)}{\sin(t)}$ (d') $\csc(t) = \dfrac{1}{\sin(t)}$

In the definition of the tangent function, no real number t such that $\cos t = 0$ is allowed. Let us determine those real numbers t such that

$\cos t = 0$. By definition, $\cos t$ is the first coordinate of the point $F(t)$ on the unit circle $\mathcal{C}$. Since there are only two points on the unit circle with first coordinate zero,

$$F(t) = P(0, 1)$$

or

$$F(t) = P(0, -1).$$

The first of these is obviously satisfied by $t = \pi/2$; and from the equation

$$F(t + 2n\pi) = F(t)$$

when $t = \pi/2$,

$$F(\pi/2 + 2n\pi) = F(\pi/2) = P(0, 1).$$

For any number t, if $F(t) = P(0, 1)$, the point $(0, 1)$ has been reached as the end point of an arc measured *from* $I(1, 0)$; and hence, $t = \pi/2 + 2n\pi$ for some integer n. Also if $F(t) = P(0, -1)$, $t = (3\pi)/2 + 2n\pi$ for some integer n. But we may write;

$$(3\pi)/2 + 2n\pi = (\pi/2 + \pi) + 2n\pi$$
$$= \pi/2 + (2n + 1)\pi.$$

Therefore, if $F(t) = P(0, \pm1)$, then

$$t = \pi/2 + 2n\pi \qquad \text{or} \qquad t = \pi/2 + (2n + 1)\pi$$

for some integer n. Notice that in the first case we add *even* multiples of π to $\pi/2$, and in the second case we add *odd* multiples of π to $\pi/2$. Since every integer is either even or odd, we have

$$F(t) = P(0, \pm1) \text{ if and only if } t = \pi/2 + k\pi$$

for some integer k. Hence,

$$\cos(t) = 0 \text{ if and only if } t \in \{\pi/2 + k\pi \mid k \text{ is an integer}\}.$$

The domain of the tangent function is then the set of all real numbers t *except* those of the set $\{\pi/2 + k\pi \mid k \text{ is an integer}\}$; *i.e.*,

$$\mathcal{D}(\tan) = \mathbf{R} - \{\pi/2 + k\pi \mid k \text{ is an integer}\}.$$

In a similar manner, one shows that

$$\mathcal{D}(\cot) = \mathbf{R} - \{k\pi \mid k \text{ is an integer}\}.$$

Let us consider now the range of the sine and cosine. If t is a real number,

$$F(t) = P(a, b)$$

for some numbers a, b. By definition

$$\sin(t) = b \quad \text{and} \quad \cos(t) = a.$$

Since $P(a, b)$ is a point on the unit circle $\mathcal{C}$,

$$a^2 + b^2 = 1$$

or

$$(\cos t)^2 + (\sin t)^2 = 1.$$

It is customary to write $\cos^2 t$ for $(\cos t)^2$ so that the last equation above may be expressed as

$$\cos^2 t + \sin^2 t = 1. \tag{5-1}$$

Hence, from Eq. (5-1) we have

$$0 \le \sin^2 t = 1 - \cos^2 t \le 1$$

and

$$0 \le \cos^2 t = 1 - \sin^2 t \le 1.$$

Therefore (see Example 4–38),

$$-1 \le \sin t \le 1$$

and

$$-1 \le \cos t \le 1.$$

We conclude from this that the range of both the sine and cosine is contained in $[-1, 1]$. As a matter of fact, the range of these functions is exactly the set $[-1, 1]$, but we cannot at this stage offer a proof of the fact that for any number $z \in [-1, 1]$ there are numbers t_1, t_2 such that

$$\sin(t_1) = z \quad \text{and} \quad \cos(t_2) = z.$$

For convenience of reference we list in the table below the domain and range of the trigonometric functions. $\mathbf{R}$ denotes the set of real numbers.

Function	Domain	Range
sine	$\mathbf{R}$	$[-1, 1]$
cosine	$\mathbf{R}$	$[-1, 1]$
tangent	$\mathbf{R} - \{\pi/2 + k\pi\}$	$\mathbf{R}$
cotangent	$\mathbf{R} - \{k\pi\}$	$\mathbf{R}$
secant	$\mathbf{R} - \{\pi/2 + k\pi\}$	$\{x \mid x \ge 1\} \cup \{x \mid x \le -1\}$
cosecant	$\mathbf{R} - \{k\pi\}$	$\{x \mid x \ge 1\} \cup \{x \mid x \le -1\}$

We have just seen that for all real numbers t,

$$\sin^2 t + \cos^2 t = 1.$$

If $\cos t \neq 0$, we may divide the above and obtain

$$\frac{\sin^2 t}{\cos^2 t} + 1 = \frac{1}{\cos^2 t}$$

or

$$\left(\frac{\sin t}{\cos t}\right)^2 + 1 = \left(\frac{1}{\cos t}\right)^2$$

Then, using the definition of tan and sec,

(5-2) $$\tan^2 t + 1 = \sec^2 t.$$

In a similar manner, if we divide Eq. **(5-1)** by $\sin^2 t$ and use the definition of csc and cot, we obtain

(5-3) $$\cot^2 t + 1 = \csc^2 t.$$

Equations **(5-1)**, **(5-2)**, **(5-3)** are true for all real numbers t for which the functions involved are defined and for this reason are referred to as *identities*.

We now derive an identity which allows one to compute $\cos(t - s)$ in terms of $\sin t$, $\sin s$, $\cot t$, $\cos s$. In fact, we have the identity

(5-4) $$\cos(t - s) = (\cos t)(\cos s) + (\sin t)(\sin s)$$

for all real numbers t, s. In order to prove this we refer to Fig. 5–10 where the point $F(s)$ is in the first quadrant and $F(t)$ is in the third quadrant. These particular positions of $F(s)$ and $F(t)$ are not necessary to the proof of **(5-4)**, but do clearly indicate the essential element in the proof, which

Figure 5–10

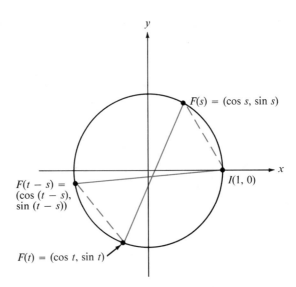

$F(s) = (\cos s, \sin s)$

$I(1, 0)$

$F(t - s) = (\cos(t - s), \sin(t - s))$

$F(t) = (\cos t, \sin t)$

is that the cords $F(t)F(s)$* and $F(t - s)I(1, 0)$ are equal in length. This being so, the distance formula gives

$$\sqrt{[\cos (t - s) - 1]^2 + [\sin (t - s) - 0]^2}$$
$$= \sqrt{(\cos t - \cos s)^2 + (\sin t - \sin s)^2}.$$

If we square the above, we get

$$\cos^2 (t - s) - 2 \cos (t - s) + 1 + \sin^2 (t - s)$$
$$= \cos^2 t - 2 \cos t \cdot \cos s + \cos^2 s + \sin^2 t - 2 \sin t \cdot \sin s + \sin^2 s.$$

This reduces to

$$2 - 2 \cos (t - s) = 2 - 2(\cos t \cdot \cos s + \sin t \cdot \sin s),$$

and this last equation is equivalent to Eq. **(5-4)**. It should be observed in the above manipulations that we have used identity Eq. **(5-1)** three times, to replace

$$\cos^2 (t - s) + \sin^2 (t - s)$$
$$\cos^2 t + \sin^2 t$$
$$\cos^2 s + \sin^2 s$$

each by 1.

Now we use identity Eq. **(5-4)** to derive several other identities. Choose $t = 0$ in Eq. **(5-4)**:

$$\cos (-s) = \cos (0 - s) = \cos 0 \cdot \cos s + \sin 0 \cdot \sin s$$
$$= 1 \cdot \cos s + 0 \cdot \sin s$$
$$= \cos s.$$

Hence, for all real numbers s,

$$\cos (-s) = \cos s. \tag{5-5}$$

In Eq. **(5-4)** take $t = \pi/2$:

$$\cos (\pi/2 - s) = \cos \pi/2 \cdot \cos s + \sin \pi/2 \cdot \sin s$$
$$= 0 \cdot \cos s + 1 \cdot \sin s$$
$$= \sin s.$$

Therefore, for all real numbers s,

$$\cos (\pi/2 - s) = \sin s. \tag{5-6}$$

---

*The notation AB means the cord or line segment connecting the points A, B.

Taking $t = -\pi/2$ in **(5-4)**, gives

$$\cos(-\pi/2 - s) = \cos(-\pi/2) \cdot \cos s + \sin(-\pi/2) \cdot \sin s$$
$$= 0 \cdot \cos s + (-1) \cdot \sin s = -\sin s.$$

This gives (a) $\cos(-\pi/2 - s) = -\sin s$. But also, by **(5-4)**,

$$\cos(\pi/2 + s) = \cos\big(\pi/2 - (-s)\big)$$
$$= \cos \pi/2 \cdot \cos(-s) + \sin \pi/2 \cdot \sin(-s)$$
$$= 0 \cdot \cos(-s) + 1 \cdot \sin(-s)$$
$$= \sin(-s).$$

Therefore, (b) $\cos(\pi/2 + s) = \sin(-s)$. From **(5-5)**,

$$\cos(-\pi/2 - s) = \cos[-(\pi/2 + s)] = \cos(\pi/2 + s).$$

We then have from (a) and (b)

(5-7) $$\sin(-s) = -\sin s$$

for all real numbers s.

By writing $t + s = t - (-s)$ and using Eq. **(5-4)** and **(5-7)**, we have that for all real numbers s, t,

(5-8) $$\cos(t + s) = \cos t \cdot \cos s - \sin t \cdot \sin s.$$

In Eq. **(5-6)** take $s = \pi/2 - t$:

$$\cos[\pi/2 - (\pi/2 - t)] = \sin(\pi/2 - t)$$

or

(5-9) $$\cos t = \sin(\pi/2 - t)$$

for all real numbers t.

Now using Equations **(5-9)**, **(5-6)** and **(5-4)**,

$$\sin(s + t) = \cos[\pi/2 - (s + t)]$$
$$= \cos[(\pi/2 - s) - t]$$
$$= \cos(\pi/2 - s) \cdot \cos t + \sin(\pi/2 - s) \cdot \sin t$$
$$= \sin s \cdot \cos t + \cos s \cdot \sin t.$$

Therefore, for all real numbers s, t,

(5-10) $$\sin(s + t) = \sin s \cdot \cos t + \cos s \cdot \sin t.$$

We list below several other important trigonometric identities, the proofs of which will be left to the student.

For all real numbers s, t for which the functions involved are defined,

$$\sin (t - s) = \sin t \cdot \cos s - \cos t \cdot \sin s \qquad \text{(5-11)}$$

$$\tan (t + s) = \frac{\tan t + \tan s}{1 - (\tan t) \cdot (\tan s)} \qquad \text{(5-12)}$$

$$\tan (t - s) = \frac{\tan t - \tan s}{1 + (\tan t) \cdot (\tan s)} \qquad \text{(5-13)}$$

$$\sin (2t) = 2 \sin t \cdot \cos t \qquad \text{(5-14)}$$

$$\cos (2t) = \cos^2 t - \sin^2 t \qquad \text{(5-15)}$$

$$= 2 \cos^2 t - 1$$

$$= 1 - 2 \sin^2 t$$

$$\sin (t/2) = \pm \sqrt{\frac{1 - \cos t}{2}} \qquad \text{(5-16)}$$

$$\cos (t/2) = \pm \sqrt{\frac{1 + \cos t}{2}} \qquad \text{(5-17)}$$

$$\sin t \cdot \cos s = (\tfrac{1}{2})[\sin (t + s) + \sin (t - s)] \qquad \text{(5-18)}$$

$$= (\tfrac{1}{2})[\sin (s + t) - \sin (s - t)]$$

$$(\tfrac{1}{2})[\sin t + \sin s] = \left[\sin \left(\frac{t + s}{2} \right) \right]\left[\cos \left(\frac{t - s}{2} \right) \right] \qquad \text{(5-19)}$$

Show that for all real numbers t for which the functions are defined,

$$\cos t + \csc t = \frac{\sin t}{1 - \cos t}.$$

Example 5–11

Solution. By definition of cot and csc,

$$\cot t + \csc t = \frac{\cos t}{\sin t} + \frac{1}{\sin t} = \frac{\cos t + 1}{\sin t} = \frac{(1 - \cos t)(1 + \cos t)}{(1 - \cos t) \cdot \sin t}$$

$$= \frac{1 - \cos^2 t}{(1 - \cos t) \cdot \sin t} = \frac{\sin^2 t}{(1 - \cos t) \cdot \sin t} \quad \text{by Eq. (5-1)}$$

$$= \frac{\sin t}{1 - \cos t}$$

This proves the identity.

Compute $\sin (5\pi/12)$.

Example 5–12

Solution. Since $5\pi/12 = \pi/6 + \pi/4$, we have by Eq. (5-10),

$$\sin (5\pi/12) = \sin (\pi/6 + \pi/4) = \sin \pi/6 \cdot \cos \pi/4 + \cos \pi/6 \cdot \sin \pi/4$$

$$= (\tfrac{1}{2})(1/\sqrt{2}) + (\tfrac{1}{2})\sqrt{3} \cdot (1/\sqrt{2}) = (\sqrt{2}/4)(1 + \sqrt{3})$$

14. Given that $\sin (\pi/3) = \frac{1}{2}\sqrt{3}$, $\cos (\pi/3) = \frac{1}{2}$, $\sin (-\pi/4) = -\frac{1}{2}\sqrt{2}$, and $\cos (-\pi/4) = \frac{1}{2}\sqrt{2}$ use appropriate identities to compute the following.

(a) $\sin (-\pi/3)$

(b) $\cos (\pi/4)$

(c) $\cos (-7\pi/12)$ [*Hint:* write $-\pi/4 - \pi/3$ for $-7\pi/12$.]

(d) $\sin (-\pi/8)$

(e) $\tan (-\pi/4)$

(f) $\tan (\pi/3)$

(g) $\sin (2\pi/3)$

(h) $\cos (2\pi/3)$

(i) $\cos (\pi/8)$

15. Given that $\sin (1.2915) = 0.9613$ and $\cos (1.2915) = 0.2756$ compute the following.

(a) $\tan (1.2915)$ (b) $\sin (-1.2915)$

(c) $\cos (-1.2915)$ (d) $\sin (2.5830)$

16. Given that $\cos (0.0116) = 0.9999$ and $\sin (0.0116) = 0.0116$ compute the following.

(a) $\tan (0.0116)$ (b) $\sin (-0.0116)$

(c) $\cos (-0.0116)$ (d) $\tan (-0.0116)$

(e) $\sin (0.0058)$

17. Given that $\sin (1.4370) = 0.9911$ and $\cos (1.4370) = 0.1334$ compute the following.

(a) $\sin (-1.4370)$ (b) $\cos (-1.4370)$

(c) $\tan (1.4370)$ (d) $\sin (2.8740)$

(e) $\cos (2.8740)$ (f) $\sin (0.7185)$

(g) $\cos (0.7185)$

18. Complete the proof of Eq. **(5-8)**.

19. Prove Eq. **(5-11)** by using Eq. **(5-10)**.

20. Prove Eq. **(5-12)** by using Eq. **(5-9)** and Eq. **(5-10)**.

21. Prove Eq. **(5-13)**.

22. Prove Eq. **(5-14)** by using Eq. **(5-10)**.

23. Prove Eq. **(5-15)**.

24. Prove Eq. **(5-16)** by taking $t/2$ for t in Eq. **(5-15)**. Also discuss the $\pm$ sign in this identity; *i.e.*, discuss the conditions under which the $+$ is appropriate and the conditions under which the sign should be $-$.

25. Prove Eq. **(5-17)**. Discuss the $\pm$ sign in this identity.

26. Prove Eq. **(5-18)** by computing $\sin(t+s)$ and $\sin(t-s)$.

27. Prove Eq. **(5-19)** by using Eq. **(5-18)**.

28. (*a*) Show that if $\sec t$ is defined for the number t, then
$$\sec t \geq 1 \qquad \text{or} \qquad \sec t \leq -1.$$

   (*b*) Show that if $\csc t$ is defined for the number t, then
$$\csc t \geq 1 \qquad \text{or} \qquad \csc t \leq -1.$$

29. Prove the following identities.

   (*a*) $\dfrac{1 - \sin u}{\cos u} = \sec u - \tan u$

   (*b*) $\dfrac{1}{2(1 - \sin x)} + \dfrac{1}{2(1 + \sin x)} = \sec^2 x$

   (*c*) $\sin y + \cos^2 y \csc y = \csc y$

   (*d*) $3 \cos^2 t + 4 \sin^2 t = 4 - \cos^2 t$

   (*e*) $\tan^2 z - \sin^2 z = \tan^2 z \sin^2 z$

30. Prove the following identities:

   (*a*) $\cos t = -\sin (3\pi/2 + t)$

   (*b*) $\tan x + \cot x = (\sec x)(\csc x)$

   (*c*) $(\sec t + \tan t)(1 - \sin t) = \cos t$

   (*d*) $(\sec t - \tan t)^{-1} = \sec t + \tan t$

   (*e*) $(1 + \cot^2 t) = (\csc^2 t - 1)(\tan^2 t + 1)$

   (*f*) $\sin^4 t - \cos^4 t = 2 \sin^2 t - 1$

31. Prove the following identities.

   (*a*) $(\tan v + 1)^2 = \sec^2 v(1 + \sin (2v))$

   (*b*) $\sin^2 w(\cot w + 1)^2 = 1 + \sin (2w)$

   (*c*) $\dfrac{\sin v + \cos v}{\tan v} = \cos v + \csc v - \sin v$

   (*d*) $\dfrac{\csc v - \cot v}{\tan v} = \dfrac{\cos v}{1 + \cos v}$

(e) $1 - \cos^4 t = \sin^2 t(2 - \sin^2 t)$

(f) $\dfrac{\tan w}{1 - \cot w} + \dfrac{\cot w}{1 - \tan w} = 1 + \tan w + \cot w$

5.4

Angles, Triangles and the Trigonometric Functions

If an angle AOB is given, we may take a coordinate system with x-axis along one of its sides and origin at the vertex as shown in Fig. 5–11. If the unit circle is centered at the origin, the angle AOB determines two arcs on this circle. The two arcs in turn each have a length which can be used as a measure of the angle. This kind of angle measure is known as *radian* measure. It is common practice to use the word "angle" ambiguously to mean both the geometric figure and one of the two arc lengths associated with this geometric figure. We shall not argue with this practice here; usually our main concern will be with the numbers (or measures) associated with the angle and we will distinguish between the two numbers by drawing small arcs as in Fig. 5–11.

Figure 5–11

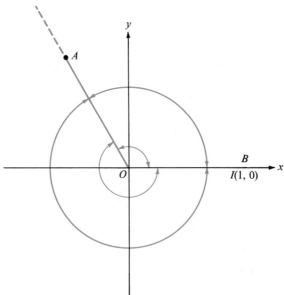

If any number t is given, a *directed angle* will be associated with t whose radian measure is t. This is accomplished by thinking of the x-axis as the *initial* side of each angle and rotating the x-axis about the origin until it coincides with the terminal side and has determined an arc on the unit circle of length $|t|$ (see Fig. 5–12). In the same way it is possible to associ-

Figure
5–12

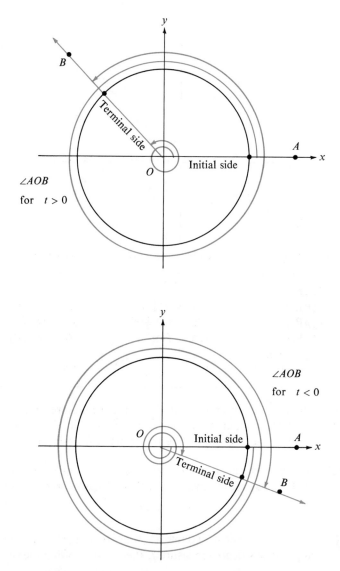

∠AOB
for t > 0

∠AOB
for t < 0

ate with any number t a directed angle whose measure is $t°$. It will be recalled that in degree measure, a right angle is 90; *i.e.*, a right angle (as in
Fig. 5–13) measures 90 degrees (90°). The radian measure of a right angle
is by our definition $\frac{1}{4}$ of the length of the circumference of the unit circle:
$\frac{1}{4}(2\pi) = (\frac{1}{2})\pi$. From this we may conclude that a straight angle measures
180° and π radians since a straight angle is twice as "large," and, hence,
has measure twice that of a right angle. In like manner, we can arrive at
the information given in the following table:

**Figure
5–13**

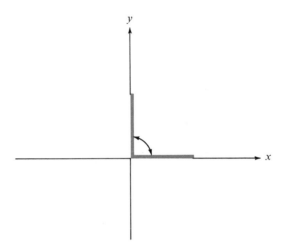

Angle	Radian measure	Degree measure
Right ⌐	$(\frac{1}{2})\pi$	90
$\frac{1}{2}$(Right)	$(\frac{1}{4})\pi[=\frac{1}{2}(\frac{1}{2})\pi]$	$45[=\frac{1}{2}(90)]$
$(\frac{1}{3})$(Right)	$\pi/6[=(\frac{1}{3})(\frac{1}{2})\pi]$	$30[=(\frac{1}{3})(90)]$
$(\frac{2}{5})$(Right)	$\pi/5[=(\frac{2}{5})(\frac{1}{2})\pi]$	$36[=(\frac{2}{5})(90)]$
$(\frac{1}{90})$(Right)	$\pi/180$	1
$(2/\pi)$(Right)	1	$180/\pi$

From the table, one observes that if an angle measures 1°, its radian measure is $\pi/180$, and if an angle measures 1 radian, its degree measure is $180/\pi$. From this fact we obtain the following conversion rules.

(I) TO CONVERT FROM RADIAN MEASURE TO DEGREE MEASURE: If an angle A has radian measure t, then A has degree measure

$$t\cdot(180/\pi).$$

(II) TO CONVERT FROM DEGREE MEASURE TO RADIAN MEASURE: If an angle A has degree measure t, then A has radian measure

$$t\cdot(\pi/180).$$

**Example
5–13**

Find the degree measure of an angle which measures 14 radians.

Solution. From the conversion rule I:

$$14(180/\pi) = 2520/\pi$$

is the degree measure.

What is the sum of the interior angles of a triangle in radian measure?

Example
5–14

Solution. From geometry, we know that the sum is as large as a straight angle; and a straight angle has radian measure π.

The use of degree measure of angles leads to new functions in terms of the trigonometric functions. If we are given a number t such that

$$0 \le t \le 360,$$

t is the degree measure of some angle; and by our conversion rule, this angle has radian measure

$$t \cdot (\pi/180).$$

Next, the number $t \cdot (\pi/180)$ measures a certain arc on the unit circle leading to the point

$$F\big(t \cdot (\pi/180)\big).$$

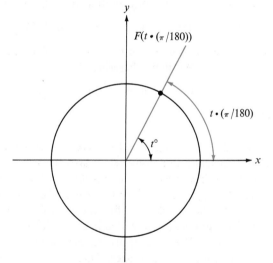

Figure
5–14

(See Fig. 5–14.) Now we may define

$$f(t) = \sin\big(t \cdot (\pi/180)\big) \qquad t \in [0, 360]$$
$$g(t) = \cos\big(t \cdot (\pi/180)\big) \qquad t \in [0, 360].$$

These functions usually are written

$$f(t) = \sin(t^\circ) \qquad g(t) = \cos(t^\circ),$$

that is,

$$\sin(t°) = \sin\left(t\cdot(\pi/180)\right)$$
$$\cos(t°) = \cos\left(t\cdot(\pi/180)\right).$$

It should be emphasized here that these two functions f, g are *not* the same functions sine and cosine as originally defined.

Example 5–15

Find $\sin(60°)$.

Solution. By definition

$$\sin(60°) = \sin(60\cdot\pi/180) = \sin(\pi/3) = \tfrac{1}{2}\sqrt{3}.$$

Example 5–16

Find $\cos(270°)$.

Solution. $\cos(270°) = \cos(270\cdot\pi/180) = \cos(3\pi/2) = 0.$

The reader has no doubt noticed that the connection the functions $\sin(t°)$ and $\cos(t°)$ have with angles is not essential since the functions

$$\sin(t\cdot\pi/180) \qquad \text{and} \qquad \cos(t\cdot\pi/180)$$

are defined for *all* real numbers t, and hence the restriction to the interval $[0, 360]$ is not necessary. We may, therefore, write

$$\sin(t°) = \sin(t\cdot\pi/180) \qquad \text{and} \qquad \cos(t°) = \cos(t\cdot\pi/180)$$

for all real numbers t.

The table on page 362 is arranged so that one may convert from degree measure to radian measure and conversely; also the values of the trigonometric functions are given for $t°$ where $0 \le t \le 90$ and for some real numbers s where $0 \le s \le \pi/2 \doteq 1.5708$ ($\doteq$ is used for "approximately"). This presents us with the problem of finding the values of the trigonometric functions for numbers not included in the table. For example, how do we find $\sin(3.2463)$? We may approximate $\sin(3.2463)$ as follows. First, $\pi \doteq 3.1416$ to four decimal places. Then

$$3.2463 = 3.1416 + 0.1047$$
$$\doteq \pi + 0.1047.$$

Next, by Eq. **(5-10)**

$$\sin(\pi + 0.1047) = \sin\pi\cos(0.1047) + \cos\pi\sin(0.1047)$$
$$= -\sin(0.1047)$$

Since $0 < 0.1047 < 1.5708$, sin (0.1047) may be found in the table. In fact, sin (0.1047) = 0.1045. Therefore, sin (3.2463) $\doteq$ -0.1045.

In finding sin (3.2463) we used a *reference number*, 0.1047; if x is any number and

$$x = n\pi \pm r \quad \text{where } n \in Z \quad \text{and} \quad 0 \leq r \leq \pi/2$$

then r is called a reference number for x. With the aid of reference numbers and our knowledge of the trigonometric functions in the interval $[0, \pi/2]$ we may compute the trigonometric functions for any number. This is made possible by the following formulas:

$$\sin (n\pi \pm r) = \pm\sin r \qquad \textbf{(5-20)}$$
$$\cos (n\pi \pm r) = \pm\cos r \qquad \textbf{(5-21)}$$
$$\tan (n\pi \pm r) = \pm\tan r \qquad \textbf{(5-22)}$$

These three results are immediate consequences of Eq. **(5-10)**, Eq. **(5-11)**, Eq. **(5-8)**, Eq. **(5-4)**. The information contained in these formulas may be summarized by saying that *the value of a trigonometric function for any number* x *is the same as the value of the trigonometric function for the reference number* r, *except possibly for sign.*

Find cos (16.2403).

Example
5–17

Solution. We first see that

$$5\pi \doteq 5(3.1416) < 16.2403 < 6(3.1416) \doteq 6\pi.$$

Then

$$16.2403 = 6(3.1416) - 2.6093$$

and

$$16.2403 = 5(3.1416) + 0.5323.$$

Since $0 < 0.5323 < 1.5708$, we see that 0.5323 is the reference number for 16.2403. Therefore, we use the table to find cos (0.5323) = 0.8616. From the above information about reference numbers we conclude that cos (16.2403) = ±0.8616. But we can see from the equation

$$16.2403 = 5(3.1416) + 0.5323$$

that $F(16.2403)$ is in quadrant III so that cos (16.2403) < 0. Hence, cos (16.2403) = -0.8616.

Equations (5-20, 5-21, 5-22) are called *reduction* formulas for the trigonometric functions. Fortunately, these reduction formulas have analogues for the functions sin ($t°$), cos ($t°$), tan ($t°$):

$$\sin [(180n)° \pm r°] = \pm \sin (r°) \qquad \textbf{(5-23)}$$
$$\cos [(180n)° \pm r°] = \pm \cos (r°) \qquad \textbf{(5-24)}$$
$$\tan [(180n)° \pm r°] = \pm \tan (r°) \qquad \textbf{(5-25)}$$

To indicate the method of proof we will show that Eq. (5-23) holds:

$$\sin [(180n)° \pm r°] = \sin \left[(180n \pm r) \cdot \frac{\pi}{180} \right]$$

by definition

$$= \sin \left(n\pi \pm \frac{r\pi}{180} \right)$$

by Eq. (5-20)

$$= \pm \sin \left(\frac{r\pi}{180} \right)$$

by definition

$$= \pm \sin (r°).$$

Example 5–18

Compute tan [($-486.5)°$].

Solution.

$$-3(180°) < -486.5° < -2(180°).$$

Then

$$-486.5° = -3(180°) + 53.5°$$

and

$$-486.5° = -2(180°) + (-126.5).$$

Since $0 < 53.5 < 90$, we see that 53.5 is the reference number and this implies that tan [($-486.5)°$] $= \pm \tan (53.5°)$. From the equation above involving the reference number the terminal side of an angle whose measure is $-486.5°$ will be in quadrant III so that tan [($-486.5)°$] is positive. From the table, tan ($53.5°$) $= 1.3514$ and, hence,

$$\tan [(-486.5)°] = 1.3514.$$

If we are given the problem of finding sin (0.2153), we find that although $0 < 0.2153 < 1.5708$, the number 0.2153 is not in Table I. Also it is not possible to find 15°20′ in Table I. This deficiency in Table I can be overcome to some extent by employing a method of approximation known as *interpolation*. This will be illustrated by finding sin (0.2153). The first step is to find the numbers in Table I which are closest to 0.2153; these are 0.2094 and 0.2182. Next, for convenience we arrange our information as has been done below:

$$0.0088 \begin{Bmatrix} 0.2094 \\ 0.2153 \\ 0.2182 \end{Bmatrix} 0.0059 \qquad x \begin{Bmatrix} \sin(0.2094) = 0.2079 \\ \sin(0.2153) = \quad ? \\ \sin(0.2182) = 0.2164 \end{Bmatrix} 0.0085$$

In the above scheme 0.0088 is the difference between the upper and lower numbers and 0.0059 is the difference between the first and second numbers; 0.0085 is the difference between sin (0.2094) and sin (0.2182) and x represents the difference between sin (0.2094) and sin (0.2153) (which we do not know). The next step in this process is to solve the following equation for x:

$$\frac{0.0059}{0.0088} = \frac{x}{0.0085}.$$

Then $x = 0.0085(59/88) \doteq 0.0057$. Now, since x was the difference between sin (0.2094) and sin (0.2153) and since sin (0.2094) < sin (0.2182), we *add* $x = 0.0057$ to sin (0.2094) = 0.2079 to obtain our approximation to sin (0.2153):

$$\sin(0.2153) \doteq 0.2079 + 0.0057 = 0.2136.$$

In this particular case the approximation obtained is correct to four decimal places.

The justification for interpolation is indicated in Fig. 5–15. The assumption is made that for "small" intervals such as [0.2094, 0.2182] the curve

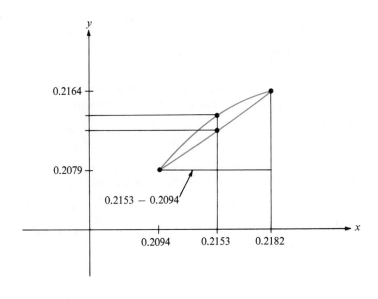

Figure
5–15

differs little from the line joining the endpoints of the curve. Then the computations are carried out on the basis of similar triangles.

Example 5–19

Compute cos (15°20′).

Solution. The information needed from Table I is arranged below:

$$30' \begin{cases} 15° \\ 15°20' \\ 15°30' \end{cases} 20' \qquad x \begin{cases} \cos{(15°)} = 0.9659 \\ \cos{(15°20')} = \quad ? \\ \cos{(15°30')} = 0.9636 \end{cases} 0.0023$$

Now the equation to be solved is

$$\frac{20}{30} = \frac{x}{0.0023}.$$

$$x = (0.0023)(2/3) \doteq 0.0015.$$

In this case we see that cos (15°) > cos (15°30′) so that we *subtract* 0.0015 from cos (15°) = 0.9659:

$$\cos{(15°20')} \doteq 0.9659 - 0.0015 = 0.9644.$$

Exercises

32. If each of the following numbers is the radian measure of an angle, find the corresponding degree measure.

    (a) 2π (b) 0   (c) $\pi/8$ (d) $11\pi/6$

33. If each of the following numbers is the degree measure of an angle, find the corresponding radian measure.

    (a) 4   (b) 0.59   (c) $\frac{13}{4}$ (d) 0.32

34. If each of the following numbers is the radian measure of an angle, use Table I to find the corresponding degree measure.

    (a) 0.1833   (b) 1.4137   (c) 0.0698
    (d) 0.9861   (e) 0.8465   (f) 0.5934

35. If each of the following numbers is the degree measure of an angle, use Table I to find the corresponding radian measure.

    (a) 1.5   (b) 83   (c) 19.5   (d) 56.5   (e) 71   (f) 38

36. Each number given below is the radian measure of some angle. Find a reference number in each case.

Example: 8.7. We wish to find a number r such that $8.7 = n\pi \pm r$ where $0 \le r \le \pi/2 \doteq 1.5708$ for some integer n. Since $8.7 > 0$ we see that n must be positive. Trying the first few positive integers we have

$$1 \cdot \pi \doteq 1 \cdot (3.1416) = 3.1416 < 8.7$$
$$2 \cdot \pi \doteq 2 \cdot (3.1416) = 6.2832 < 8.7$$
$$3 \cdot \pi \doteq 3 \cdot (3.1416) = 9.4248 > 8.7$$

Then we see that $2 \cdot (3.1416) < 8.7 < 3 \cdot (3.1416)$. Also

$$8.7 = 2 \cdot (3.1416) + 2.1468$$
$$8.7 = 3 \cdot (3.1416) - 0.7248$$

We have now found the number we desire, namely 0.7248, since $0 < 0.7248 < 1.5708$.

(*a*) 3.2715   (*b*) 12.25   (*c*) −1.4
(*d*) −13.0083   (*e*) 33.1111

37. Each number given below is the degree measure of some angle. Find a reference number in each case.

(*a*) 135°   (*b*) 256°   (*c*) 413.5°   (*d*) −8.5°
(*e*) −29°   (*f*) −16.5°

38. Use reference numbers and Table I to find sin, cos, tan of each of the following numbers.

(*a*) 9.4859   (*b*) 8.5783   (*c*) −4.9829
(*d*) −7.2780   (*e*) 32.9344   (*f*) −30.0110

39. Use reference numbers and Table I to find sin, cos, tan of each of the following:

(*a*) 95°   (*b*) 458°   (*c*) −383°
(*d*) 181.5°   (*e*) −10.5°   (*f*) −724.5°

40. Compute each of the following:

(*a*) tan (38°10′)   (*b*) cos (22°40′)
(*c*) sin (0.5340)   (*d*) cot (1.2702)
(*e*) sin (29.2104)   (*f*) cos (324°20′)
(*g*) tan (−19.3864)   (*h*) cos (−422°35′)

41. Use Table I and solve each of the equations below (see Exercise 5).

(*a*) cos ($x°$) = 0.9426   (*b*) sin ($x°$) = 0.8988
    cos (x) = 0.9426   sin (x) = 0.8988
(*c*) sin ($x°$) = 0.1800
    sin (x) = 0.1800

Solution. In Table I there is no entry of 0.1800. But we see that 0.1800 lies between sin $(10°) = 0.1736$ and sin $(10.5°) = 0.1822$. Therefore, we use interpolation as before:

$$0.5° \begin{bmatrix} 10° \\ x° \\ 10.5° \end{bmatrix} x° - 10° \quad 0.0064 \begin{cases} \sin(10°) = 0.1736 \\ \sin(x°) = 0.1800 \\ \sin(10.5°) = 0.1822 \end{cases} 0.0086$$

$$\frac{x - 10}{0.5} = \frac{0.0064}{0.0086} = \frac{64}{86} = \frac{32}{43}$$

$$x - 10 = (0.5)\left(\frac{32}{43}\right) \doteq 0.372$$

$$x = 10 + 0.372 = 10.37$$

We may convert $0.37°$ into minutes: $60 \cdot (0.37) = 22.2$. Then $x° = 10.37° = 10°22.2'$. To obtain sin $(x) = 0.1800$ we proceed in the same way:

$$0.0088 \begin{bmatrix} 0.1745 \\ x \\ 0.1833 \end{bmatrix} x - 0.1745 \quad 0.0064 \begin{cases} \sin(0.1745) = 0.1736 \\ \sin(x) = 0.1800 \\ \sin(0.1833) = 0.1822 \end{cases} 0.0086$$

$$\frac{x - 0.1745}{0.0088} = \frac{0.0064}{0.0086} = \frac{64}{86} = \frac{32}{43} \doteq 0.744$$

$$x - 0.1745 = (0.0088) \cdot (0.744) \doteq 0.0065$$

$$x = 0.0065 + 0.1745 = 0.1810$$

(d) $\tan(x°) = 0.6615$ (e) $\cos(x°) = 0.6075$
     $\tan(x) = 0.6615$ $\cos(x) = 0.6075$
(f) $\sin(x°) = 0.9231$ (g) $\cos(x°) = 0.1723$
     $\sin(x) = 0.9231$ $\cos(x) = 0.1723$

5.5
The Law of Sines, the Law of Cosines, Solving Triangles

Let us now consider a point $A(a, b)$ as in Fig. 5–16(a). We let t be a measure of the angle AOI as shown. By the distance formula,

$$OA = \sqrt{a^2 + b^2}.$$

For convenience we let $\sqrt{a^2 + b^2} = r$. From the fact that the triangles ABO and PCO are similar, we have

$$\frac{b}{r} = \frac{\sin t}{1} \quad \text{and} \quad \frac{-a}{r} = \frac{-\cos t}{1}$$

or

$$b = r \sin t \quad \text{and} \quad a = r \cos t. \tag{5-26}$$

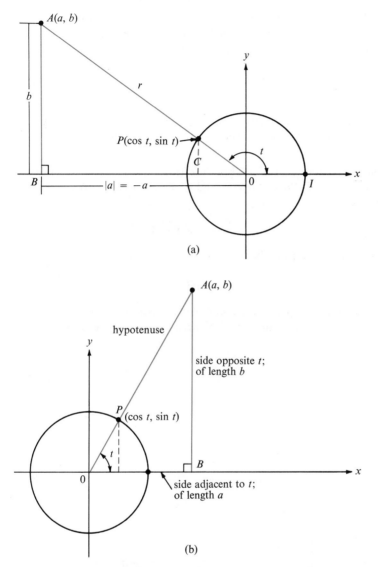

(a)

(b)

This says that the coordinates of a point A may be given in terms of the distance the point A is from the origin O and a measure t of the angle AOI. Fig. 5-16(b) illustrates this situation with A in the first quadrant. In this case we may express $\sin t$ and $\cos t$ in terms of the length of the sides of the right triangle AOB:

$$\sin t = \frac{L(\text{opposite})}{L(\text{hypotenuse})} \qquad \cos t = \frac{L(\text{adjacent})}{L(\text{hypotenuse})}$$

We have used $L(\ )$ here to mean "length of." This shows that the definitions we gave of the trigonometric functions for triangles in Section 5.1 coincides with the more general definition given in Section 5.2.

Theorem 24

The area of any triangle is $\frac{1}{2}$ the product of the length of any two sides and the sine of the angle included by the two sides.

Proof. Consider $\triangle ABC$ of Fig. 5–17(a). Choose a coordinate system as shown in Fig. 5–17(b). Then using Eq. **(5-26)**, with $h = b$ and $c = r$ in this case, we have

$$h = c \cdot \sin \measuredangle A.$$

Also, we know that the area of $\triangle ABC$ is $\frac{1}{2}bh$. Therefore, the area is

$$\tfrac{1}{2}bh = \tfrac{1}{2}bc \cdot \sin \measuredangle A$$

by substituting for h. This completes the proof for sides b, c and in the same way we could show that the area is $\frac{1}{2}ab \cdot \sin \measuredangle C$ and $\frac{1}{2}ac \cdot \sin \measuredangle B$.

Figure 5–17

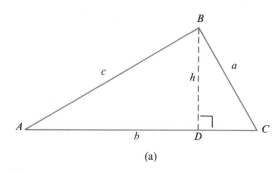

(a)

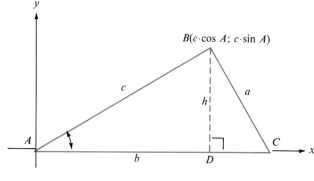

(b)

If $\triangle ABC$ is labeled as in Fig. 5-17(a), then

$$\frac{\sin \angle A}{a} = \frac{\sin \angle B}{b} = \frac{\sin \angle C}{c}$$

Proof. By Theorem 24 we have

$$\tfrac{1}{2}bc \cdot \sin \angle A = \tfrac{1}{2}ac \cdot \sin \angle B = \tfrac{1}{2}ab \cdot \sin \angle C.$$

Then dividing by $\tfrac{1}{2}abc$ completes the proof.

If a triangle ABC has $A = 43°$, $C = 49°$ and $b = 1$, find B, a, c.

Example
5–20

Solution. First of all, since the sum of the interior angles of ABC is $180°$,

$$B = 180° - (43° + 49°)$$
$$= 180° - 92° = 88°.$$

By the law of sines,

$$\frac{\sin (43°)}{a} = \frac{\sin (88°)}{1} = \frac{\sin (49°)}{c}.$$

Then, using Table I of trigonometric functions, we have

$$a = (\sin 43°)/(\sin 88°) \doteq (.6820)/(.9994) = 0.6824$$
$$c = (\sin 49°)/(\sin 88°) \doteq (.7547)/(.9994) = 0.7552.$$

Let $\triangle ABC$ be labeled as in Fig. 5–18. Then the following equations hold:

(a) $a^2 = b^2 + c^2 - 2bc \cdot \cos \angle A$

(b) $b^2 = a^2 + c^2 - 2ac \cdot \cos \angle B$

(c) $c^2 = a^2 + b^2 - 2ab \cdot \cos \angle C$.

Proof. Let a coordinate system be set up as indicated in Fig. 5–18. Then by Eq. **(5-26)** C has coordinates $(b \cdot \cos \angle A, b \cdot \sin \angle A)$; the coordinates of B are $(c, 0)$. Using the distance formula for the distance between B and C we have

$$a = \sqrt{(b \cdot \cos \angle A - c)^2 + (b \cdot \sin \angle A - 0)^2}$$

or

$$a^2 = b^2 \cos^2 \measuredangle A - 2bc \cdot \cos \measuredangle A + c^2 + b^2 \sin^2 \measuredangle A$$
$$= b^2(\cos^2 \measuredangle A + \sin^2 \measuredangle A) + c^2 - 2bc \cdot \cos \measuredangle A$$
$$= b^2 \cdot 1 + c^2 - 2bc \cdot \cos \measuredangle A$$
$$= b^2 + c^2 - 2bc \cdot \cos \measuredangle A.$$

This establishes (*a*); (*b*) and (*c*) are proved in a similar manner.

**Figure
5–18**

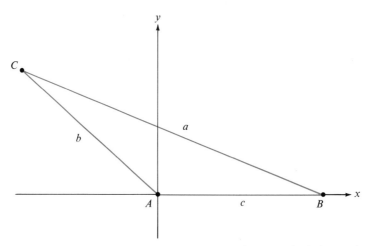

When we speak of *solving a triangle* we mean that certain information is given about the measures of sides and angles of the triangle and that we wish to determine the measures of the remaining parts, if possible. In Example 5–20 we solved the triangle ABC in which b, $m \measuredangle C$ and $m \measuredangle A$ were given. In order to obtain a unique solution in solving a triangle we

**Figure
5–19**

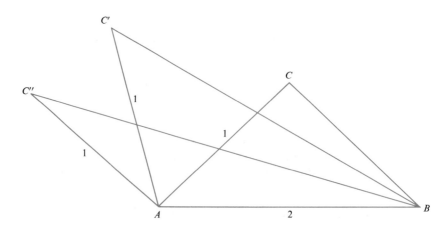

must be given sufficient information. There are, for example, many different, non-congruent triangles having two sides measuring 1 and 2 inches respectively (see Fig. 5–19). The reader may recall from high school geometry that in order to uniquely determine the measures of the six parts of a triangle we need to know *at least* three parts. Although three parts are not always sufficient, fewer than three are never sufficient. We list below all possible cases of three given parts of a triangle and follow this list with an analysis.

GIVEN INFORMATION IN SOLVING

TRIANGLES	SOLUTIONS
(*a*) Three sides (*SSS*)	If a solution exists, it is always unique.
(*b*) Two sides and an angle	
(i) Angle included between sides (*SAS*)	A solution always exists and is unique.
(ii) Angle not included between sides (*SSA*)	If a solution exists, it may not be unique.
(*c*) Two angles and side (*ASA* and *AAS*)	If a solution exists, it is unique.
(*d*) Three angles (*AAA*)	A solution, if it exists, is never unique.

(*a*) *Case SSS.* In this case a solution is always unique. However, for given *a*, *b*, *c* a solution may not exist. For example, there is no solution for $a = 1$, $b = 1$ and $c = 2$. The reason for this is that $a + b = c$ and in any triangle the sum of the lengths of two sides is greater than the third. A solution that does exist may be found by using the law of Cosines and the Law of Sines.

Analysis of Solution Cases

(*b*) *Two sides and an angle.*

(i) *Case SAS.* A solution always exists and is unique; it may be found by using the Law of Cosines.

(ii) *Case SSA.* Here we may have none, one or two solutions as illustrated in Fig. 5–20. In Fig. 5–20 we are given $m \angle A$, *a* and *c*; Fig. 5–20(*a*) has no solution because *a* is not long enough; (*b*) has two solutions as indicated; (*c*) has exactly one solution. The solutions may be found, if they exist, by the Law of Sines.

**Figure
5–20**

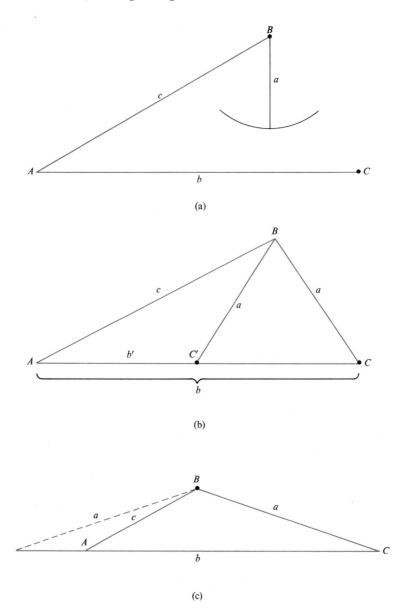

(a)

(b)

(c)

(c) *Case ASA and AAS.* A solution may not exist in this case. One example for which no solution exists is $m \angle A = 95°$, $m \angle B = 100°$, $c = 14$. The reason a solution does not exist for this information is that a triangle can have at most one obtuse angle. When a solution exists it is unique and may be found by the Law of Sines.

(d) *Case AAA.* No solution exists if $m \angle A + m \angle B + m \angle C \neq 180°$. Otherwise $m \angle A + m \angle B + m \angle C = 180°$ and infinitely many solutions exist, all of which lead to similar triangles.

Example 5–20 was the case *ASA*. The case (b) (ii) is the only troublesome situation. In Fig. 5–20(b) $\triangle ABC$ and $\triangle ABC'$ both have the given measures $m \angle A$, a, c but $b \neq b'$, $m \angle ABC \neq m \angle ABC'$, $m \angle ACB \neq m \angle AC'B$. Hence, from the given information, $m \angle A$, a, c, we would not obtain a unique solution. When this case arises it is necessary to be alert for two possible solutions. In Fig. 5–20(b) $\triangle BCC'$ is isosceles so that $m \angle BCC' = m \angle C'CB$ and consequently $\angle ACB$, $\angle AC'B$ are supplementary. Thus *one of our solutions contains an acute angle and the other an obtuse angle.* We may conclude that for this case, when a solution exists, the supplement of certain angles should be checked for a possible second solution (see Example 5–21).

Solve the triangle *ABC* when $a = 3$, $c = 4.9740$ and $m \angle A = 30°$.

Example
5–21

Solution. This is the *SSA* case and by the Law of Sines

$$\frac{\sin \angle A}{a} = \frac{\sin \angle C}{c}$$

or

$$\frac{\sin 30°}{3} = \frac{\sin \angle C}{4.9740}.$$

Then $\sin \angle C = \dfrac{4.9740}{3} \cdot (0.5000) \doteq 0.8290$. One solution for the equation $\sin \angle C = 0.8290$ is $m \angle C = 56°$; another solution is the supplement of 56°: $180° - 56° = 124°$ (that 124° is a solution follows from Eq. **(5-23)**). Then if $m \angle C = 56°$, $m \angle B = 180° - (m \angle A + m \angle C) = 180° - 86° = 94°$; and if $m \angle C = 124°$, $m \angle B = 180° - (m \angle A + m \angle C) = 180° - 154° = 26°$. If $m \angle B = 94°$, then

$$\frac{\sin 30°}{3} = \frac{\sin 94°}{b}$$

and since $\sin 94° = \sin(180° - 94°) = \sin 86° = 0.9976$,

$$b = \frac{3 \cdot \sin 94°}{\sin 30°} = \frac{3 \cdot (0.9976)}{0.5000} \doteq 5.9856.$$

For $m \angle B = 26°$

$$b = \frac{3 \cdot \sin 26°}{\sin 30°} = \frac{3 \cdot (0.4384)}{0.5000} \doteq 2.6304.$$

Therefore, the two solutions are $b = 5.9856$, $m \angle B = 94°$, $m \angle C = 56°$ and $b = 2.6304$, $m \angle B = 26°$, $m \angle C = 124°$. These two solutions are illustrated in Fig. 5–21.

Figure
5–21

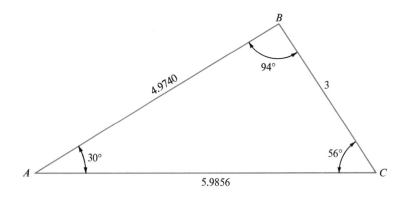

(a)

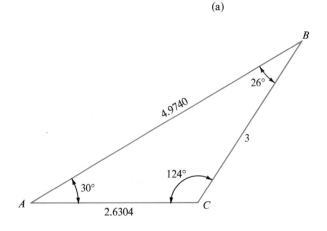

(b)

Solve $\triangle ABC$ if $m \angle A = 30°$, $a = 1$, $c = 4.9740$.

Example
5-22

Solution. This is the *SSA* case. Using the Law of Sines

$$\frac{\sin 30°}{1} = \frac{\sin \angle C}{4.9740}$$

and $\sin \angle C = (4.9740) \cdot (0.5000) \doteq 2.4870$. Since there is no $\angle C$ such that $\sin \angle C = 2.4870$, we see that no solution exists.

Solve $\triangle ABC$ where $a = 6$, $c = 4.9740$ and $m \angle A = 30°$.

Example
5-23

Solution. This is the *SSA* case.

$$\frac{\sin 30°}{6} = \frac{\sin \angle C}{4.9740}$$

$$\sin \angle C = (1/6) \cdot (0.5000) \cdot (4.9740) \doteq 0.4145.$$

Then $m \angle C = 24.5°$ is one possibility and $180° - 24.5° = 155.5°$ is another.

Case 1 ($m \angle C = 24.5°$). Then

$$m \angle B = 180° - (30° + 24.5°) = 180° - 54.5° = 125.5°$$

and

$$\frac{\sin 30°}{6} = \frac{\sin \angle B}{b} = \frac{\sin 125.5°}{b}$$

$$b = \frac{6 \cdot \sin 125.5°}{\sin 30°} = 12 \cdot \sin 125.5° = 12 \cdot (0.8141)$$

$$\doteq 9.7692.$$

Hence, one solution is $b = 9.7692$, $m \angle B = 125.5°$ and $m \angle C = 24.5°$.

Case 2 ($m \angle C = 155.5°$). In this case

$$m \angle B = 180° - (30° + 155.5°) = 180° - 185.5° < 0.$$

Since $m \angle B$ cannot be negative, this case cannot occur. Therefore, the solution given by Case 1 is unique.

Find two solutions for $\triangle ABC$ if $m \angle A = 10°$, $m \angle B = 50°$, $m \angle C = 120°$.

Example
5-24

Solution. We may choose $a = 1$ and then find b and c. We have by the Law of Sines

$$\frac{\sin 10°}{1} = \frac{\sin 50°}{b} = \frac{\sin 120°}{c}$$

$$b = \frac{\sin 50°}{\sin 10°} = \frac{0.7660}{0.1736} \doteq 4.412$$

$$c = \frac{\sin 120°}{\sin 10°} = \frac{\sin (180° - 60°)}{\sin 10°} = \frac{\sin 60°}{\sin 10°} = \frac{0.8660}{0.1736} \doteq 4.988$$

Therefore, one solution is $a = 1$, $b = 4.412$, $c = 4.988$. Since any other solution will necessarily produce a triangle similar to this one, we may obtain another solution as $2a$, $2b$, $2c$, *i.e.*, $a' = 2$, $b' = 8.824$, $c' = 9.976$.

42. Find the area of each triangle given below by using Theorem 24.
    (a) $m \angle A = m \angle B = m \angle C = 60°$, $a = b = c = 1$.
    (b) $m \angle A = 30°$, $m \angle B = 60°$, $m \angle C = 90°$, $a = \frac{1}{2}$, $b = \frac{1}{2}\sqrt{3}$, $c = 1$.
    (c) $m \angle A = m \angle B = 45°$, $m \angle C = 90°$, $a = 10$, $b = 10$, $c = 10\sqrt{2}$.
    (d) $m \angle A = 22°$, $m \angle B = 58°$, $m \angle C = 100°$, $a = 0.3746$, $b = 0.8480$, $c = 0.9848$.
    (e) $m \angle A = 47°$, $m \angle B = 56°$, $m \angle C = 77°$, $a = 1.4628$, $b = 1.6580$, $c = 1.9488$.
    (f) $m \angle A = 121°$, $m \angle B = 26°$, $m \angle C = 33°$, $a = 8.572$, $b = 4.384$, $c = 5.446$.

43. Solve the following triangles if there is a solution.
    (a) $m \angle A = 28°$, $m \angle B = 61°$, $a = 0.4695$.
    (b) $m \angle A = 56°$, $m \angle C = 94°$, $c = 0.9976$.
    (c) $m \angle B = 103.5°$, $m \angle C = 124°$, $a = 14.9$.
    (d) $m \angle B = 103.5°$, $m \angle C = 12°$, $a = 14.9$.
    (e) $m \angle A = 29.5°$, $m \angle B = 54°$, $c = 22.1$.
    (f) $m \angle A = 32.5°$, $m \angle C = 44.5°$, $b = 3.786$.

44. Solve the following triangles if there is a solution.
    (a) $m \angle A = 28°$, $b = 1$, $c = 1$.
    (b) $m \angle A = 96°$, $b = 1$, $c = 2$.
    (c) $m \angle B = 45°$, $a = 2.1$, $c = 0.3$.
    (d) $m \angle B = 112°$, $a = 1.25$, $c = 3.1$.
    (e) $m \angle C = 63.5°$, $a = 2.4$, $b = 5$.
    (f) $m \angle C = 135°$, $a = 4$, $b = 1$.

45. Solve the following triangles if a solution exists.
    (a) $a = 1, b = 2, c = 2$.
    (b) $a = 0.5, b = 1, c = 1.25$.
    (c) $a = 10, b = 4, c = 15$.
    (d) $a = 3, b = 4, c = 5$.
    (e) $a = 2, b = 3, c = 4$.
    (f) $a = 0.13, b = 1.5, c = 1.41$.

46. Solve the following triangles if a solution exists; if two solutions exist, find both.
    (a) $a = 1, b = 1, m \angle A = 30°$.
    (b) $a = 2, b = 5, m \angle A = 90°$.
    (c) $b = 10, c = 6, m \angle C = 10°$.
    (d) $b = 7.6, c = 0.765, m \angle B = 49.5°$.
    (e) $a = 34, c = 8.5, m \angle A = 15°$.
    (f) $a = 1.13, c = 2.25, m \angle C = 8.5°$.

47. At a point C on the west bank of a river it is determined that $\angle ACB = 60°$ and $\angle ABC = 90°$, where A, B are two points on the east bank (see the figure below). Points A and B are known to be 2 miles apart. What is the width of the river at C?

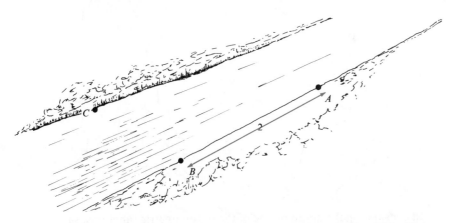

48. In solving triangles can you explain why solutions in the *SSS, ASA, AAS,* and *SAS* cases are unique?

49. A man 6′ tall stands 20′ away from a flagpole and makes a sighting on the top of the pole. He finds that his line of sight makes a 55° angle with the horizontal (see the figure below). How tall is the flagpole?

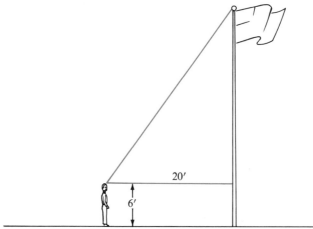

50. Two points A and B one mile apart lie in a valley below a mountain peak. It is found that the line of sight from A to the peak makes a 30° angle with AB and that the line of sight from B makes a 40° angle. How high is the peak above the line AB?

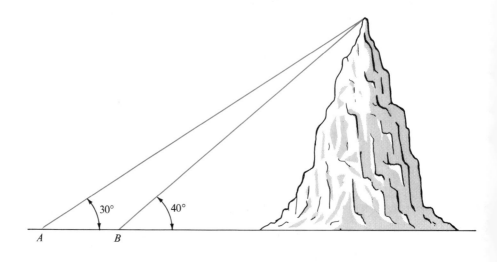

In considering the graph of one of the trigonometric functions, say cosine for example, we naturally need information about its domain and range. Knowing that the range of cosine is $[-1, 1]$ tells us that its graph will lie between the lines $y = 1$ and $y = -1$. Just as important, however, is the fact that the cosine is a periodic function. If we graph the cosine function on the interval $[-\pi, \pi]$ for instance, the graph at *any* other number t can be obtained. If we wish to obtain the graph of cosine on the interval $[\pi, 3\pi]$, we reason that $t \in [-\pi, \pi]$ if and only if $t + 2\pi \in [\pi, 3\pi]$. But

$\cos(t + 2\pi) = \cos t$. Consequently, the graph of cosine on $[\pi, 3\pi]$ is just a repetition of its graph on $[-\pi, \pi]$. In a similar fashion we can extend the graph of the cosine function along the x-axis in either direction from the interval $[-\pi, \pi]$ as far as we like.

To construct the graph of cosine on the interval $[-\pi, \pi]$, we first plot some points $(t, \cos t)$ for a few convenient values of t. These numbers t and their corresponding numbers $\cos t$ we may obtain from Exercise 8. The following table summarizes this information. In the fourth column of this table we use 3.14 as an approximation for π. Having plotted the points indicated, we connect these with a smooth curve to obtain the graph of cosine on $[-\pi, \pi]$ as given in Fig. 5–22.

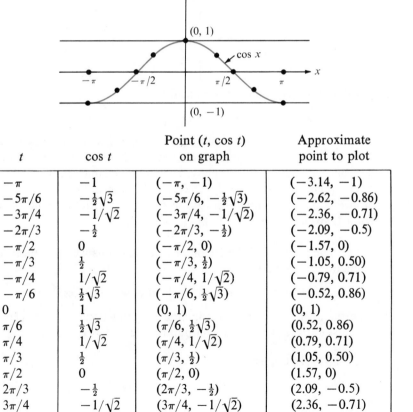

Figure 5–22

t	$\cos t$	Point $(t, \cos t)$ on graph	Approximate point to plot
$-\pi$	-1	$(-\pi, -1)$	$(-3.14, -1)$
$-5\pi/6$	$-\frac{1}{2}\sqrt{3}$	$(-5\pi/6, -\frac{1}{2}\sqrt{3})$	$(-2.62, -0.86)$
$-3\pi/4$	$-1/\sqrt{2}$	$(-3\pi/4, -1/\sqrt{2})$	$(-2.36, -0.71)$
$-2\pi/3$	$-\frac{1}{2}$	$(-2\pi/3, -\frac{1}{2})$	$(-2.09, -0.5)$
$-\pi/2$	0	$(-\pi/2, 0)$	$(-1.57, 0)$
$-\pi/3$	$\frac{1}{2}$	$(-\pi/3, \frac{1}{2})$	$(-1.05, 0.50)$
$-\pi/4$	$1/\sqrt{2}$	$(-\pi/4, 1/\sqrt{2})$	$(-0.79, 0.71)$
$-\pi/6$	$\frac{1}{2}\sqrt{3}$	$(-\pi/6, \frac{1}{2}\sqrt{3})$	$(-0.52, 0.86)$
0	1	$(0, 1)$	$(0, 1)$
$\pi/6$	$\frac{1}{2}\sqrt{3}$	$(\pi/6, \frac{1}{2}\sqrt{3})$	$(0.52, 0.86)$
$\pi/4$	$1/\sqrt{2}$	$(\pi/4, 1/\sqrt{2})$	$(0.79, 0.71)$
$\pi/3$	$\frac{1}{2}$	$(\pi/3, \frac{1}{2})$	$(1.05, 0.50)$
$\pi/2$	0	$(\pi/2, 0)$	$(1.57, 0)$
$2\pi/3$	$-\frac{1}{2}$	$(2\pi/3, -\frac{1}{2})$	$(2.09, -0.5)$
$3\pi/4$	$-1/\sqrt{2}$	$(3\pi/4, -1/\sqrt{2})$	$(2.36, -0.71)$
$5\pi/6$	$-\frac{1}{2}\sqrt{3}$	$(5\pi/6, -\frac{1}{2}\sqrt{3})$	$(2.62, -0.86)$
π	-1	$(\pi, -1)$	$(3.14, -1)$

To obtain the graph of sine on $[-\pi, \pi]$, we may use the equation

$$\sin t = \cos (t - \pi/2)$$

which is immediately derivable from Eq. **(5-4)**. This equation allows us to compute $\sin t$ for any number t simply by computing $\cos (t - \pi/2)$. Since we already have the graph of cosine, the graph of sine is obtained by shifting to the right by $(\frac{1}{2})\pi$. (See Fig. 5–23.)

Figure
5–23

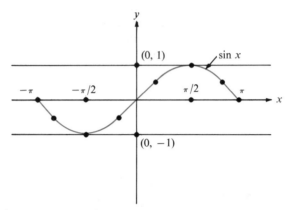

To graph the tangent function we recall that

$$\tan (t + \pi) = \tan t$$

for all numbers in the domain of tangent. Using this fact we may graph tangent on $(-\pi/2, \pi/2)$ for example, and make extensions from this. Notice that the interval chosen above is open since tangent is not defined at the end points of $(-\pi/2, \pi/2)$. The graph of tangent on $(-\pi/2, \pi/2)$ is given in Fig. 5–24. [Compare the graph of tangent with the graph of the function L given in Exercise 13(e).]

In certain applications of the trigonometric functions we are concerned with functions of the type

$$f(x) = A \sin (bx + c) \qquad \text{and} \qquad g(x) = A \cos (bx + c).$$

These are known as *generalized* trigonometric functions and are seen to be composite. Here A, b, c are real numbers and $A \neq 0, b \neq 0$.

Since for all real numbers x

$$-1 \leq \sin (bx + c) \leq 1,$$

we have, if $A > 0$,

$$-A \leq A \sin (bx + c) \leq A.$$

Figure
5–24

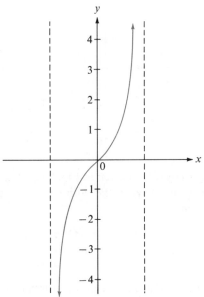

It is easy to see, then, that the range of both f and g is $[-A, A]$ if $A > 0$. This shows that the graph of f and g will lie between the lines

$$y = A \quad \text{and} \quad y = -A.$$

The number $|A|$ is called the *amplitude* of these functions.

The function f is periodic as the following computation shows:

$$
\begin{aligned}
f(x + 2\pi/|b|) &= A \sin [b(x + 2\pi/|b|) + c] \\
&= A \sin (bx \pm 2\pi + c) \\
&= A \sin [(bx + c) \pm 2\pi] \\
&= A \sin (bx + c) \\
&= f(x).
\end{aligned}
$$

And similarly,

$$g(x + 2\pi/|b|) = g(x) \quad \text{for all } x.$$

The number $2\pi/|b|$ is called the *period* of these functions. To study the graph of these functions a few special cases may be instructive.

The graph of $f(x) = \cos (x + 1)$ is given in Fig. 5–25. The effect of adding 1 is evidently to "shift" the graph of cosine one unit to the left. Generally, for any $c > 0$ the graph of $\cos (x + c)$ will be "shifted" c units to the left from the graph of cosine. The shift will be to the right if $c < 0$.

Example
5–25

**Figure
5–25**

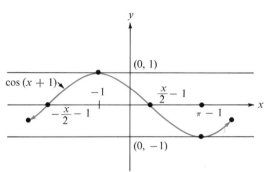

**Example
5–26**

The graph of $f(x) = \cos\left(\frac{1}{2}\right)x$ is given in Fig. 5–26. The effect here of the $\frac{1}{2}$ is to "stretch" the cosine curve in the x-direction. This effect is also indicated by the equation

$$f(x + 4\pi) = \cos\left[\tfrac{1}{2}(x + 4\pi)\right]$$
$$= \cos\left[\left(\left(\tfrac{1}{2}\right)x + 2\pi\right)\right]$$
$$= \cos\left(\tfrac{1}{2}\right)x = f(x).$$

This suggests that the distance between the numbers at which $\cos\left(\frac{1}{2}\right)x$ repeats is $2(2\pi)$ rather than 2π as with $\cos x$. This situation is typical: if b is a positive number, then the graph of the function

$$f(x) = \cos(bx)$$

is stretched in the x-direction from the position of $\cos x$ if $b < 1$ and is "compressed" in the x-direction from the position of $\cos x$ if $b > 1$.

**Figure
5–26**

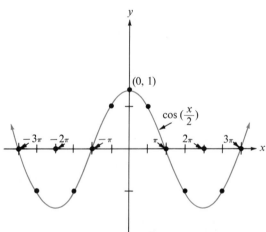

**Example
5–27**

As indicated in Examples 5–25 and 5–26, the graph of

$$f(x) = \cos\left[\left(\tfrac{1}{2}\right)x + 1\right]$$

will be changed from that of cos x in two respects: there is both a stretching (from the $\frac{1}{2}$) and a shifting (from the 1). This graph is given in Fig. 5–27. It should also be noted that the $\frac{1}{2}$ has played a part in the shift. To see why this is so, we only have to remember that the graph of $f(x) = \cos(x + 1)$ meets the line $y = 1$ at $x = -1$, *i.e.*, $f(-1) = \cos[(-1) + 1] = \cos 0 = 1$. But $\cos[\frac{1}{2}(-1) + 1] = \cos(\frac{1}{2}) \neq 1$. However, $\cos[\frac{1}{2}(-2) + 1] = \cos 0 = 1$. This indicates that the $\frac{1}{2}$ has contributed to the shift and the graph of $f(x) = \cos[(\frac{1}{2})x + 1]$ is shifted 2 units from the graph of $f(x) = \cos(\frac{1}{2})x$ instead of only 1 unit as we might at first expect.

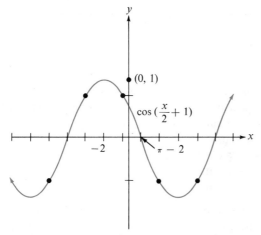

Figure
5–27

The effect of multiplying the function $f(x) = \cos[(\frac{1}{2})x + 1]$ by a number $A > 0$ if a stretch in the y-direction if $A > 1$ and a compression in the y-direction if $A < 1$. This is illustrated in Fig. 5–28 which gives the graph of

Example
5–28

$$f(x) = 3\cos[(\tfrac{1}{2})x + 1].$$

If we wish to graph one of the generalized trigonometric functions, it is helpful to begin by determining the amplitude and the period. If the amplitude is $|A|$, we have seen that the graph of the function lies between the lines

$$y = A \quad \text{and} \quad y = -A.$$

Next, if the graph is obtained over an interval which has length equal to the function's period, we may determine the behavior of the function over any other interval.

**Figure
5–28**

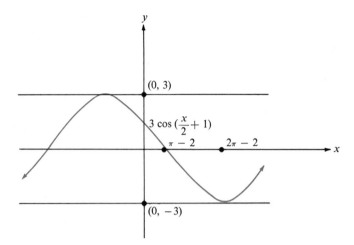

Exercises

51. For each of the functions f given below find the amplitude, the period, at least one zero and tell what lines the graph lies between.
    (a) $f(x) = -3 \sin (2x + 1)$
    (b) $f(x) = 5 \cos (5x - 1)$
    (c) $f(x) = -\frac{2}{3} \cos (-3x + 7)$
    (d) $f(x) = -8 \sin (\pi x - 7)$
    (e) $f(x) = 1.24 \cos (-3\pi + 1.8)$

52. Determine the amplitude and period of each of the following functions and graph each over an interval having a length equal to the period of the function.

    (a) $f(x) = (\frac{1}{3}) \sin x$ (b) $g(x) = \cos (\pi x)$
    (c) $f(x) = \sin (7x)$ (d) $g(x) = -2 \cos x$
    (e) $f(x) = \sin (2x - 1)$ (f) $g(x) = -2 \cos (\pi x + \pi/2)$

53. Graph each of the following.

    (a) $f(x) = \sin x - \cos x$ (b) $g(x) = \tan x + 1$
    (c) $h(x) = \sec x$ (d) $j(x) = \sec x - 1$

54. Is the function $k(x) = \sin (x^2)$ periodic?

55. Is the function $p(x) = \cos |x|$ periodic? Graph this function and compare with the graph of the cosine function.

56. Let the function q be defined as follows:

$$q(x) = \sin ([x] + 1 - x).$$

Construct the graph of this function.

The trigonometric equations which we shall consider here are easily solved by using the following results.

(I) $\sin t = \sin s$ if and only if

$$\begin{cases} t = s + 2n\pi, \\ \text{or} \\ t = -s + (2n + 1)\pi. \end{cases}$$
for some integer n

(II) $\sin t = -\sin s$ if and only if

$$\begin{cases} t = -s + 2n\pi, \\ \text{or} \\ t = s + (2n + 1)\pi. \end{cases}$$
for some integer n

(III) $\sin t = \cos s$ if and only if

$$\begin{cases} t = s + (4n + 1)\pi/2, \\ \text{or} \\ t = -s + (4n + 1)\pi/2. \end{cases}$$
for some integer n

(IV) $\cos t = \cos s$ if and only if

$$\begin{cases} t = s + 2n\pi, \\ \text{or} \\ t = -s + 2n\pi. \end{cases}$$
for some integer n

We will prove the first of these and leave the others for exercises. First, if $t = s + 2n\pi$ for some integer n, it is clear that $\sin t = \sin s$. Also if $t = -s + (2n + 1)\pi$,

$$\begin{aligned} \sin t &= \sin \left(-s + (2n + 1)\pi\right) \\ &= \sin \left(-s + 2n\pi + \pi\right) \\ &= \sin \left(-s + \pi\right) \\ &= \sin \left(\pi - s\right) \\ &= \sin \left(\pi\right) \cos s - \cos \left(\pi\right) \sin s \\ &= \sin s. \end{aligned}$$

Thus the first part of (I) is proved. To prove the second part, we start with the following identity [see Eq. **(5-18)**]:

$$\sin (x + y) - \sin (x - y) = 2 \sin y \cdot \cos x.$$

In this identity, if we take

$$t = x + y \qquad \text{and} \qquad s = x - y,$$

we have

$$x = \tfrac{1}{2}(s + t) \qquad \text{and} \qquad y = \tfrac{1}{2}(t - s)$$

which gives the identity

$$\sin t - \sin s = 2 \sin \left(\frac{t - s}{2}\right) \cdot \cos \left(\frac{t + s}{2}\right).$$

Now if $\sin t = \sin s$, $\sin t - \sin s = 0$ and this implies that

$$2 \sin \left[\tfrac{1}{2}(t - s)\right] \cdot \cos \left[\tfrac{1}{2}(t + s)\right] = 0.$$

From this we conclude that

$$\sin \left[\tfrac{1}{2}(t - s)\right] = 0 \qquad \text{or} \qquad \cos \left[\tfrac{1}{2}(t + s)\right] = 0.$$

If $\sin \left[\tfrac{1}{2}(t - s)\right] = 0$, then $\tfrac{1}{2}(t - s) = n\pi$, for some integer n, and $t - s = 2n\pi$. This then gives

$$t = s + 2n\pi.$$

If $\cos \left[\tfrac{1}{2}(t + s)\right] = 0$, $\tfrac{1}{2}(t + s) = (2n + 1)\pi/2$ and this gives $t = -s + (2n + 1)\pi$, for some integer n.

<table>
<tr><td>

**Example
5–29**

</td><td>

Solve the equation $2 \cos x - 1 = 0$; *i.e.*, find all real numbers x such that $2 \cos x - 1 = 0$.

Solution. If x is a number such that $2 \cos x - 1 = 0$, then

$$\cos x = \tfrac{1}{2}.$$

But, $\cos (\pi/3) = \tfrac{1}{2}$, and so,

$$\cos x = \cos (\pi/3).$$

Then, from (IV) above,

$$x = \pi/3 + 2n\pi \qquad \text{or} \qquad x = -\pi/3 + 2n\pi$$

where n is an integer. It is clear that either choice of x is a solution to the equation. Therefore, the solution set of this equation is

$$\{\pi/3 + 2n\pi \mid n \text{ is an integer}\} \; \cup \; \{-\pi/3 + 2n\pi \mid n \text{ is an integer}\}.$$

</td></tr>
<tr><td>

**Example
5–30**

</td><td>

Solve the equation $\sin^2 x - 2 \sin x = 3$.

Solution. If x is a number such that $\sin^2 x - 2 \sin x = 3$, then

$$\sin^2 x - 2 \sin x - 3 = 0$$

or

$$(\sin x - 3)(\sin x + 1) = 0.$$

</td></tr>
</table>

From this,

$$\sin x = 3 \quad \text{or} \quad \sin x = -1.$$

Since there is no number such that $\sin x = 3$, we conclude that $\sin x = -1$. But $\sin(-\frac{1}{2}\pi) = -1$ and by (I) above

$$x = -(\tfrac{1}{2})\pi + 2n\pi \quad \text{or} \quad x = (\tfrac{1}{2})\pi + (2n + 1)\pi$$
$$= (\tfrac{1}{2})\pi + \pi + 2n\pi$$
$$= \tfrac{1}{2}(3\pi) + 2n\pi.$$

In this case,

$$\{-(\tfrac{1}{2})\pi + 2n\pi\} = \{\tfrac{1}{2}(3\pi) + 2n\pi\}$$

and hence, the solution set for the equation is

$$\{-(\tfrac{1}{2})\pi + 2n\pi\} \cup \{\tfrac{1}{2}(3\pi) + 2n\pi\} = \{-(\tfrac{1}{2})\pi + 2n\pi\}.$$

Solve the equation $\sin(3t) = \cos t$.

**Example
5–31**

Solution. From (III)

$$3t = t + (4n + 1)\pi/2 \quad \text{or} \quad 3t = -t + (4n + 1)\pi/2.$$

Thus,

$$t = (4n + 1)\pi/4 = \pi/4 + n\pi$$

or

$$t = (4n + 1)\pi/8 = \pi/8 + \tfrac{1}{2}(n\pi).$$

The solution set is, therefore,

$$\{(\tfrac{1}{4})\pi + n\pi\} \cup \{(\tfrac{1}{8})\pi + \tfrac{1}{2}(n\pi)\}.$$

57. In each of the following you are to draw a conclusion from the given information.

Exercises

 (a) Given: $\sin(x) = \sin(3\pi/2)$
 (b) Given: $\sin(y - 1) = -1$
 (c) Given: $\sin(w) = \cos(4w)$
 (d) Given: $\sin(x + y) = \cos(x - y)$
 (e) Given: $\cos(3v) = \cos(2v)$
 (f) Given: $\cos(4x + 1) = \cos(3x + 2)$
 (g) Given: $\sin(\pi t) = \sin(2\pi s)$

58. Prove that $\sin t = -\sin s$ if and only if for some integer n
$$t = -s + 2n\pi \quad \text{or} \quad t = s + (2n + 1)\pi.$$

59. Prove that $\sin t = \cos s$ if and only if for some integer n
$$t = s + (4n + 1)\pi/2 \quad \text{or} \quad t = -s + (4n + 1)\pi/2.$$
[*Hint:* Use the identity $\cos s = \sin [(\tfrac{1}{2})\pi - s]$ and (I).]

60. Prove that $\cos t = \cos s$ if, and only if, for some integer n
$$t = s + 2n\pi \quad \text{or} \quad t = -s + 2n\pi.$$

61. Solve the following equations:
    (a) $2 \sin x - 1 = 0$
    (b) $2 \sin (x - 1) = 0$
    (c) $\cos^2 x + 3 \sin x - 3 = 0$ [*Hint:* Write $\cos^2 x$ in terms of $\sin^2 x$ and factor — or use the quadratic formula.]
    (d) $2 \cos x = -1$
    (e) $2 \sin x \cdot \cos x + \sin x = 0$
    (f) $\cos (2x) = -1$

62. Show that there is no real number x such that
$$2 \sin^2 x + 4 \sin x + 3 = 0.$$

63. Solve the equations:

    (a) $\sin (x/3) = \cos x$ (b) $\cos (2x) = -\sin x$
    (c) $-\sin (2x) = \cos x$ (d) $\sin (7t) = \sin (-5t)$

64. Solve the equations:

    (a) $\tan x \cdot \csc x - 1 = 0$ (b) $2 \tan (x/2) = \cot (x/2)$
    (c) $\sin^4 y - \cos^4 y = 1$ (d) $\sin s + \cos s = 1$

5.8
The Inverse Trigonometric Functions

Consider a function f defined as follows:

$$f(x) = \sin (x), \quad x \in [a, b]$$

where a, b are real numbers and $a < b$. An examination of the graph of the sine function on the set of *all* real numbers reveals that whether or not f is 1-1 depends upon the interval $[a, b]$. In Fig. 5–29(a), the interval chosen is $[-\pi, \pi/4]$ and in this case f is not 1-1. In Fig. 5–29(b), the interval chosen is $[-\pi/2, \pi/4]$ and here f is 1-1. This implies, of course, that in the first case f^{-1} is not a function, and in the second case f^{-1} is a function. Due to the fact that sine is periodic of period 2π, f^{-1} is not a function if $[a, b]$ is chosen to have a length greater than 2π.

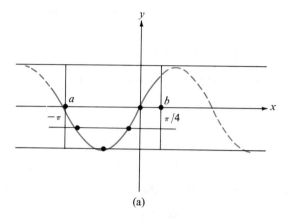

(a)

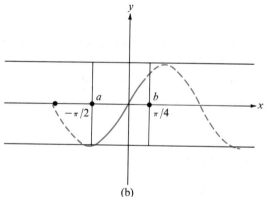

(b)

We obviously may make an infinite number of choices for $[a, b]$ each of which will give us a 1-1 function f. All the functions listed below are 1-1, for example:

$$f_1(x) = \sin (x), \quad x \in [-5\pi/2, -3\pi/2].$$

$$f_2(x) = \sin (x), \quad x \in [-3\pi/2, -\pi/2].$$

$$f_3(x) = \sin (x), \quad x \in [-\pi/2, \pi/2].$$

$$f_4(x) = \sin (x), \quad x \in [\pi/2, 3\pi/2].$$

$$f_5(x) = \sin (x), \quad x \in [3\pi/2, 5\pi/2].$$

Then each of the relations $f_1^{-1}, f_2^{-1}, f_3^{-1}, f_4^{-1}, f_5^{-1}$ is a function. The domain of each is $[-1, 1]$ since $\sin (x)$ takes as values all numbers of the set $[-1, 1]$ for $x \in [a, b]$, where $[a, b]$ is any one of the intervals above. It is customary to call $f_3{}^{-1}$ *the* inverse sine function and denote it by "Arc sin." Thus,

$$f_3{}^{-1}(x) = \text{Arc sin } (x) \quad \text{for} \quad x \in [-1, 1].$$

In a similar way we define Arc cos and Arc tan.

Definition

(a) If $f(x) = \sin (x)$, $x \in [-\pi/2, \pi/2]$, then

$$\text{Arc sin } = f^{-1}.$$

(b) If $g(x) = \cos (x)$, $x \in [0, \pi]$, then

$$\text{Arc cos } = g^{-1}.$$

(c) If $h(x) = \tan (x)$, $x \in (-\pi/2, \pi/2)$, then

$$\text{Arc tan } = h^{-1}.$$

From the definition, we have

$$\mathfrak{D}(\text{Arc sin}) = [-1, 1], \qquad \mathfrak{R}(\text{Arc sin}) = [-\pi/2, \pi/2].$$
$$\mathfrak{D}(\text{Arc cos}) = [-1, 1], \qquad \mathfrak{R}(\text{Arc cos}) = [0, \pi].$$
$$\mathfrak{D}(\text{Arc tan}) = \mathbf{R}, \qquad \mathfrak{R}(\text{Arc tan}) = (-\pi/2, \pi/2).$$

The graphs of the functions Arc sin, Arc cos and Arc tan are given in Fig. 5–30.

Example 5–32

Show that $\cos (\text{Arc sin } x) = \sqrt{1 - x^2}$.

Solution. We use the identity

$$\cos^2 t + \sin^2 t = 1$$

and the fact that $\sin (\text{Arc sin } x) = x$. Substitution of $t = \text{Arc sin } x$ in the identity gives

$$\cos^2 (\text{Arc sin } x) + \sin^2 (\text{Arc sin } x) = 1$$

or

$$[\cos (\text{Arc sin } x)]^2 + [\sin (\text{Arc sin } x)]^2 = 1.$$

Then

$$[\cos (\text{Arc sin } x)]^2 = 1 - x^2$$

and

$$\cos (\text{Arc sin } x) = \pm\sqrt{1 - x^2}.$$

Figure
5-30

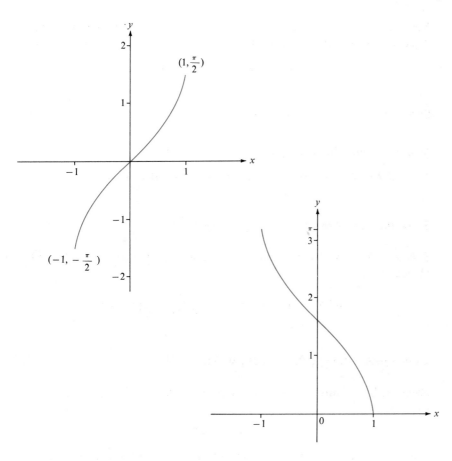

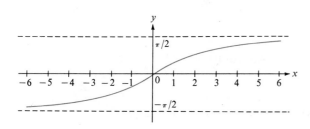

But since the range of Arc sin x is $[-\pi/2, \pi/2]$ and the cosine is not negative on this set, we must have

$$\cos (\text{Arc sin } x) = \sqrt{1 - x^2}.$$

Example 5–33

Show that Arc sin $(\cos x) = x + \pi/2$.

Solution. Using the identity $\sin (x + \pi/2) = \cos x$

$$\text{Arc sin } (\cos x) = \text{Arc sin } \left(\sin (x + \pi/2)\right) = x + \pi/2.$$

Example 5–34

Show that $\tan (\text{Arc sin } x) = x/\sqrt{1 - x^2}$.

Solution. We have from Example 5–32 that $\cos (\text{Arc sin } x) = \sqrt{1 - x^2}$. Hence,

$$\tan (\text{Arc sin } x) = \frac{\sin (\text{Arc sin } x)}{\cos (\text{Arc sin } x)} = \frac{x}{\sqrt{1 - x^2}}.$$

Example 5–35

Compute Arc sin $(\frac{1}{2}\sqrt{2})$, Arc cos (0), Arc tan (1).

Solution. By definition of Arc sin, we look for a number $x \in [-\pi/2, \pi/2]$ such that

$$\sin (x) = \tfrac{1}{2}\sqrt{2}.$$

The number $x = \pi/4$ is the only one meeting these requirements. Hence,

$$\text{Arc sin } (\tfrac{1}{2}\sqrt{2}) = \pi/4.$$

Similarly,

$$\text{Arc cos } (0) = \pi/2$$

since

$$\cos (\pi/2) = 0 \quad \text{and} \quad \pi/2 \in [0, \pi].$$

Also,

$$\text{Arc tan } (1) = \pi/4$$

since

$$\tan (\pi/4) = 1 \quad \text{and} \quad \pi/4 \in (-\pi/2, \pi/2).$$

65. Use Definitions and Table I to compute each of the following.

(a) Arc cos (0.9962) (b) Arc tan (1.0355)
(c) Arc sin (0.5592) (d) Arc cos (0.9636)
(e) Arc sin (0.9997) (f) Arc tan (2.9887)
(g) Arc cos (0.9998) (h) Arc sin (0.8704)

66. Use Table I to solve each of the following for x, where $1.5708 \doteq \pi/2$ and $-\pi/2 \leq x \leq 0$. For the cosine solve in $[\pi/2, \pi]$.

(a) $\sin (x) = -0.1736$. *Solution.* By Eq. (5-7) $\sin (-x) = -\sin (x)$. Now solve $\sin (y) = 0.1736$; from Table I, $y = 0.1745$. Then $\sin(-0.1745), = -\sin(0.1745) = -0.1736$. Hence, $x = -0.1745$.

(b) $\cos (x) = 0.9455$ (c) $\tan (x) = -8.7769$
(d) $\sin (x) = -0.8339$ (e) $\cos (x) = 0.4462$
(f) $\tan (x) = -12.7062$

67. Use the results of Exercise 66 above to compute the following.

(a) Arc sin (-0.1736) (b) Arc cos (0.9455)
(c) Arc tan (-8.7769) (d) Arc sin (-0.8339)
(e) Arc cos (0.4462) (f) Arc tan (-12.7062)

68. Compute each of the following.

(a) Arc sin (0.1800) (b) Arc tan (0.6615)
    Arc sin (-0.1800) Arc tan (-0.6615)
(c) Arc cos (0.6075) (d) Arc sin (0.9231)
    Arc cos (-0.6075) Arc sin (-0.9231)
(e) Arc cos (0.1723)
    Arc cos (-0.1723)

69. Compute the following:

(a) Arc sin (0) (b) Arc sin (1)
    Arc cos (0)   Arc cos (1)
    Arc tan (0)   Arc tan (1)
(c) Arc sin (-1) (d) Arc sin $(\frac{1}{2}\sqrt{2})$
    Arc cos (-1) Arc cos $(-\frac{1}{2}\sqrt{2})$
    Arc tan (-1)
(e) Arc sin $(-\frac{1}{2})$ (f) Arc sin $(\frac{1}{2}\sqrt{3})$
    Arc cos $(\frac{1}{2})$ Arc cos $(-\frac{1}{2}\sqrt{3})$

70. Prove each of the following:

(a) $\cos (2 \text{ Arc sin } x) = 1 - 2x^2$
(b) $\sin (2 \text{ Arc cos } x) = 2x\sqrt{1 - x^2}$

(c) $\sin (\text{Arc tan } x) = x/\sqrt{1 + x^2}$

(d) $\sin (2 \text{ Arc sin } x) = 2x\sqrt{1 - x^2}$

(e) $\cos (\text{Arc sin } t + \text{Arc cos } s) = s\sqrt{1 - t^2} - t\sqrt{1 - s^2}$

(f) $\tan (x + \text{Arc tan } x) = \dfrac{x + \tan x}{1 - x \tan x}$

71. Use the method of composites (see Sec. 1.5) to compute f^{-1} for each of the functions below:

(a) $f(x) = \text{Arc cos } \sqrt{x^2 - 1}, \; x \in [1, \sqrt{2}]$

(b) $f(x) = 1 + \sin (2x), \; x \in [-\pi/4, \pi/4]$

INVERSE FUNCTIONS, EXPONENTIAL AND LOGARITHMIC FUNCTIONS

If R is a relation, the inverse of R, R^{-1}, has been defined as follows:

$$R^{-1} = \{(a, b) \mid (b, a) \in R\}.$$

In other words, a relation R^{-1} is obtained from R by interchanging first and second elements in the ordered pairs of R. We will be concerned here with the real functions R which have the property that R^{-1} is also a function. It will be recalled from Chapter 1 that R^{-1} is a function if and only if R is a 1-1 function. By definition, R is a 1-1 function if and only if

$$(x_1, y) \in R \text{ and } (x_2, y) \in R \text{ implies that } x_1 = x_2.$$

In Fig. 6–1 the situation illustrated is that in which $x_1 \neq x_2$ and both (x_1, y) and (x_2, y) are elements of R. Fig. 6–1 thus gives the graph of a function R which is not 1-1. It is evident from this illustration that the property of being 1-1 may be interpreted geometrically as follows:

The function R is 1-1 if and only if every line which is parallel to the x-axis intersects the graph of R at most once.

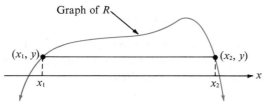

Graph of R

(x_1, y) (x_2, y)

x_1 x_2

x

Figure 6–1

The graph of $f(x) = x^2$ on $[-1, 1]$ is given in Fig. 6–2. It is clear from this graph that f is not a 1-1 function since any line parallel to the x-axis and lying between the lines $y = 0$ and $y = 1$ will intersect the graph exactly twice. It also is evident from the definition that f is not 1-1 since $(1, 1) \in f$ and $(-1, 1) \in f$.

Example 6–1

**Figure
6–2**

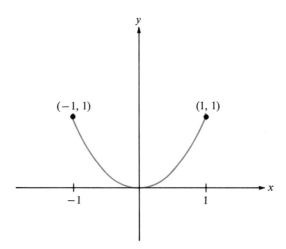

**Example
6–2**

Show that a linear function

$$f(x) = ax + b, a \neq 0$$

is 1-1.

Solution. Suppose that (x_1, y), (x_2, y) are both members of f; *i.e.*,

$$y = f(x_1) = ax_1 + b$$

and

$$y = f(x_2) = ax_2 + b.$$

Then

$$ax_1 + b = ax_2 + b.$$

Hence,

$$ax_1 = ax_2,$$

and since $a \neq 0$, $x_1 = x_2$. Therefore, by definition f is 1-1. It will be recalled that the graphs of linear functions are straight lines and so the result just obtained is obvious geometrically.

**Example
6–3**

Show that the function

$$Q(x) = x^3$$

is 1-1.

Solution. Assume that $(x_1, y) \in Q$ and $(x_2, y) \in Q$. Then $x_1^3 = x_2^3$. Since x_1, x_2 cannot have different signs in this situation, we have

$$x_1 \geq 0 \quad \text{and} \quad x_2 \geq 0$$

or

$$x_1 < 0 \quad \text{and} \quad x_2 < 0.$$

Clearly, if either x_1 or x_2 is zero, so is the other — and in that case, $x_1 = x_2$. Hence, suppose that $x_1 \neq 0$ and $x_2 \neq 0$. Therefore,

$$x_1 > 0 \quad \text{and} \quad x_2 > 0$$

or

$$x_1 < 0 \quad \text{and} \quad x_2 < 0.$$

In either case $x_1 x_2 > 0$. Now from

$$x_1^3 = x_2^3, \quad x_1^3 - x_2^3 = 0;$$

and

$$(x_1 - x_2)(x_1^2 + x_1 x_2 + x_2^2) = 0.$$

From this,

$$x_1 - x_2 = 0$$

or

$$x_1^2 + x_1 x_2 + x_2^2 = 0.$$

We have just seen that $x_1 x_2 > 0$, and hence,

$$x_1^2 + x_1 x_2 + x_2^2 > 0,$$

so that

$$x_1 - x_2 = 0 \quad \text{and} \quad x_1 = x_2.$$

Therefore, $Q(x) = x^3$ is a 1-1 function by definition.

Show that the function

$$f(x) = |x| \quad \text{for} \quad x \in [-1, 1]$$

is not 1-1.

Example
6–4

Solution. $f(-\frac{1}{2}) = |-\frac{1}{2}| = \frac{1}{2}$ and $f(\frac{1}{2}) = |\frac{1}{2}| = \frac{1}{2}$. Hence, f is not 1-1.

Example 6–5

Show that the function given below is 1-1:
$$g(x) = x^2 \quad \text{for} \quad x \geq 0.$$

Solution. Suppose that $g(x_1) = g(x_2)$. Then $x_1{}^2 = x_2{}^2$ and
$$x_1{}^2 - x_2{}^2 = 0,$$

or
$$(x_1 - x_2)(x_1 + x_2) = 0.$$

Therefore,
$$x_1 - x_2 = 0 \quad \text{or} \quad x_1 + x_2 = 0.$$

Since in this case $x_1 \geq 0$ and $x_2 \geq 0$,
$$x_1 + x_2 \neq 0$$

unless both x_1 and x_2 are zero — and in that case $x_1 = x_2$. So $x_1 - x_2 = 0$ and $x_1 = x_2$ which shows that g is 1-1.

Example 6–6

Describe the inverse of the function g above.

Solution. Since g is 1-1, g^{-1} is a function. We then may use composites to compute g^{-1}:
$$x = g(g^{-1}(x)) = (g^{-1}(x))^2.$$

Therefore, since $x \geq 0$,
$$g^{-1}(x) = \sqrt{x}.$$

Example 6–7

Show that the function
$$g(x) = \frac{x - 1}{x + 1}$$

is 1-1 and describe its inverse.

Solution. If x_1, x_2 are numbers such that
$$(x_1 - 1)/(x_1 + 1) = (x_2 - 1)/(x_2 + 1),$$

then
$$(x_1 - 1)(x_2 + 1) = (x_1 + 1)(x_2 - 1)$$

or

$$x_1 x_2 - x_2 + x_1 - 1 = x_1 x_2 + x_2 - x_1 - 1.$$

Therefore,

$$-x_2 + x_1 = x_2 - x_1$$
$$2(x_2 - x_1) = 0$$
$$x_2 - x_1 = 0$$
$$x_1 = x_2$$

and g is 1-1. To compute g^{-1}:

$$x = g(g^{-1}(x)) = (g^{-1}(x) - 1)/(g^{-1}(x) + 1).$$

Hence,

$$x(g^{-1}(x) + 1) = g^{-1}(x) - 1$$
$$xg^{-1}(x) - g^{-1}(x) = -x - 1$$
$$(x - 1)g^{-1}(x) = -x - 1$$
$$g^{-1}(x) = (-x - 1)/(x - 1) = (1 + x)/(1 - x).$$

Note that the domain of g is the set of all real numbers except -1 and that of g^{-1} is the set of all real numbers except 1.

Now, assuming that $a \neq b$, compare the relative positions for the points on the coordinate system corresponding to the pairs (a, b) and (b, a).

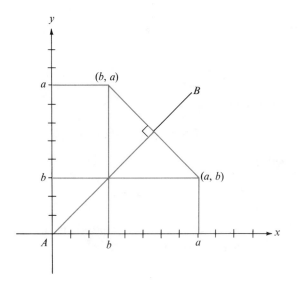

Figure
6–3

In Fig. 6–3 these pairs have been plotted. Observe that in Fig. 6–3 if a line is drawn through the point representing (a, b) perpendicular to the line AB, this line passes through the point representing (b, a). *The points representing* (a, b) *and* (b, a) *are said to be symmetric with respect to the line* AB.

The graph of T^{-1}, for a relation T, may then be constructed as follows:

For each point P of the graph of T plot the point Q which is symmetric to P with respect to the line $y = x$.

This means that through each point P of the graph of T we draw a line L perpendicular to the line D which bisects quadrants I and III. If I is the point of D where L intersects, we choose Q on L opposite P such that $PI = IQ$. (See Fig. 6–4.)

Figure 6–4

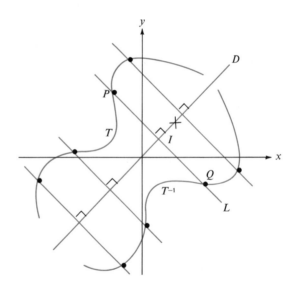

Example 6–8

Construct the graph of

$$T^{-1}(x) = \tfrac{1}{2}(x - 3)$$

from the graph of

$$T(x) = 2x + 3.$$

Solution. Since T^{-1} is a linear function its graph will be determined by any two points on it. The graph is given in Fig. 6–5.

**Figure
6–5**

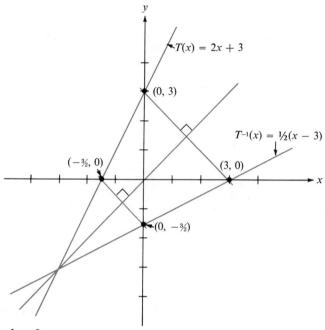

The graphs of

$$f(x) = x^2, \quad \text{for} \quad x \geq 0$$

and

$$f^{-1}(x) = \sqrt{x}, \quad \text{for} \quad x \geq 0$$

have been constructed in Fig. 6–6.

**Example
6–9**

**Figure
6–6**

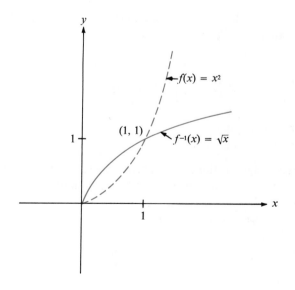

1. In Fig. 6–7, which of the graphs are graphs of 1-1 functions?

Figure
6–7

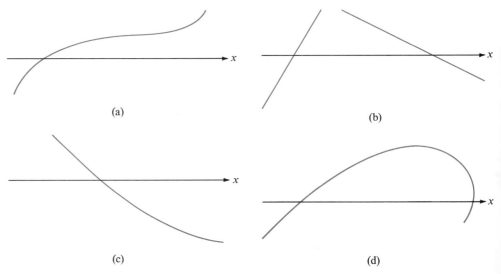

(a)

(b)

(c)

(d)

2. Use the graph of each of the following functions to determine if it is 1-1.
   (a) $f(x) = [x]$
   (b) $g(x) = x - [x]$
   (c) $T(x) = \cos x$
   (d) $L(x) = 3 \sin (x + 1)$, $x \in [0, 2\pi]$
   (e) $g(x) = -x^2 + 1$, $x \in [-1, 0]$
   (f) $g(x) = -x^2 + 1$, $x \in [-1, 1]$

3. Graph each of the following functions. Then, construct the graph of the function's inverse geometrically.
   (a) $T(x) = -x^2$, $x \leq 0$
   (b) $Q(x) = x^3 + 1$, $x \in [-1, 2]$
   (c) $C(x) = \sqrt{2 - x^2}$, $x \in [0, \sqrt{2}]$
   (d) $C(x) = \sqrt{2 - x^2}$, $x \in [-\sqrt{2}, \sqrt{2}]$
   (e) $L(x) = x + 1$
   (f) $K(x) = -x + 1$
   (g) $M(x) = -(\frac{1}{3})x - 6$

4. For each of the functions f given below give a description of f^{-1} using the method of composites as described in Example 6–6.
   (a) $f(x) = 7 - 2x$
   (b) $f(x) = \sqrt{1 - x^2}$, $x \in [-1, 0]$
   (c) $f(x) = 1/x$, $x > 0$
   (d) $f(x) = 3/(1 - x)$, $x > 1$

(e) $f(x) = x/(x + 1)$, $x < -1$

(f) $f(x) = x + \sqrt{x}$. [*Hint:* Solve the quadratic

$$x = (\sqrt{f^{-1}(x)})^2 + \sqrt{f^{-1}(x)} \quad \text{for} \quad \sqrt{f^{-1}(x)}.]$$

(g) $f(x) = \dfrac{ax + b}{cx + d}$, for $ad \neq bc$.

The functions that we shall examine now are based upon the meaning of a^t, where t and a denote numbers and $a > 0$. We assume that the reader is to some extent familiar with this notation. For example, if $a = 2$ and $t = 3$, the usual meaning of 2^3 is 8, *i.e.*, $2 \cdot 2 \cdot 2$. Generally, if t is a positive integer, a^t means

$$a \cdot a \cdot a \cdots a$$

with t factors, each of which is a. From this starting point, when t is a positive integer, it is possible to go further and give meaning to a^t for any real number t. This is accomplished in such a way that the following statements are true for all real numbers t, s:

$$a^t \cdot a^s = a^{t+s} \qquad \text{(6-1)}$$
$$a^t / a^s = a^{t-s} \qquad \text{(6-2)}$$
$$(a^t)^s = a^{ts} \qquad \text{(6-3)}$$
$$(ab)^t = a^t \cdot b^t \qquad \text{(6-4)}$$

We shall not at this point pursue the definition of a^t further. It is enough for our purposes to know that when $a > 0$, a^t means a number for each real number t such that Equations (6-1), (6-2), (6-3), (6-4) above hold.*

If $a > 0$, a^t is a real number for each real number t and we may define a real function

$$\{(t, a^t) \mid t \text{ is a real number}\}.$$

This function we denote by E_a and call it the *exponential function of base* a. In our usual functional notation we have

$$E_a(t) = a^t \quad \text{for all real } t.$$

---

*The reader who is interested in the precise definition of a^t may consult *The Number System* by B. K. Youse, Dickenson Pub. Co., 1965.

As examples of members of E_a, we have

$$(0, a^0) = (0, 1) \in E_a \qquad \text{or} \qquad E_a(0) = 1.$$
$$(1, a^1) = (1, a) \in E_a \qquad \text{or} \qquad E_a(1) = a.$$
$$(-1, a^{-1}) = (-1, 1/a) \in E_a \qquad \text{or} \qquad E_a(-1) = 1/a.$$
$$(\sqrt[3]{5}, a^{\sqrt[3]{5}}) \in E_a \qquad \text{or} \qquad E_a(\sqrt[3]{5}) = a^{\sqrt[3]{5}}.$$

From Equations **(6-1)**, **(6-2)**, and **(6-3)**, we have the following rules:

(6-5) $\qquad\qquad\qquad E_a(t + s) = E_a(t) \cdot E_a(s)$

(6-6) $\qquad\qquad\qquad E_a(t - s) = E_a(t)/E_a(s)$

(6-7) $\qquad\qquad\qquad E_a(ts) = [E_a(t)]^s = [E_a(s)]^t$

We illustrate the graph of E_a for $a = 3$. Some convenient pairs belonging to E_3 are the following:

$$(0, 1) \in E_3.$$
$$(1, 3) \in E_3.$$
$$(-1, \tfrac{1}{3}) \in E_3.$$
$$(2, 9) \in E_3.$$
$$(-2, \tfrac{1}{9}) \in E_3, \text{ etc.}$$

Then, assuming that the graph of E_a is a smooth curve, we may obtain the graph of E_3 as in Fig. 6–8.

Figure 6–8

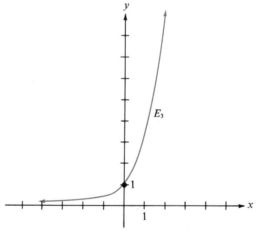

Now consider the graph of $E_{1/3}$. We have for any real number t

$$E_{1/3}(t) = (\tfrac{1}{3})^t = 3^{-t} = E_3(-t).$$

Suppose now that (x, y) is a point on the graph of $E_{1/3}$, i.e., $(x, y) \in E_{1/3}$. Then $y = E_{1/3}(x) = E_3(-x)$ as above. But this says that $(-x, y) \in E_3$.

And in like manner we can show that if $(z, w) \in E_3$, then $(-z, w) \in E_{1/3}$. Hence,

$$(x, y) \in E_{1/3} \text{ if and only if } (-x, y) \in E_3. \qquad \textbf{(6-8)}$$

This is illustrated in Fig. 6–9 in which the graphs of both E_3 and $E_{1/3}$ have been given on the same coordinate-axis system.

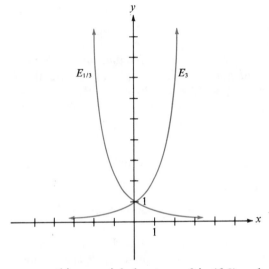

**Figure
6–9**

There is, of course, nothing special about $a = 3$ in **(6-8)** and we may have for any real number $a (> 0)$

$$(x, y) \in E_{1/a} \text{ if and only if } (-x, y) \in E_a. \qquad \textbf{(6-9)}$$

Notice that in Fig. 6–9 the graph of E_3 rises while the graph of $E_{1/3}$ drops (assuming that the curve is traversed in a direction of increasing first co-ordinates). This illustrates a general property of the exponential functions which we may state as follows:

$$\text{If } 0 < a < 1 \quad \text{and} \quad x < y, \quad \text{then} \quad E_a(y) < E_a(x). \qquad \textbf{(6-10)}$$
$$\text{If } 1 < a \qquad \text{and} \quad x < y, \quad \text{then} \quad E_a(x) < E_a(y). \qquad \textbf{(6-11)}$$

Let f be a function.

(*a*) For every pair of numbers x, y in the domain of f such that $x < y$, if

$$f(x) < f(y),$$

then f is called a *strictly increasing* function.

(b) For every pair of numbers x, y in the domain of f such that $x < y$, if

$$f(x) > f(y),$$

then f is called a *strictly decreasing* function.

From these definitions, and Equations **(6-10)** and **(6-11)** above, it is clear that

(6-12) E_a is a strictly decreasing function for $0 < a < 1$.

(6-13) E_a is a strictly increasing function for $a > 1$.

It will be obvious to the reader that *if a function is either strictly increasing or strictly decreasing, it is 1-1*. This implies, of course, that f^{-1}, for such a function f, is a function. Of immediate importance for us is the fact that for each $a > 0$ such that $a \neq 1$, E_a^{-1} is a function. The function E_a^{-1} is called the *logarithm function of base* a. E_a^{-1} is usually denoted by "$\log_a$." Thus,

$$w = E_a(z) = a^z \quad \text{if and only if} \quad z = \log_a(w).$$

Therefore, for all z and w,

$$\log_a(a^z) = z \quad \text{and} \quad a^{\log_a(w)} = w.$$

The properties of the function E_a stated in Equations **(6-5)**, **(6-6)**, and **(6-7)** may be used to derive corresponding properties of $\log_a$. For example, Eq. **(6-5)** may be used to prove the following:

(6-14) $\log_a(t \cdot s) = \log_a(t) + \log_a(s)$

where t, s are positive numbers. To prove this, let $w = \log_a(t)$ and $z = \log_a(s)$. Then by definition $E_a(w) = t$ and $E_a(z) = s$.

Now according to Eq. **(6-5)**,

$$E_a(w + z) = E_a(w) \cdot E_a(z)$$

or substituting for $E_a(w)$ and $E_a(z)$,

$$E_a(w + z) = t \cdot s.$$

This gives

$$\log_a[E_a(w + z)] = \log_a(t \cdot s)$$

or

$$w + z = \log_a(t \cdot s)$$

which is the same as **(6-14)**.

In a similar way, we may prove the following:

$$\log_a \left(\frac{s}{t}\right) = \log_a (s) - \log_a (t) \qquad \text{(6-15)}$$

$$\log_a (s^t) = t \cdot \log_a (s) \qquad \text{(6-16)}$$

where s, t are positive real numbers.

Since $a^t > 0$ for all real numbers t, the range of E_a is a subset of the positive real numbers. In fact, if $a \neq 1$,

$$\Re(E_a) = \{x \mid x > 0\}.$$

Because of the relationship between the domain and range of functions and their inverses,

$$\mathfrak{D}(\log_a) = \{x \mid x > 0\}$$

and

$$\Re(\log_a) = \{t \mid t \text{ is a real number}\}.$$

Show that $\log_a$ is a 1-1 function.

Example
6–10

Solution. Suppose that s, t are positive real numbers such that

$$\log_a (s) = \log_a (t).$$

Then

$$s = E_a[\log_a (s)] = E_a[\log_a (t)] = t.$$

Hence, by definition $\log_a$ is a 1-1 function.

Compute the inverse f^{-1} of the function

Example
6–11

$$f(x) = E_a(\sqrt{x}), x \geq 0, a \neq 1.$$

Solution. We know that $f(f^{-1}(x)) = x$. Thus

$$x = f(f^{-1}(x)) = E_a(\sqrt{f^{-1}(x)})$$

[How do we know here that $f^{-1}(x) \geq 0$?] Therefore,

$$\log_a (x) = \log_a [E_a(\sqrt{f^{-1}(x)})] = \sqrt{f^{-1}(x)}$$

and

$$f^{-1}(x) = [\log_a(x)]^2.$$

We may construct the graph of the functions $\log_a$ in the usual way by making use of the fact that $\log_a$ is the inverse of E_a. The graphs of $\log_3$ and $\log_{1/3}$ are given in Fig. 6–10.

Figure 6–10

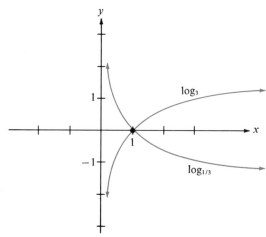

The graphs of $\log_{1/3}$ and $\log_3$ of Fig. 6–10 indicate that the functions $\log_{1/3}$ and $\log_3$ are strictly decreasing and strictly increasing respectively. These are properties which the functions $\log_a$ inherit from E_a.

(6-17)   If $0 < a < 1$, $\log_a$ is a strictly decreasing function.

(6-18)   If $a > 1$, $\log_a$ is a strictly increasing function.

To prove Eq. **(6-17)**, suppose that $\log_a$ is *not* a strictly decreasing function for $0 < a < 1$. Then by definition, there are two numbers x, y in the domain of $\log_a$ such that $x < y$ and

$$\log_a (x) \not> \log_a (y).$$

From this, we may conclude that

$$\log_a (x) \leq \log_a (y).$$

If $\log_a (x) < \log_a (y)$, we know that $E_a[\log_a (x)] > E_a[\log_a (y)]$ since E_a is a strictly decreasing function. But this implies that $x > y$, contrary to the fact that $x < y$. Therefore, we cannot have $\log_a (x) < \log_a (y)$. On the other hand, if $\log_a (x) = \log_a (y)$, then $x = y$ (since $\log_a$ is 1-1) and this again contradicts the fact that $x < y$. This then proves **(6-17)**. A similar proof for **(6-18)** may be given.

5. Plot the graph of each of the following functions:

(a) E_2 (b) E_5 (c) E_{10}

6. (a) Using only the graph of E_2, plot on the same axis the graph of $E_{1/2}$.
   (b) Using only the graph of E_{10}, plot on the same axis the graph of $E_{1/10}$.

7. Prove the statements in Equations **(6-5)**, **(6-6)**, **(6-7)**.

8. Using the fact that E_a, $(a \neq 1)$, and $\log_a$ are 1-1 functions, solve the following equations.
   (a) $2^x = 2^{x^2-1}$
   (b) $a^{7x} = 1$ [Hint: $a^0 = 1$.]
   (c) $3^{7x} = 2$ [Hint: $2 = 3^{\log_3 2}$.]
   (d) $9(3^{x^2}) = 1$
   (e) $\log_a (x^2 - 1) = 0$
   (f) $\log_5 (2x - 1) + \log_5 (2x + 1) = 1$
   (g) $\log_{1/2} (x^2 - 7) - \log_{1/2} (2x^2 + 1) = 1$.

9. Use information about the functions E_a to prove the following properties of $\log_a$.
   (a) If $a > 1$, $\log_a$ is strictly increasing.
   (b) $\log_a \left(\dfrac{t}{s} \right) = \log_a (t) - \log_a (s)$ for positive numbers s, t.
   (c) $\log_a (s^t) = t \log_a (s)$ for all numbers t and positive numbers s.
   (d) $\log_a (t) = \dfrac{\log_b (t)}{\log_b (a)}$ for all positive numbers t, a, b such that $a \neq 1, b \neq 1$.
   (e) If $\log_a (t) = \log_b (s) = w$, then $\log_{ab} (ts) = w$ for all positive numbers t, s, a, b such that $a \neq 1, b \neq 1$.

10. Compute the inverse of each of the following functions:

   (a) $f(x) = 4^x$ (b) $g(x) = \frac{1}{7}(2^{x-1})$
   (c) $h(x) = E_a(1/x) + 1$ (d) $f(x) = \log_2 (x)$
   (e) $g(x) = \log_3 (2x) + \log_3 (x)$ (f) $h(x) = \log_a (7x + 2)$

11. Obtain the solution sets for each of the following inequalities:
   (a) $2^x \leq 1$
   (b) $5^{x^2} < 5^x$
   (c) $3^{x-1} \leq 9^{1-x^2}$
   (d) $(\frac{1}{2})^x \leq 2^{x^2}$
   (e) $E_{1/3} (\sin x) \leq E_{1/3} (\cos x)$, $x \in [0, 2\pi]$
   (f) $\log_2 (1 - x) < \log_2 (2x)$
   (g) $\log_{1/7} (x^3) < \log_{1/7} (2x)$

6.3
Common Logarithms
and Computations

The function $\log_a$ is called the *common logarithm* function when $a = 10$. This function is often used as a tool to simplify numerical calculations which involve multiplications and divisions. The basis for this usefulness in computation is the following theorem.

Theorem
27

If t, s are positive real numbers, then

$$(a)\ \log_{10}(t \cdot s) = \log_{10}(t) + \log_{10}(s)$$

$$(b)\ \log_{10}\left(\frac{t}{s}\right) = \log_{10}(t) - \log_{10}(s)$$

$$(c)\ \log_{10}(s^t) = t \cdot \log_{10}(s)$$

Proof. Part (a) has already been proved as Eq. **(6-14)** on p. 228. For part (b) let $w = \log_{10}(t)$ and $z = \log_{10}(s)$. By Eq. **(6-6)** on p. 226

$$E_{10}(w - z) = E_{10}(w)/E_{10}(z) = t/s.$$

Then

$$\log_{10}(t/s) = w - z = \log_{10}(t) - \log_{10}(s).$$

For part (c), if $z = \log_{10}(s)$, then $t \cdot z = t \cdot \log_{10}(s)$. Now use Eq. **(6-7)** on p. 226:

$$E_{10}(t \cdot z) = E_{10}(t \cdot \log_{10}(s) = [E_{10}))\log_{10}(s))]^t = s^t$$

or

$$t \cdot z = \log_{10}(s^t)$$

i.e.

$$t \cdot \log_{10}(s) = \log(s^t).$$

From this point we will write log for $\log_{10}$.

Before we can use Theorem 27 for carrying out numerical calculations, it will be necessary to discuss Table II found on pp. 364–365. This is a table of common logarithms of numbers between 1 and 10 in increments of 0.01. For example, we find from the table that

$$\log(8.27) \doteq 0.9175$$

(to four decimal places of accuracy). If $x \in [1, 10)$, then

$$1 \leq x < 10$$

and by Eq. **(6-18)**

$$\log (1) \leq \log (x) < \log (10)$$

or

$$0 \leq \log (x) < 1.$$

We therefore see that all the entries in Table II lie in the interval [0, 1). Since Table II contains only the common logarithms of numbers in the interval [1, 10), we will not find log (82.7), log (0.827), or log (0.000827). It is possible to compute these common logarithms, however, by using *scientific notation* and Theorem 27:

$$82.7 = 8.27 \times 10$$
$$0.827 = 8.27 \times 10^{-1}$$
$$0.000827 = 8.27 \times 10^{-4}$$

Then using Theorem 27 (*a*) and (*c*)

$$
\begin{aligned}
\log (82.7) &= \log (8.27 \times 10) \\
&= \log (8.27) + \log (10) && \text{by } (a) \\
&= \log (8.27) + 1 && \text{(since } \log (10) = 1)
\end{aligned}
$$

$$
\begin{aligned}
\log (0.827) &= \log (8.27 \times 10^{-1}) \\
&= \log (8.27) + \log (10^{-1}) && \text{by } (a) \\
&= \log (8.27) + (-1) \cdot \log (10) && \text{by } (c) \\
&= \log (8.27) - 1
\end{aligned}
$$

$$
\begin{aligned}
\log (0.000827) &= \log (8.27 \times 10^{-4}) \\
&= \log (8.27) + \log (10^{-4}) && \text{by } (a) \\
&= \log (8.27) + (-4) \cdot \log (10) && \text{by } (c) \\
&= \log (8.27) - 4
\end{aligned}
$$

The important part of these calculations is that each of the numbers log (82.7), log (0.827), log (0.000827) is expressed in terms of log (8.27). The number log (8.27) is called the *mantissa* for each of the numbers log (8.27), log (82.7), log (0.827), log (0.000827). The integer -4 is called the *characteristic* of log (0.000827) because

$$\log (0.000827) = \log (8.27) - 4.$$

Similarly, log (82.7) has characteristic 1 because

$$\log (82.7) = \log (8.27) + 1$$

and log (0.827) has characteristic -1 because

$$\log (0.827) = \log (8.27) - 1.$$

Using scientific notation we can express any positive number x as

$$x = a \cdot 10^n$$

where $a \in [1, 10)$ and n is an integer. Then

$$
\begin{aligned}
\log (x) &= \log (a \cdot 10^n) \\
&= \log (a) + \log (10^n) && \text{by Theorem 27 } (a) \\
&= \log (a) + n \cdot \log (10) && \text{by Theorem 27 } (c) \\
&= \log (a) + n
\end{aligned}
$$

The mantissa of $\log (x)$ is $\log (a)$ and n is the characteristic of $\log (x)$.

Example 6–12

Find the mantissa and characteristic of $\log (24500)$.

Solution. $24500 = 2.45 \times 10^4$. Therefore,

$$\log (24500) = \log (2.45) + 4$$

so that 4 is the characteristic of $\log (24500)$. Using Table II the mantissa is $\log (2.45) \doteq 0.3892$.

It will be observed that the finding of the characteristic is nothing more than moving the decimal point. Consider 24500 above. The characteristic of $\log (24500)$ is 4 and 4 is the number of places we need to move the decimal place to the left in 24500 in order to obtain 2.45. If the decimal point is moved to the right, then the characteristic is negative. For the number 0.00762 we need to move the decimal point 3 places to the right to obtain 7.62 (a number in $[1, 10)$) so that the characteristic of $\log (0.00762)$ is -3.

Suppose we wish to find $\log (3.102)$. Table II contains no entry for 3.102 although this is a number in $[1, 10)$. There is, however, an entry for 3.10 and one for 3.11. Since

$$3.10 < 3.102 < 3.11$$

and since log is an increasing function we know that

$$\log (3.10) < \log (3.102) < \log (3.11)$$

To find an approximation for $\log (3.102)$ we use interpolation in much the same way as we have done in Chapter 5 for trigonometric functions.

$$
0.01 \left\{ \begin{array}{l} 3.10 \\ 3.102 \\ 3.11 \end{array} \right\} 0.002 \qquad
0.0014 \left\{ \begin{array}{l} \log (3.10) = 0.4914 \\ \log (3.102) = x \\ \log (3.11) = 0.4928 \end{array} \right\} x - 0.4914
$$

Then

$$\frac{x - 0.4914}{0.0014} = \frac{0.002}{0.01} = 0.2$$

$$x - 0.4914 = 0.2(0.0014) = 0.00028 \doteq 0.0003$$

$$x \doteq 0.4914 + 0.0003 = 0.4917$$

Find log (8295).

Example
6–13

Solution. log (8295) = log (8.295) + 3. Then log (8.295) is found by interpolation:

$$0.01 \begin{Bmatrix} 8.29 \\ 8.295 \\ 8.30 \end{Bmatrix} 0.005 \qquad 0.0005 \begin{Bmatrix} \log (8.29) = 0.9186 \\ \log (8.295) = x \\ \log (8.30) = 0.9191 \end{Bmatrix} x - 0.9186$$

$$\frac{x - 0.9186}{0.0005} = \frac{0.005}{0.01} = 0.5$$

$$x - 0.09186 = (0.5)(0.0005) = 0.00025 \doteq 0.0003$$

$$x \doteq 0.9186 + 0.0003 = 0.9189$$

Thus

$$\log (8295) = \log (8.295) + 3$$
$$\doteq 0.9189 + 3$$
$$= 3.9189$$

Let us now consider the reverse problem. For example, suppose we are given the number 0.7604 and asked to solve the equation log $(x) = 0.7604$. Using Table II we find that log (5.76) = 0.7604. Hence, the solution is $x = 5.76$. If the given number had been 0.7608, Table II would not give us a solution directly. But again we may interpolate.

$$0.01 \begin{Bmatrix} 5.76 \\ x \\ 5.77 \end{Bmatrix} x - 5.76 \qquad 0.0008 \begin{Bmatrix} \log (5.76) = 0.7604 \\ \log (x) = 0.7608 \\ \log (5.77) = 0.7612 \end{Bmatrix} 0.0004$$

$$\frac{x - 5.76}{0.01} = \frac{0.0004}{0.0008} = 0.5$$

$$x - 5.76 = (0.5)(0.01) = 0.005$$

$$x = 5.76 + 0.005 = 5.765$$

Example 6–14

Solve the equation $\log(x) = 13.4721$.

Solution. The number 13.4721 is larger than 1 so that Table II cannot be used directly. However,

$$\log(x) = 0.4721 + 13$$

shows that the mantissa of $\log(x)$ is 0.4721 and the characteristic is 13. Since the characteristic of $\log(x)$ gives information only about the position of the decimal point in x, we investigate the equation $\log(y) = 0.4721$. Interpolation is required:

$$0.01\left\{\begin{bmatrix}2.96\\ y\\ 2.97\end{bmatrix}y - 2.96\right. \qquad 0.0015\left\{\begin{bmatrix}\log(2.96) = 0.4713\\ \log(y) = 0.4721\\ \log(2.97) = 0.4728\end{bmatrix}0.0008\right.$$

$$\frac{y - 2.96}{0.01} = \frac{0.0008}{0.0015} = \frac{8}{15}$$

$$y - 2.96 = (8/15)(0.01) \doteq 0.005$$

$$y \doteq 2.96 + 0.005 = 2.965$$

To complete the solution we see that the equation

$$\log(x) = 0.4721 + 13$$

is equivalent to

$$E_{10}\big(\log(x)\big) = E_{10}(0.4721 + 13)$$

or

$$x = 10^{0.4721+13}$$
$$= 10^{0.4721} \times 10^{13}$$

Also, the equation

$$\log(y) = 0.4721$$

is equivalent to

$$y = 10^{0.4721}$$

Since we know that $y \doteq 2.965$, we have

$$x \doteq 10^{0.4721} \times 10^{13}$$
$$= 2.965 \times 10^{13}$$
$$= 29,650,000,000,000$$

In Example 6–14 it was possible to solve the equation $\log(y) = 0.4721$ to obtain $y = 2.965$ and then obtain x simply by shifting the decimal in

2.965 to the right 13 places. If the original equation had been $\log (x) = 0.4721 - 13$, we would have proceeded in the same way except that the decimal would have been shifted 13 places to the *left*. To handle an equation of the type

$$\log (x) = -4.7621$$

we write $-4.7621 = (5 - 4.7621) - 5 = 0.2379 - 5$. Now the equation

$$\log (x) = 0.2379 - 5$$

may be solved as in Example 6–14.

Compute $(0.0837) \cdot (12.6)$ by using common logarithms. Example
6–15

Solution. We let $x = (0.0837) \cdot (12.6)$. Then

$$\begin{aligned}
\log (x) &= \log [(0.0837) \times (12.6)] \\
&= \log (0.0837) + \log (12.6) \qquad \text{by Theorem 27 } (a) \\
&= \log (8.37) - 2 + \log (1.26) + 1 \\
&= \log (8.37) + \log (1.26) - 1
\end{aligned}$$

From Table II we find that

$$\log (8.37) \doteq 0.9227$$
$$\log (1.26) \doteq 0.1004$$

Therefore,

$$\begin{aligned}
\log (x) &= \log (8.37) + \log (1.26) - 1 \\
&= (0.9227 + 0.1004) - 1 \\
&= 1.0231 - 1 \\
&= 0.0231
\end{aligned}$$

Now the equation $\log (x) = 0.0231$ must be solved and interpolation is needed:

$$0.01\left\{\begin{matrix}1.05 \\ x \\ 1.06\end{matrix}\right]x - 1.05 \qquad 0.0041\left\{\begin{matrix}\log (1.05) = 0.0212 \\ \log (x) \quad= 0.0231 \\ \log (1.06) = 0.0253\end{matrix}\right] 0.0019$$

$$\frac{x - 1.05}{0.01} = \frac{0.0019}{0.0041} = \frac{19}{41} \doteq 0.4634$$
$$x - 1.05 \doteq (0.01)(0.4634) \doteq 0.0046$$
$$x \doteq 1.05 + 0.0046 = 1.0546$$

Thus $(0.0837) \cdot (12.6) \doteq 1.0546$.

Example 6–16

Use common logarithms to approximate the number

$$x = \sqrt{\frac{490}{93.2}}$$

Solution.

$$\log(x) = \log\left(\frac{490}{93.2}\right)^{1/2}$$

$$= \tfrac{1}{2}\cdot\log\left(\frac{490}{93.2}\right) \quad \text{by Theorem 27 }(c)$$

$$= \tfrac{1}{2}(\log(490) - \log(93.2)) \quad \text{by Theorem 27 }(b)$$

$$= \tfrac{1}{2}(\log(4.90) + 2 - (\log(9.32) + 1))$$

$$\doteq \tfrac{1}{2}(0.6902 + 2 - (0.9694 + 1)) \quad \text{by Table II}$$

$$= \tfrac{1}{2}(-0.2792 + 1)$$

$$= \tfrac{1}{2}(0.7208)$$

$$= 0.3604$$

Next, x is obtained by interpolation:

$$0.01\begin{Bmatrix}2.29 \\ x \\ 2.30\end{Bmatrix}x - 2.29 \qquad 0.0019\begin{Bmatrix}\log(2.29) = 0.3598 \\ \log(x) = 0.3604 \\ \log(2.30) = 0.3617\end{Bmatrix}0.0006$$

$$\frac{x - 2.29}{0.01} = \frac{0.0006}{0.0019} = \frac{6}{19} \doteq 0.3158$$

$$x - 2.29 \doteq (0.01)(0.3158) \doteq 0.0032$$

$$x \doteq 2.29 + 0.0032 = 2.2932$$

Therefore,

$$\sqrt{\frac{490}{93.2}} \doteq 2.2932$$

Exercises

12. Find the mantissa and characteristic of each of the following.
    Example: log (0.000925). Since log (0.000925) = log (9.25) − 4 the mantissa is log (9.25) and the characteristic is −4.

    (a) log (221)   (b) log (3459)   (c) log (0.010)
    (d) log (55)   (e) log (5,280)   (f) log (0.000012)

13. Use Table II and interpolation, if necessary, to approximate each of the following:

    (a) log (7.12)   (b) log (618)   (c) log (0.0031)
    (d) log (4.756)   (e) log (4756)   (f) log (0.05718)
    (g) log (22.37)   (h) log (0.00001212)   (i) log (3623)
    (j) log (0.007944)

14. Use Table II and interpolation, if necessary, to solve each of the following:

    (a) log (x) = 0.6096   (b) log (x) = 0.9212
    (c) log (x) = 1.7193   (d) log (x) = 0.3560 − 2
    (e) log (x) = −3.5884

    Solution. Write −3.5884 = (4 − 3.5884) − 4 = 0.4116 − 4. Then by Table II log (2.58) = 0.4116 so that x = 0.000258 (the decimal point in 2.58 was moved 4 places to the left because the characteristic is −4).

    (f) log (x) = −2.1124   (g) log (x) = −0.0250
    (h) log (x) = 0.7431   (i) log (x) = 0.9471
    (j) log (x) = 0.6370   (k) log (x) = 0.3515
    (l) log (x) = 1.3515   (m) log (x) = 0.3515 − 2
    (n) log (x) = 3.2262   (o) log (x) = −0.5233
    (p) log (x) = −1.6109   (q) log (x) = −11.2403

15. Use common logarithms to approximate each of the following.
    (a) 7.32 × 1.5
    (b) 240 × 122
    (c) 0.59 × 3.5
    (d) 2,040 × 128
    (e) 24.7 ÷ 0.29
    (f) 0.8 ÷ 0.325
    (g) $\sqrt{265}$
    (h) $\sqrt{0.375}$
    (i) $\sqrt[3]{14}$
    (j) $\sqrt{35} \times \sqrt[3]{50}$
    (k) $\sqrt{(411)/(0.82)}$
    (l) $\sqrt{(0.5276)/(1,422)}$
    (m) $\sqrt[3]{8,042} \times (1.2073)/(0.00258)$
    (n) $(0.0072) \times [(1,162)/(482)] \times \sqrt{2,029}$

16. Suppose x is any real positive number. Show that the characteristic of log (x) is [log (x)] (the greatest integer in log (x)).

7

THE
BINOMIAL
THEOREM

Before discussing the binomial theorem, we take up two very useful notational devices. Both notations will be helpful in our discussion of the binomial theorem.

7.1
The Σ-notation

Let n be a natural number and suppose that $a_1, a_2, a_3, \ldots, a_n$ are numbers. Then, we mean by the symbol $\sum_{i=1}^{n} a_i$ the number

$$a_1 + a_2 + a_3 + \cdots + a_n.$$

The Greek letter "Σ" is supposed to suggest "sum" and we read $\sum_{i=1}^{n} a_i$ as "the sum of a_i from $i = 1$ to $i = n$."

**Example
7–1**

If for each natural number i, $a_i = i$ we have

$$\sum_{i=1}^{n} a_i = a_1 + a_2 + a_3 + \cdots + a_n$$
$$= 1 + 2 + 3 + \cdots + n$$
$$= (\tfrac{1}{2})(n)(n + 1) \text{ by Exercise 29}(a), \text{ Chap. 2}$$

**Example
7–2**

Suppose that for each natural number i we have $a_i = 1$. Then

$$\sum_{i=1}^{n} a_i = \sum_{i=1}^{n} 1 = a_1 + a_2 + \cdots + a_n$$
$$= 1 + 1 + \cdots + 1$$
$$= n.$$

If $P(x) = c_0 + c_1x + c_2x^2 + \cdots + c_mx^m$, $c_m \neq 0$, is a polynomial function, we may express P by using the Σ-notation as follows. Let $a_i = c_ix^i$ for each i. Then

Example
7–3

$$\sum_{i=0}^{m} a_i = \sum_{i=0}^{m} c_ix^i = c_0x^0 + c_1x + c_2x^2 + \cdots + c_mx^m$$

or

$$P(x) = \sum_{i=0}^{m} c_ix^i.$$

Let $a_k = (1 + k)$ for each positive integer k. Then

Example
7–4

$$\begin{aligned}
\sum_{k=1}^{n} a_k &= \sum_{k=1}^{n} (1 + k) \\
&= a_1 + a_2 + a_3 + \cdots + a_n \\
&= (1 + 1) + (1 + 2) + (1 + 3) + \cdots + (1 + n) \\
&= (1 + 1 + 1 + \cdots + 1) + (1 + 2 + 3 + \cdots + n) \\
&= n + (\tfrac{1}{2})(n)(n + 1) \\
&= \sum_{k=1}^{n} 1 + \sum_{k=1}^{n} k.
\end{aligned}$$

In the following theorem we list some of the properties of the Σ-notation.

If n, m are natural numbers with $m < n$, and c, a_i, b_i are any real numbers, then

(a) $\displaystyle\sum_{i=1}^{n} ca_i = c \sum_{i=1}^{n} a_i.$

(b) $\displaystyle\sum_{i=1}^{n} (a_i + b_i) = \sum_{i=1}^{n} a_i + \sum_{i=1}^{n} b_i.$

(c) $\displaystyle\sum_{i=1}^{n} (a_i - a_{i-1}) = a_n - a_0.$

(d) $\displaystyle\sum_{i=m}^{n} a_i = \sum_{i=m+k}^{n+k} a_{i-k}.$

Proof:

(a) This will be proved by applying the mathematical induction principle to the statement

$$P(n): \sum_{i=1}^{n} ca_i = c \sum_{i=1}^{n} a_i.$$

By the definition of Σ we have:

$$\sum_{i=1}^{1} ca_i = ca_1 \quad \text{and} \quad \sum_{i=1}^{1} a_i = a_1,$$

and so

$$\sum_{i=1}^{1} ca_i = c \sum_{i=1}^{1} a_i$$

which shows that $P(1)$ is true. Next, the statements $P(k)$ and $P(k+1)$ are

$$P(k): \sum_{i=1}^{k} ca_i = c \sum_{i=1}^{k} a_i$$

and

$$P(k+1): \sum_{i=1}^{k+1} ca_i = c \sum_{i=1}^{k+1} a_i.$$

Now, if $P(k)$ is true, we have

$$\sum_{i=1}^{k+1} ca_i = \sum_{i=1}^{k} ca_i + ca_{k+1}$$
$$= c \sum_{i=1}^{k} a_i + ca_{k+1}$$
$$= c\left(\sum_{i=1}^{k} a_i + a_{k+1} \right)$$
$$= c \sum_{i=1}^{k+1} a_i.$$

Hence, if $P(k)$ is true, then $P(k+1)$ is true. We conclude that $P(n)$ is true for all natural numbers n.

(d) By definition of Σ, we have

$$\sum_{i=m}^{n} a_i = a_m + a_{m+1} + a_{m+2} + \cdots + a_n$$

and,

$$\sum_{i=n+k}^{n+k} a_{i-k} = a_{(m+k)-k} + a_{(m+k+1)-k} + \cdots + a_{(n+k)-k}$$
$$= a_m + a_{m+1} + a_{m+2} + \cdots + a_n.$$

The proofs of (b) and (c) will be left to the reader. (See Exercise 2.)

The reader will have noticed that the subscript i in Σa_i is not important in the sense that any other letter would serve just as well. For example,

$$\sum_{i=1}^{n} a_i = \sum_{r=1}^{n} a_r = \sum_{j=1}^{n} a_j = \sum_{x=1}^{n} a_x, \text{ etc.}$$

Show that $\displaystyle\sum_{k=1}^{n} a_k - \sum_{k=1}^{n} b_k = \sum_{k=1}^{n} (a_k - b_k)$.

Example
7-5

Solution. From (*a*) of Theorem 28 with $c = -1$,

$$-\sum_{k=1}^{n} b_k = -1 \cdot \sum_{k=1}^{n} b_k = \sum_{k=1}^{n} -1 \cdot b_k = \sum_{k=1}^{n} (-b_k).$$

Then using (*b*) of Theorem 28,

$$\sum_{k=1}^{n} a_k - \sum_{k=1}^{n} b_k = \sum_{k=1}^{n} a_k + \left(-\sum_{k=1}^{n} b_k \right)$$

$$= \sum_{k=1}^{n} a_k + \sum_{k=1}^{n} (-b_k)$$

$$= \sum_{k=1}^{n} [a_k + (-b_k)] \qquad \text{by (b)}$$

$$= \sum_{k=1}^{n} (a_k - b_k).$$

The results of Theorem 28 may be used to successively develop formulas for

Example
7-6

$$\sum_{i=1}^{n} i, \sum_{i=1}^{n} i^2, \sum_{i=1}^{n} i^3, \text{ etc.}$$

We already know that $\displaystyle\sum_{i=1}^{n} i = (\tfrac{1}{2})(n)(n+1)$. Take $a_i = i^3$ in part (*c*) of Theorem 28. This results in

$$\sum_{i=1}^{n} (a_i - a_{i-1}) = a_n - a_o = n^3 - 0^3 = n^3.$$

Now, using the Theorem, we have

$$n^3 = \sum_{i=1}^{n} (a_i - a_{i-1})$$

$$= \sum_{i=1}^{n} [i^3 - (i-1)^3]$$

$$= \sum_{i=1}^{n} i^3 - \sum_{i=1}^{n} (i-1)^3 \qquad \text{(by Example 7–5)}$$

$$= \sum_{i=1}^{n} i^3 - \sum_{i=1}^{n} (i^3 - 3i^2 + 3i - 1)$$

$$= \sum_{i=1}^{n} i^3 - \left(\sum_{i=1}^{n} i^3 - \sum_{i=1}^{n} 3i^2 + \sum_{i=1}^{n} 3i - \sum_{i=1}^{n} 1 \right) \qquad \begin{array}{l} \text{(by parts } (a) \text{ and} \\ (b) \text{ of Theorem 28)} \end{array}$$

$$= \sum_{i=1}^{n} i^3 - \sum_{i=1}^{n} i^3 + 3\sum_{i=1}^{n} i^2 - 3\sum_{i=1}^{n} i + \sum_{i=1}^{n} 1 \qquad \begin{array}{l} \text{(by part } (a) \text{ of} \\ \text{Theorem 28)} \end{array}$$

$$= 3 \sum_{i=1}^{n} i^2 - 3(\tfrac{1}{2})(n)(n+1) + n$$

Therefore,

$$3 \sum_{i=1}^{n} i^2 = n^3 + \tfrac{3}{2}(n)(n+1) - n$$

$$= n(n^2 + \tfrac{3}{2}n + \tfrac{1}{2})$$

$$= \frac{n}{2}(2n^2 + 3n + 1)$$

and

$$\sum_{i=1}^{n} i^2 = \frac{n}{6}(2n^2 + 3n + 1).$$

Exercises

1. Use the definition of Σ-notation to write the following in unabbreviated form.

(a) $\displaystyle\sum_{t=1}^{5} 7t$ (b) $\displaystyle\sum_{j=5}^{10} (j - \tfrac{1}{2})$

(c) $\displaystyle\sum_{s=1}^{8} a^s$ (d) $\displaystyle\sum_{n=1}^{4} 7$

2. Prove parts (b) and (c) of Theorem 28.

3. Prove each of the following by mathematical induction.

(a) If, for each i, $a_i \neq 0$, then $\displaystyle\sum_{i=1}^{n} |a_i| > 0$

(b) $\displaystyle\sum_{t=1}^{n} t^3 = \left(\sum_{t=1}^{n} t \right)^2$

(c) $\displaystyle\sum_{t=1}^{m-1} t^3 < \frac{m^4}{4} < \sum_{t=1}^{m} t^3$

(d) $\displaystyle\sum_{i=1}^{n} (a + id) = an + (\tfrac{1}{2})dn(n + 1)$

4. Use induction to prove that $\displaystyle\sum_{i=1}^{n} at^{i-1} = a(1 - t^n)/(1 - t)$, where a, t are numbers and $t \neq 1$. Use this result to compute the following:

(a) $\displaystyle\sum_{i=1}^{6} 2^{i-1}$ (b) $\displaystyle\sum_{i=1}^{5} (\tfrac{1}{2})2^i$

(c) $\displaystyle\sum_{t=1}^{3} 8^{(t-1)/3}$ (d) $\displaystyle\sum_{t=0}^{3} \frac{1}{3^t}$

(e) $\displaystyle\sum_{k=1}^{n-1} \frac{4}{5^k}$ (f) $\displaystyle\sum_{k=1}^{n-1} (\tfrac{1}{2})5^k$

5. In a manner similar to Example 7–6, derive a formula for $\displaystyle\sum_{i=1}^{n} i^3$.

6. Prove that if n is a natural number and $a_1, a_2, a_3, \ldots, a_n$ are real numbers, then

$$\left| \sum_{i=1}^{n} a_i \right| \leq \sum_{i=1}^{n} |a_i|.$$

Suppose that a motorist wishes to travel from town A to town C by passing through town B, and that there are two roads connecting A to B and three roads connecting B to C. In how many different ways may the motorist make the trip from A to C? (See Fig. 7–1.) If the motorist takes road 1 from A to B, then he may travel from B to C by any one of the three roads connecting these towns; *i.e.*, he may make the trip from A to C in three ways by taking road 1. Similarly, he may make the trip in three ways if he takes road 2. We conclude that it is possible to make the trip in a total of six different ways.

This example illustrates the following general counting principle: *if a task* T_1 *can be performed in* t_1 *different ways and a task* T_2 *can be performed in* t_2 *ways, then the tasks* T_1, T_2 *can be performed successively in* $t_1 \cdot t_2$ *ways.*

The above counting principle is stated only for two tasks T_1, T_2. Suppose that we have n tasks $T_1, T_2, T_3, \ldots, T_n$ and T_1 can be performed in t_1 ways; T_2 can be performed in t_2 ways; etc. We claim that the sequence of

Figure
7–1

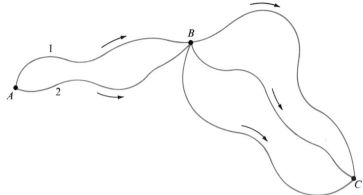

tasks $T_1, T_2, T_3, \ldots, T_n$ can be performed in $t_1 \cdot t_2 \cdot t_3 \cdots t_n$ ways. To prove this we may apply mathematical induction to the statement:

$P(n)$: the sequence $T_1, T_2, T_3, \ldots, T_n$ can be performed in
$t_1 \cdot t_2 \cdot t_3 \cdots t_n$ ways.

The statement $P(1)$ is simply "T_1 can be performed in t_1 ways." Since we are given this information, $P(1)$ is true. The sequence $T_1, T_2, T_3, \ldots,$ T_k, T_{k+1} of tasks may be performed by first performing $T_1, T_2, T_3, \ldots, T_k$ and then performing T_{k+1}. If $P(k)$ is true, the sequence $T_1, T_2, T_3, \ldots, T_k$ may be performed in $t_1 \cdot t_2 \cdot t_3 \cdots t_k$ ways; and by the counting principle, sequence $T_1, T_2, T_3, \ldots, T_k, T_{k+1}$ may be performed $(t_1 \cdot t_2 \cdot t_3 \cdots t_k) \cdot t_{k+1}$ many ways. But this says that $P(k + 1)$ is true if $P(k)$ is true. By mathematical induction we conclude that the following is true.

Theorem
29

Let n be a natural number and suppose that T_i is a task for each $i = 1, 2, 3, \ldots, n$ which can be performed in t_i ways. Then the sequence of tasks $T_1, T_2, T_3, \ldots, T_n$ can be performed in $t_1 \cdot t_2 \cdot t_3 \cdots t_n$ ways.

Example
7–7

How many 4-digit numbers are there which involve only the digits 1, 2, 3?

Solution. Any such number may be obtained by writing down four digits $a_1 a_2 a_3 a_4$ where a_i is 1, 2 or 3 for each $i = 1, 2, 3, 4$. For example, 2132 is such a number. Each of the "tasks" of writing down a_i can be performed in three ways. Hence, there are $3 \cdot 3 \cdot 3 \cdot 3 = 81$ such numbers.

Example
7–8

If S is a set of n elements and T is a set of m elements, how many elements has the Cartesian product $S \times T$?

Solution. The set $S \times T$ consists of all ordered pairs (x, y) such that $x \in S$ and $y \in T$. Since the x may be chosen in n ways and the y in m ways, we may choose x, y in $n \cdot m$ ways. Hence, the set $S \times T$ contains nm elements.

Let S be a set having $n > 0$ number of elements, and suppose that $a_1, a_2, a_3, \ldots, a_n$ are the elements of S. If T is a subset of S, we may indicate that an element a_i is or is not a member of T by writing 1 or 0 respectively. For example, if $S = \{a_1, a_2, a_3, a_4\}$, then we could indicate that T is the set $\{a_2, a_4\}$ by writing

$$
\begin{array}{cccc}
a_1 & a_2 & a_3 & a_4 \\
| & | & | & | \\
0 & 1 & 0 & 1
\end{array}
$$

In this way, every subset of S corresponds uniquely to a sequence of n zeros and ones. In other words, every sequence of n zeros and ones represents a subset of S and conversely every subset of S has such a representation. The sequence $000 \cdots 0$ (all zeros) represents the null subset of S while $111 \cdots 1$ (all ones) represents S itself. The question of how many subsets of S there are may be answered by determining the number of sequences of n zeros and ones. Every such sequence may be obtained as

$$z_1 z_2 z_3 \cdots z_n$$

where each z_i is 0 or 1. Since there are 2 choices for each z_i, there are $2 \cdot 2 \cdots 2 = 2^n$ such sequences. Therefore, *a set S of n elements has 2^n subsets.*

If $S = \{a_1, a_2, a_3, \ldots, a_n\}$ is a set, and we impose an order on its elements — as we have done already for $n = 2$ — we obtain an *ordered n-tuple* which is denoted by $(a_1, a_2, a_3, \ldots, a_n)$. If, for each i, j such that $i \neq j$, we have $a_i \neq a_j$, then the n-tuple $(a_1, a_2, a_3, \ldots, a_n)$ is called a *permutation* of n elements.

How many ordered n-tuples can be formed from a set of m elements?

Solution. Since each a_i in an n-tuple $(a_1, a_2, a_3, \ldots, a_n)$ may be chosen in m different ways, there are $m \cdot m \cdot m \cdots m$ or m^n ordered n-tuples. From this it is evident that from a set of 10 elements we may construct 10^2 or 100 ordered pairs.

Example 7–9

How many permutations of n elements can be formed from a set of $m \geq n$ elements?

Example 7–10

Solution. To construct a permutation $(a_1, a_2, a_3, \ldots, a_n)$, we must make certain that if $i \neq j$, then $a_i \neq a_j$. Since each a_i belongs to a set of m elements, we may reason as follows: a_1 may be chosen in m ways; a_2 may be chosen in $(m - 1)$ ways (since the choice for a_1 cannot be repeated); a_3 may be chosen in $(m - 2)$ ways (since the choice for a_1 and a_2 cannot be repeated), etc.; finally, a_n may be chosen in $m - (n - 1)$ ways or in $m - n + 1$ ways. Therefore, there are $m \cdot (m - 1) \cdot (m - 2) \cdot (m - 3) \cdot \cdots \cdot (m - n + 1)$ permutations of n elements from a set of m elements. This number is sometimes denoted by $P(m, n)$; *i.e.*, by definition,

$$P(m, n) = m(m - 1) \cdot (m - 2) \cdot (m - 3) \cdot \cdots \cdot (m - n + 1).$$

From Example 7–10, if $m = n$,

$$P(n, n) = n(n - 1)(n - 2)(n - 3) \cdots 2 \cdot 1.$$

The product $n(n - 1)(n - 2)(n - 3) \cdots 2 \cdot 1$ is usually denoted by $n!$ (Define $0! = 1$). Hence, $P(n, n) = n!$ and *the number of permutations of* n *elements is* n!

The symbol $\binom{n}{k}$, where n, k are non-negative integers and $k \leq n$, denotes the total number of subsets of k elements from a set of n elements. We just have seen that each set $\{b_1, b_2, b_3, \ldots, b_k\}$ of k elements gives rise to $k!$ permutations; and since each permutation of k elements from a set of n elements is obtained from a subset of k elements, we have

$$\binom{n}{k} \cdot k! = P(n, k) = n(n - 1)(n - 2)(n - 3) \cdots (n - k + 1)$$

or

$$\binom{n}{k} = \frac{n(n - 1)(n - 2)(n - 3) \cdots (n - k + 1)}{k!}.$$

Since

$$n! = [n(n - 1)(n - 2) \cdots (n - k + 1)][(n - k)(n - k - 1) \cdots 2 \cdot 1]$$
$$= [n(n - 1)(n - 2) \cdots (n - k + 1)][(n - k)!],$$

we may write

$$\binom{n}{k} = \frac{[n(n - 1)(n - 2) \cdots (n - k + 1)][(n - k)!]}{k!(n - k)!} = \frac{n!}{k!(n - k)!}.$$

Example 7–11

A set of 10 elements contains $\binom{10}{5}$ 5-element subsets; *i.e.*,

$$\frac{10!}{5! \, 5!} = \frac{6 \cdot 7 \cdot 8 \cdot 9 \cdot 10}{1 \cdot 2 \cdot 3 \cdot 4 \cdot 5} = 252.$$

A set of 10 elements contains 10 one-element subsets. This agrees with the results just obtained since

$$\binom{10}{1} = \frac{10!}{1! \, (9!)} = 10.$$

Example
7–12

For a set of 4 elements there are $4! = 24$ permutations.

Example
7–13

From an ordinary pack of playing cards, how many bridge hands are possible?

Example
7–14

Solution. An ordinary pack of playing cards contains 52 cards and a bridge hand consists of 13 cards. Therefore, the question is simply: "How many subsets are there of 13 elements from a set of 52?" Hence, the total number of bridge hands from a bridge deck of cards is

$$\binom{52}{13} = \frac{52!}{13! \, (52 - 13)!} = \frac{52!}{13! \, 39!} = \frac{40 \cdot 41 \cdot 42 \ldots 52}{13!}.$$

This comes to 635,013,559,600.

Suppose that license plates are to be made as follows:

Example
7–15

$$A\text{-}B\text{-}CDEF$$

where A, C, D, E, F are digits $0, 1, 2, 3, 4, \cdots, 9$ except that $A \neq 0, C \neq 0$ and B can be any letter of the alphabet except "O". How many plates can be made subject to these restrictions?

Solution. A, C may each be chosen in nine ways; D, E, F each in 10 ways; and B in 25 ways. Thus, there are

$$9 \cdot 25 \cdot 9 \cdot 10 \cdot 10 \cdot 10 = 2,025,000$$

possible plates.

7. Let $B = \{1, 3, 5, 7\}$.

   (*a*) Using numbers in B, how many 1 digit numbers can be formed?
   (*b*) Using numbers in B, how many 2 digit numbers can be formed?
   (*c*) Using numbers in B, how many 3 digit numbers can be formed?
   (*d*) Using numbers in B, how many numbers of n digits can be formed?

8. Call letters for radio stations consist of either 3 or 4 letter Latin alphabet sequences; each sequence begins with the letter W.
   (*a*) How many 3 letter sequences are there which begin with W?
   (*b*) How many 4 letter sequences are there which begin with W?
   (*c*) How many radio stations can there be which have 3 or 4 call letters?

9. A telephone number can be any sequence of seven digits from the set 0, 1, 2, 3, 4, 5, 6, 7, 8, 9 except that the number may not begin with 0. How many telephone numbers are there?

10. A true-false test is given with 10 questions; each question can be answered true or false. In how many different ways can the test be completed?

11. A certain state in the U.S. decides to design its auto license plates as follows:

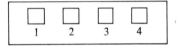

where the first "block" is to be a letter of the alphabet not "O" and the other "blocks" are to be digits from the set {0, 1, 2, 3, 4, 5, 6, 7, 8, 9} except that block 2 cannot be "0." How many plates can this state make with this design?

12. The state referred to in Exercise 11 above discovered one year — purely by accident — that with the tag design then in existence there were not enough license plates for each auto to have one! With some outside help, it was determined that there were at least 75,000 and no more than 100,000 autos in the state. The Governor of the state solved the problem by affixing four additional blocks to the tag design as shown below:

The Governor argued that by adding four more blocks for numbers the number of tags would be increased by a factor of 4 : $4 \cdot 22500 = 90000$. What was the smallest number of blocks which could have been added to solve the problem? How many tags could have been made with just one additional block? How many tags were possible by the Governor's method? What was the excess number of tags that resulted by the Governor's method? (Assume that 100,000 were needed.) The year following the Great Tag Surplus the I.C.C. paid the state *not* to produce auto tags. This action greatly endeared the Governor to the legislature and in later years he became affectionately known as Old Blockhead.

13. Let A be a set with 5 elements. (*a*) How many pairs (a, b) are there with $a \in A$, $b \in A$? (*b*) How many pairs (a, b) are there with $a \in A$, $b \in A$, and $a \neq b$?

    Let A be a set with k elements. (*a*) How many pairs (a, b) are there with $a \in A$ and $b \in A$? (*b*) How many pairs (a, b) are there with $a \in A$, $b \in A$, and $a \neq b$?

    In Exercises 14–19 express each problem in the proper notation and then make the computation requested. The notation and computations refer to the formula $P(n, k) = n \cdot (n - 1) \cdot (n - 2) \cdots (n - k + 1)$.

14. How many permutations are there of 15 things taken 10 at a time? Notation: $P(15, 10)$.

    Solution: The $n - k + 1$ factor is $15 - 10 + 1 = 6$. $P(15, 10) = 15 \cdot 14 \cdot 13 \cdot 12 \cdot 11 \cdot 10 \cdot 9 \cdot 8 \cdot 7 \cdot 6$.

15. Compute the number of permutations of 5 things taken 2 at a time.

16. Compute the number of permutations of 12 things taken 3 at a time.

17. Compute the number of permutations of 99 things taken 1 at a time.

18. Compute the number of permutations of m things taken 5 at a time, where 5 is no larger than m.

19. Compute the number of permutations of 35 things taken t at a time, where t is no larger than 35.

    In Exercises 20–24, express the problem in the appropriate notation and compute.

20. The number of 5 element subsets of a set of 7 elements.

    Solution: $\binom{7}{5} = \frac{7!}{5!(7 - 5)!} = \frac{7!}{5!2!} = 21$

21. The number of 3-subsets of a set of 10 elements.

22. The number of 1-subsets of a set of 100 elements.

23. The number of 100-subsets of a set of 100 elements.

24. The number of 48-subsets of a set of 50 elements.

In Exercises 25–28 make the computations requested.

25. Express the number 12! in terms of the number 10!.

Solution: $12! = 12 \cdot 11 \cdot 10 \cdot 9 \cdots 2 \cdot 1$
$= 12 \cdot 11 \cdot (10 \cdot 9 \cdots 2 \cdot 1)$
$= 132 \cdot (10!)$

26. Express the number 22! in terms of 21!.

27. Express the number 38! in terms of 35!.

28. Express the number 38! as the product of 35! and $P(38, k)$ for some k.

29. Express 100! as a product of 50! and $P(100, k)$ for some k.

30. Express 432! as a product of 111! and $P(432, k)$ for some k.

31. Compute each of the following:

(a) $P(3, 2)$ (b) $P(8, 5)$ (c) $\binom{5}{2}$

(d) $\binom{5}{4}$ (e) $P(m, n - 2)$ (f) $P(t - 1, t - 1)$

32. Find m if

(a) $P(m, 2) = 12$ (b) $P(m, 3) = 120$

(c) $\binom{m}{2} = 6$ (d) $\binom{m}{3} = 20$

33. In how many ways can a committee of 4 be chosen from a group of 12 persons?

34. In an organization consisting of 8 members in how many ways can the offices of 1st, 2nd, and 3rd vice president be filled?

35. (a) How many 5-card hands are there in a deck of 52 playing cards?
    (b) How many 5-card hands are there which contain the Queen of Diamonds?
    (c) How many 5-card hands are there having all clubs?

36. How many 4-digit numbers are there which involve only 8, 5?

37. A student wishes to schedule his three classes at 8:30, 9:30, and 10:30. If there are three courses he can take at 8:30, two courses he can take at 9:30, and five courses he can take at 10:30, how many schedules are possible?

38. The student of Problem 37 discovers that the department of his major forbids taking Basketweaving 432 and Social Maladjustment 82.3 concurrently. One of these is taught at 8:30 and the other at 10:30. With this added restriction how many schedules are possible?

39. The chairman of a 10-man committee wishes to appoint a 4-man sub-committee. In how many ways can he do this?

40. The chairman of a 10-man committee wishes to appoint two 4-man sub-committees with himself as chairman of each. In how many ways can he appoint these two sub-committees if he is to be the only common member?

41. In how many ways can the chairman of an 8-man committee appoint a 3-man subcommittee if he allows himself to be a member of the sub-committee? In how many ways if he demands to always be a member of the subcommittee? In how many ways if he is never a member of the subcommittee?

42. Let $S = \{a, b, c, d\}$ be a 4-element set.
    (a) Using the technique described above of associating a sequence of 0's and 1's with subsets, write the appropriate sequence for each of the following subsets of S:

(i) $\{a\}$	(ix) $\{b, d\}$
(ii) $\{b\}$	(x) $\{c, d\}$
(iii) $\{c\}$	(xi) $\{a, b, c\}$
(iv) $\{d\}$	(xii) $\{a, b, d\}$
(v) $\{a, b\}$	(xiii) $\{b, c, d\}$
(vi) $\{a, c\}$	(xiv) $\{a, c, d\}$
(vii) $\{a, d\}$	(xv) $\{a, b, c, d\}$
(viii) $\{b, c\}$	(xvi) $\varnothing$

    (b) Give the appropriate subset of S that goes with the following sequences:

(i) 0101	(ix) 0100
(ii) 0111	(x) 0010
(iii) 1010	(xi) 1100
(iv) 1110	(xii) 0110
(v) 0001	(xiii) 0011
(vi) 1001	(xiv) 0000
(vii) 1011	(xv) 1111
(viii) 1000	(xvi) 1101

43. List all the permutations of the set $\{a, b, c\}$.

The Binomial Theorem

If n, k are non-negative integers and $n \geq k$, the symbol $\binom{n}{k}$ denotes the number

$$\frac{n!}{k!\,(n-k!)}.$$

By definition $n! = n(n-1)(n-2)(n-3)\cdots 2 \cdot 1$ if $n \geq 1$ and $0! = 1$.

The numbers $\binom{n}{k}$ are called the *binomial coefficients*.

Example 7–16

$$\binom{6}{3} = \frac{6!}{3!\,(6-3)!} = \frac{6!}{3!\,3!} = \frac{1\cdot2\cdot3\cdot4\cdot5\cdot6}{(1\cdot2\cdot3)(1\cdot2\cdot3)} = 20.$$

Example 7–17

If n is any non-negative integer, then

$$\binom{n}{0} = \frac{n!}{0!\,(n-0)!} = \frac{n!}{1\cdot(n)!} = 1.$$

Also

$$\binom{n}{n} = \frac{n!}{n!\,(n-n)!} = 1.$$

Example 7–18

Prove that if m, k are integers with $m \geq 1$, $k \geq 1$ and $m > k$, then

$$\binom{m+1}{k} = \binom{m}{k} + \binom{m}{k-1}.$$

Solution. Using the definition,

$$\binom{m}{k} + \binom{m}{k-1} = \frac{m!}{k!\,(m-k)!} + \frac{m!}{(k-1)!\,(m-k+1)!}$$

$$= \frac{(m!)(m-k+1)}{k!\,(m-k)!\,(m-k+1)}$$

$$+ \frac{k(m!)}{k(k-1)!\,(m-k+1)!}$$

$$= \frac{(m!)(m-k+1)}{k!\,(m-k+1)!} + \frac{k(m!)}{k!\,(m-k+1)!}$$

$$= \frac{(m!)(m-k+1) + k(m!)}{k!\,(m-k+1)!}$$

$$= \frac{(m+1)!}{k!\,[(m+1)-k]!} = \binom{m+1}{k}.$$

This result will be used below in the proof of the binomial theorem.

If a, b are numbers and n is an integer such that $n \geq 1$, then

$$(a + b)^n = \sum_{i=0}^{n} \binom{n}{i} a^{n-i} b^i.$$

Before proving the binomial theorem, a few illustrations will be given.

$(a + b)^2 = a^2 + 2ab + b^2$ by multiplying. Using the binomial theorem,

$$(a + b)^2 = \sum_{i=0}^{2} \binom{2}{i} a^{2-i} b^i$$

$$= \binom{2}{0} a^{2-0} b^0 + \binom{2}{1} a^{2-1} b^1 + \binom{2}{2} a^{2-2} b^2$$

$$= a^2 + 2ab + b^2.$$

Example
7–19

Using the binomial theorem to compute $(x - y)^5$, we have

$(x - y)^5 = [x + (-y)]^5$

$$= \sum_{i=0}^{5} \binom{5}{i} x^{5-i} (-y)^i$$

$$= \binom{5}{0} x^5 (-y)^0 + \binom{5}{1} x^4 (-y)^1 + \binom{5}{2} x^3 (-y)^2 + \binom{5}{3} x^2 (-y)^3$$

$$+ \binom{5}{4} x^1 (-y)^4 + \binom{5}{5} x^0 (-y)^5$$

$$= x^5 - 5x^4 y + 10x^3 y^2 - 10x^2 y^3 + 5xy^4 - y^5.$$

Example
7–20

Proof (of Theorem 30): The proof is by induction. Apply the induction principle to the statement

$$P(n) : (a + b)^n = \sum_{i=0}^{n} \binom{n}{i} a^{n-i} b^i.$$

The statement $P(1)$ is

$$(a + b)^1 = \sum_{i=0}^{1} \binom{1}{i} a^{1-i} b^i$$

and this is true since $(a + b)^1 = a + b$ and

$$\sum_{i=0}^{1} \binom{1}{i} a^{1-i} b^i = \binom{1}{0} a^{1-0} b^0 + \binom{1}{1} a^{1-1} b^1$$

$$= a + b.$$

Now suppose that $P(k)$ is true; *i.e.*, that

$$(a + b)^k = \sum_{i=0}^{k} \binom{k}{i} a^{k-i} b^i.$$

Then

$$(a + b)^{k+1} = (a + b)(a + b)^k$$

$$= (a + b) \sum_{i=0}^{k} \binom{k}{i} a^{k-i} b^i \quad \text{(if } P(k) \text{ is true)}$$

$$= a \sum_{i=0}^{k} \binom{k}{i} a^{k-i} b^i + b \sum_{i=0}^{k} \binom{k}{i} a^{k-i} b^i$$

$$= \sum_{i=0}^{k} \binom{k}{i} a^{k-i+1} b^i + \sum_{i=0}^{k} \binom{k}{i} a^{k-i} b^{i+1}$$

$$= \sum_{i=0}^{k} \binom{k}{i} a^{k-i+1} b^i + \sum_{i=1}^{k+1} \binom{k}{i-1} a^{k-i+1} b^i$$

$$= \binom{k}{0} a^{k+1} b^0 + \sum_{i=1}^{k} \binom{k}{i} a^{k-i+1} b^i + \sum_{i=1}^{k} \binom{k}{i-1} a^{k-i+1} b^i$$
$$+ \binom{k}{k} a^0 b^{k+1}$$

$$= \binom{k+1}{0} a^{k+1} b^0 + \sum_{i=1}^{k} \left[\binom{k}{i} + \binom{k}{i-1} \right] a^{k-i+1} b^i$$
$$+ \binom{k+1}{k+1} a^0 b^{k+1}$$

$$= \binom{k+1}{0} a^{k+1} b^0 + \sum_{i=1}^{k} \binom{k+1}{i} a^{(k+1)-i} b^i$$
$$+ \binom{k+1}{k+1} a^0 b^{k+1}$$

$$= \sum_{i=0}^{k+1} \binom{k+1}{i} a^{(k+1)-i} b^i.$$

This completes the proof since

$$(a + b)^{k+1} = \sum_{i=0}^{k+1} \binom{k+1}{i} a^{k+1-i} b^i$$

is the statement $P(k + 1)$, and we have shown that this is true if $P(k)$ is true.

Example 7–21

In the binomial expansion of $(2x - 1)^{10}$, what is the fifth term?

Solution. By the theorem,

$$(2x - 1)^{10} = \sum_{i=0}^{10} \binom{10}{i} (2x)^{10-i} (-1)^i.$$

Notice that the number of a term $\binom{10}{i} \cdot (2x)^{10-i}(-1)^i$ can be determined by the i. Since i starts at 0, the term $\binom{10}{i} \cdot (2x)^{10-i}(-1)^i$ is term number $(i + 1)$. Hence, the fifth term is

$$\binom{10}{4} \cdot (2x)^{10-4}(-1)^4 = 210(2x)^6 = 13{,}440x^6.$$

For an excellent discussion of the binomial theorem and related matters the reader may consult I. Niven, *Mathematics of Choice*, Random House.

44. Use the binomial theorem to expand each of the following: Exercises

    (a) $(a - b)^3$ (b) $(-3x + 1)^4$

    (c) $(x - 1)^6$ (d) $(x^{1/3} + y^{1/3})^6$

    (e) $(2/x + x/2)^5$ (f) $(a^2 - bx + x^2)^3$

45. Compute $(a + \sqrt{3})^4 + (a - \sqrt{3})^4$.

46. Compute $(\sqrt{2} + 1)^6 - (\sqrt{2} - 1)^6$.

47. Prove the following:

    (a) $\binom{n}{k} = \binom{n}{n - k}$ (b) $\binom{n}{k + 1} = \dfrac{n - k}{k + 1} \cdot \binom{n}{k}$

    (c) $\displaystyle\sum_{i=0}^{n} \binom{n}{i} = 2^n$ (d) $\displaystyle\sum_{i=0}^{n} (-1)^i \binom{n}{i} = 0$

48. (a) In the expansion of $(x - 1)^{15}$ give the 8th term.

    (b) In the expansion of $\left(\dfrac{2}{c} + \dfrac{d}{2}\right)^{10}$ give the 5th term.

    (c) Compute the coefficient of x^{18} in the expansion of $(x^2 + 5/x)^{15}$.

8

ANALYTIC GEOMETRY

In Chapter 3, the concepts of coordinate systems, the graphs of relations, etc., were discussed to some extent. These concepts are basic to a study of analytic geometry. The guiding principle behind analytic geometry is that of making questions of a geometric nature depend upon the real number system. This dependence is initiated in the case of *plane* analytic geometry with the correspondence between ordered pairs of real numbers and points of the plane. The reason that such a dependency is desirable is that it helps our understanding of geometry and aids in the solving of geometric problems.

8.1
Lines

We have seen (in Chapter 3) that the graph of a linear relation $\{(x, y) \mid Ax + By = C\}$ is a line and that every line is the graph of some linear relation. If $B \neq 0$, we may write

$$\{(x, y) \mid Ax + By = C\} = \{(x, y) \mid y = [(-A)/B]x + C/B\}.$$

Now let $m = (-A)/B$ and $b = C/B$. The number m is called the *slope* of the line L which is the graph of

$$\{(x, y) \mid y = mx + b\}.$$

If $B = 0$, the graph of $\{(x, y) \mid Ax = C\} = \{(x, y) \mid x = C/A\}$ is a vertical line; and, in this case, we say that *the line has no slope*. The significance of the slope m is that it measures the "steepness" of the line L with respect to the x-axis. (See Fig. 8–1.) In Fig. 8–1, φ is an angle that line L makes with the x-axis and $\tan \varphi$ can be computed by using triangle ABD. We have

$$\tan \varphi = (-b)/(-b/m) = m.$$

Figure
8–1

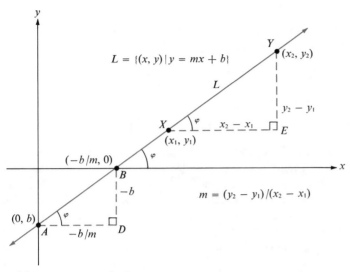

From this we may conclude:

$$0 \le \varphi < \pi/2 \text{ if and only if } m > 0 \quad \text{(see Fig. 8–2(a))} \quad \textbf{(8-1)}$$
$$\pi/2 < \varphi < \pi \text{ if and only if } m < 0 \quad \text{(see Fig. 8–2(b)).} \quad \textbf{(8-2)}$$

Figure
8–2

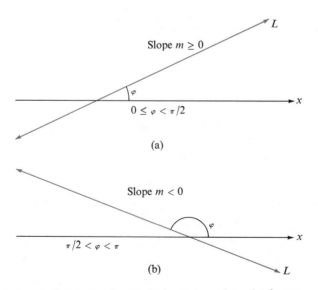

It should be noted that the line L of Fig. 8–1 used to obtain $\tan \varphi = m$ is a special case in that it does not pass through the origin. The result still holds for lines which pass through the origin (see Exercise 4 in this section).

As indicated in Fig. 8–1, the slope m of L may be computed by using *any* pair of points on L. For if $X(x_1, y_1)$ and $Y(x_2, y_2)$ are two points of L, then, triangle XYE is similar to triangle ABD so that

$$\frac{y_2 - y_1}{x_2 - x_1} = (-b)/((-b)/m) = m.$$

The reader should check to see that this result still holds for points X, Y of L other than those in quadrant I.

The reader will recall that there is one and only one line on two points P, Q. Given the points P, $Q(P \neq Q)$, is it possible to determine a linear relation whose graph is the line on P, Q? To answer this, suppose that (a, b), (c, d) are the coordinates of P, Q respectively. The distances that P, Q are respectively from the y-axis is given by $|a|$, $|c|$. Hence, if $a = c$, P, Q are each $|a|$ units from the y-axis; so that the line on P, Q is parallel to the y-axis. Therefore, a linear relation having this line as its graph is

$$\{(x, y) \mid x = a\} = \{(a, y) \mid y \text{ is a real number}\}.$$

If $a \neq c$, we may use the coordinates of P, Q as above to compute the slope of the line on P, Q:

$$\frac{d - b}{c - a} = m.$$

Now if the relation $L = \{(x, y) \mid y = mx + p\}$ is to have the line on P, Q as its graph, we must have

$$(a, b) \in L \qquad \text{and} \qquad (c, d) \in L.$$

Hence,

$$b = ma + p \qquad \text{and} \qquad d = mc + p.$$

Thus, $p = b - ma = d - mc$. This shows that we may use the two points to compute the slope m (if the line is not vertical) and either of the two points to compute p. We then know that the relation

$$L = \{(x, y) \mid y = mx + p\},$$

where m, p are given as above, has a graph which is the line on P, Q; for the graph of *every* linear relation is a line and in particular the graph of L is a line — the unique line on P, Q.

Example 8–1

What is a linear relation whose graph is the line on the points $(1, 0)$, $(\frac{1}{2}, 3)$?

Solution. The slope is

$$\frac{3 - 0}{\frac{1}{2} - 1} = \frac{3}{-\frac{1}{2}} = -6.$$

The relation is then $\{(x, y) \mid y = -6x + p\}$ where p is to be determined. Since $(1, 0) \in \{(x, y) \mid y = -6x + p\}$, we have

$$0 = -6 \cdot 1 + p$$

or $p = 6$. The desired relation is thus

$$\{(x, y) \mid y = -6x + 6\}.$$

Example
8–2

What is the slope and y-intercept of the graph of the relation

$$\{(x, y) \mid 3x - 7y = 2\}?$$

Solution. To determine the slope and y-intercept, we express the "connection" between x and y in the form $y = mx + p$:

$$\{(x, y) \mid 3x - 7y = 2\} = \{(x, y) \mid y = (\tfrac{3}{7})x - \tfrac{2}{7}\}.$$

Therefore, the slope is $\tfrac{3}{7}$ and the y-intercept is $(0, -\tfrac{2}{7})$.

If two lines k_1 and k_2 are parallel, then they make equal angles with the x-axis. If each of k_1, k_2 is perpendicular to the x-axis, then they are each graphs of linear relations of the form

$$\{(x, y) \mid x = a\}.$$

If k_1, k_2 are not vertical lines, they have equal slopes. Thus k_1, k_2 are graphs of linear relations as follows:

$$k_1: \{(x, y) \mid y = mx + p_1\}$$
$$k_2: \{(x, y) \mid y = mx + p_2\}.$$

(See Fig. 8–3.)

Figure
8–3

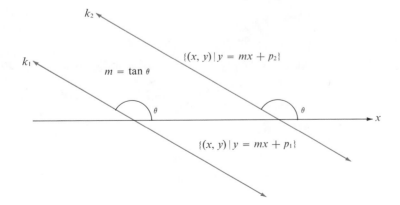

k_2

k_1

$m = \tan \theta$

$\{(x, y) \mid y = mx + p_2\}$

θ

θ

x

$\{(x, y) \mid y = mx + p_1\}$

Conversely, two relations

$$\{(x, y) \mid y = mx + p_1\} \qquad \text{and} \qquad \{(x, y) \mid y = mx + p_2\}$$

have graphs which are parallel lines.

From this point we shall not always clearly distinguish between a relation and its graph.

Let k_1, k_2 be two lines which are perpendicular to each other and neither is parallel to the x-axis. (See Fig. 8–4.) If m_1 is the slope of k_1 and m_2 is the slope of k_2, we have $\tan \varphi_1 = m_1$ and $\tan \varphi_2 = m_2$ where φ_1, φ_2 are the angles as shown in Fig. 8–4. Since φ_1 is an exterior angle of the triangle in Fig. 8–4, it is the sum of the opposite interior angles: $\varphi_1 = \pi/2 + \varphi_2$ or $\varphi_1 - \varphi_2 = \pi/2$. Then

$$0 = \cos(\pi/2) = \cos(\varphi_1 - \varphi_2)$$
$$= \cos \varphi_1 \cos \varphi_2 + \sin \varphi_1 \sin \varphi_2.$$

Figure 8–4

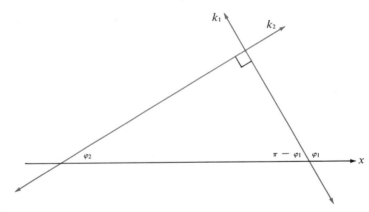

Hence,

(8-3) $$\sin \varphi_1 \sin \varphi_2 = -\cos \varphi_1 \cos \varphi_2$$

and since $\cos \varphi_1 \cos \varphi_2 \neq 0$ (why?), we may divide in Eq. **(8-3)** and obtain

$$\frac{\sin \varphi_1 \sin \varphi_2}{\cos \varphi_1 \cos \varphi_2} = -1.$$

But from this last equation,

$$\tan \varphi_1 \cdot \tan \varphi_2 = -1 \qquad \text{or} \qquad m_1 m_2 = -1.$$

Conversely, if k_1, k_2 are not perpendicular, then $m_1 m_2 \neq -1$.

If one of the two lines k_1, k_2 is parallel to the x-axis, it makes an angle 0 with the x-axis so that its slope, tan 0, is 0. Then if the other line is perpendicular to the first, it is vertical and has no slope. To summarize: Two lines k_1, k_2 are perpendicular if and only if

(a) one line has 0 slope and the other no slope, or

(b) the product of their slopes is -1.

Example 8–3

The lines $\{(x, y) \mid 2x + y = \frac{1}{2}\}$, $\{(x, y) \mid 2x + y = 10\}$ are parallel since each has slope -2. The lines $\{(x, y) \mid y = 8x\}$, $\{(x, y) \mid y = -\frac{1}{8}x - 7\}$ are perpendicular since the slope of the first is 8 and that of the second is $-\frac{1}{8}$: $8(-\frac{1}{8}) = -1$.

Example 8–4

Let (a, b) be a point not on the line

$$k = \{(x, y) \mid y = mx + q\}.$$

Develop a formula for the perpendicular distance from (a, b) to k, if $m \neq 0$.

Solution. If j is the line on (a, b) and perpendicular to k, then the distance in question is that between (a, b) and the intersection of k and j. (See Fig. 8–5.) Then the slope of j must be $-1/m$. Hence, $j = \{(x, y) \mid y = (-1/m)x + p\}$ for some number p. If (c, d) is the intersection point of j, k we have

$$(c, d) \in j \quad \text{and} \quad (c, d) \in k.$$

Hence,

$$d = mc + q \quad \text{and} \quad d = (-1/m)c + p.$$

Figure 8–5

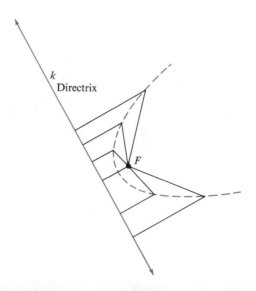

k
Directrix

F

Also, we must have $b = (-1/m)a + p$ since $(a, b) \in j$. Then $p = b + a/m$ and substituting this for p in the equation $d = (-1/m)c + p$:

(8-4)
$$d = (-1/m)c + p$$
$$= (-1/m)c + b + a/m = (1/m)(a - c) + b.$$

Then from $d = mc + q$ and Eq. (8-4)

$$mc + q = (1/m)(a - c) + b$$

and multiplying by m gives

$$m^2c + mq = a - c + bm.$$

From this equation we may solve for c:

(8-5)
$$c = \frac{a + m(b - q)}{m^2 + 1}.$$

If t is the distance between (a, b) and (c, d), the distance formula may be applied:

$$t^2 = (a - c)^2 + (b - d)^2$$
$$t^2 = (a - c)^2 + [1/m(a - c)]^2 \quad \text{from Eq. (8-4)}$$
$$= (a - c)^2[1 + (1/m)^2]$$
$$= \left(\frac{m(b - q) - am^2}{m^2 + 1}\right)^2 \cdot \frac{m^2 + 1}{m^2} \quad \text{from Eq. (8-5)}$$
$$= \frac{(b - q - am)^2}{m^2 + 1} = \frac{(am + q - b)^2}{m^2 + 1}.$$

Therefore,

$$t = \frac{|am + q - b|}{\sqrt{m^2 + 1}}.$$

This formula gives the distance between the point (a, b) and the line $\{(x, y) \mid y = mx + q\}$ in terms of the numbers a, b, m, q.

Example 8–5

What is the distance between the point $(1, 1)$ and the line

$$\{(x, y) \mid y = 7x - 1\}?$$

Solution. From Example 8–4 the distance t is

$$t = \frac{|1 \cdot 7 + (-1) - 1|}{\sqrt{7^2 + 1}} = \frac{|7 - 2|}{\sqrt{50}} = \frac{5}{5\sqrt{2}} = \frac{1}{\sqrt{2}}.$$

1. In each of the following, write a linear relation whose graph is the line indicated:
   (a) the line on the points $(0, 0)$ and $(4, -5)$.
   (b) the line on the points $(0, -5)$ and $(98, -5)$.
   (c) the line on the points $(\frac{1}{2}, 4)$ and $(\frac{1}{2}, \pi)$.
   (d) the line on the points $(1, \sqrt{2})$ and $(\frac{2}{5}, -1)$.
   (e) the line with slope 1 and on the point $(0, 10)$.
   (f) the line with slope $\frac{1}{2}$ and on the point $(1, 0)$.
   (g) the line with slope -1 and on the point $(0, 0)$.
   (h) the line with slope -2 and on the point $(-6, -1)$.
   (i) the line with slope 0 and on the point $(-5, \pi)$.
   (j) the line with no slope and on the point $(4, 1)$.

2. Graph each of the lines of Exercise 1.

3. In each of the following, write a linear relation whose graph is tangent to the given circle at the point indicated:
   (a) $\{(x, y) \mid x^2 + y^2 = 1\}$; $(-1, 0)$.
   (b) $\{(x, y) \mid x^2 + y^2 = 1\}$; $(0, 1)$.
   (c) $\{(x, y) \mid x^2 + y^2 = 1\}$; $(\frac{1}{2}\sqrt{2}, \frac{1}{2}\sqrt{2})$.
   (d) $\{(x, y) \mid (x - 1)^2 + (y + 1)^2 = 1\}$; $(2, -1)$.
   (e) $\{(x, y) \mid (x - 1)^2 + (y + 1)^2 = 1\}$; $(1, 0)$.
   (f) $\{(x, y) \mid (x - 1)^2 + (y + 1)^2 = 1\}$; $\left(\frac{1}{2}, \frac{\sqrt{3}}{2} - 1\right)$.

4. In Section 8–1 the formula

$$(y_2 - y_1)/(x_2 - x_1) = m$$

   was derived. This gives the slope m of a line in terms of two points $X(x_1, y_1)$, $Y(x_2, y_2)$ on the line. The formula was shown to hold only in one case. List the other possibilities and show that the formula holds in each case.

5. If $A(a, b)$ and $B(c, d)$ are points such that $a \neq c$, show that the line on A, B is the graph of $\{(x, y) \mid y - b = m(x - a)\}$ where $m = (d - b)/(c - a)$. This is called the *point-slope* form of the line.

6. Let $A(a, b)$, $B(c, d)$ be two distinct points. Show that the relation $\{(x, y) \mid$ the distance from (x, y) to A is the same as the distance from (x, y) to $B\}$ is a linear relation. [*Hint:* Use the distance formula.]

7. Show that the point $M\left(\dfrac{a + b}{2}, \dfrac{c + d}{2}\right)$ is the midpoint of the line segment connecting the points $A(a, c)$, $B(b, d)$.

8. Show that the diagonals of a parallelogram bisect each other.

9. In each of the following, write a linear relation having a graph parallel to the given line and passing through the given point.
(a) $\{(x, y) \mid 2x - (\tfrac{1}{2})y = 1\}$; $(1, 1)$.
(b) $\{(x, y) \mid 7 - 5y + x = 0\}$; $(6, 2)$.

10. In each of the following, write a linear relation having a graph perpendicular to the given line and passing through the given point.
(a) $\{(x, y) \mid x - y = 0\}$; $(-1, -1)$.
(b) $\{(x, y) \mid y + 1 = (\tfrac{1}{3})x\}$; $(2, \tfrac{5}{4})$.

11. Show that the graphs of the linear relations

$$\{(x, y) \mid Ax + By = C\} \qquad \text{and} \qquad \{(x, y) \mid Dx + Ey = F\}$$

intersect in exactly one point if and only if $AE - BD \neq 0$.

12. In each of the following, compute the distance between the given line and point:
(a) $\{(x, y) \mid x + y - 1 = 0\}$; $(2, 5)$.
(b) $\{(x, y) \mid 2x - 8 = y\}$; $(0, 0)$.

13. Show that a line k which intersects the x-axis at a distance $|a|$ from the origin and intersects the y-axis at a distance $|b|$ from the origin is the graph of the relation $\{(x, y) \mid (x/a) + (y/b) = 1\}$.

8.2
The Parabola

It is convenient to denote the distance between two points X, Y by $d(X, Y)$; also denote by $d(X, k)$ the perpendicular distance between the point X and the line k.

Figure
8–6

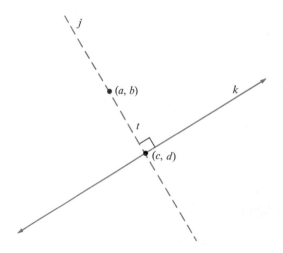

Let F be a point and let k be a line such that F is not on k. Then a relation

$$\{(x, y) \mid d((x, y), F) = d((x, y), k)\}$$

is called a *parabola*. The point F is called the *focus* of the parabola and k is called the *directrix*.

In Fig. 8–6 the definition is illustrated.

For the present, let us consider the two following cases: (1) the focus F is on the x-axis and the directrix k is perpendicular to the x-axis; and (2) the focus F is on the y-axis and the directrix k is perpendicular to the y-axis.

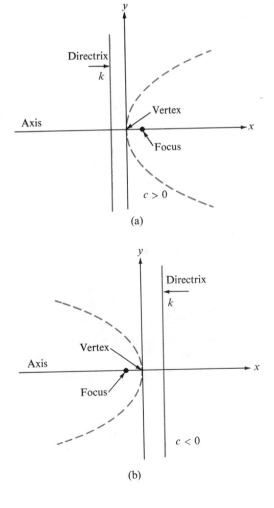

Figure
8–7

(a)

(b)

Consider case (1) with $(c, 0)$ the coordinates of F and $k = \{(x, y) \mid x = -c\}$. (See Fig. 8–7.) The expressions in the definition are

$$d((x, y), (c, 0)) = \sqrt{(x - c)^2 + y^2} \quad \text{and} \quad d((x, y), k) = |x + c|.$$

These are equal if and only if

$$(x - c)^2 + y^2 = (x + c)^2,$$

and hence if and only if

$$y^2 = 4cx.$$

The parabola in this case is

$$\{(x, y) \mid y^2 = 4cx\}.$$

Figure 8–8

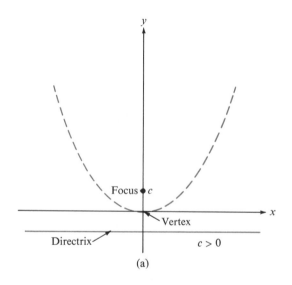

(a)

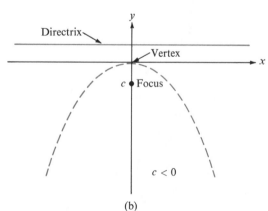

(b)

The graphs of these parabolas are indicated in Fig. 8–7. Note that the graph is nowhere to the left of the y-axis if $c > 0$ and nowhere to the right of the y-axis if $c < 0$.

If F is taken to be $(0, c)$ and k is $\{(x, y) \mid y = -c\}$, the parabola is

$$\{(x, y) \mid x^2 = 4cy\}.$$

These are illustrated in Fig. 8–8.

The parabola with focus $(1, 0)$ and directrix $\{(-1, y)\}$ is $\{(x, y) \mid y^2 = 4x\}$. Here $c = 1$.

Example 8–6

The parabola with focus $(0, \frac{1}{4})$ and directrix $\{(x, -\frac{1}{4})\}$ is $\{(x, y) \mid y = x^2\}$. This, of course, is the function $f(x) = x^2$.

Example 8–7

The line on the focus of a parabola which is perpendicular to the directrix is called the *axis* of the parabola; the point on the axis and midway between the focus and directrix is called the *vertex* of the parabola. Note that the vertex is a point of the parabola. (See Fig. 8–7 and 8–8.)

Describe the parabola $\{(x, y) \mid y = 2x^2\}$.

Example 8–8

Solution. We may write the equation $y = 2x^2$ as $x^2 = 4(\frac{1}{8})y$. Then,

$$\{(x, y) \mid y = 2x^2\} = \{(x, y) \mid x^2 = 4(\frac{1}{8})y\}.$$

Now this is seen to be the parabola with focus $(0, \frac{1}{8})$, directrix $\{(x, -\frac{1}{8})\}$; the axis of the parabola is the y-axis and vertex is the origin.

Consider now a parabola with focus F at the point $(a + c, b)$ and directrix $k: \{(a - c, y)\}$. We will show that such a parabola is a relation as follows:

$$\{(x, y) \mid (y - b)^2 = 4c(x - a)\}.$$

We have

$$d((x, y), (a + c, b)) = \sqrt{[x - (a + c)]^2 + (y - b)^2}$$

and

$$d((x, y), k) = |x - (a - c)|.$$

These numbers are equal if and only if

$$[x - (a + c)]^2 + (y - b)^2 = [x - (a - c)]^2$$

or if and only if

$$(y - b)^2 = 4c(x - a).$$

By definition the axis of the parabola in this case is the line $\{(x, b)\}$ and the vertex is the midpoint of the segment connecting $(a - c, b)$ and $(a + c, b)$; *i.e.*, (a, b) is the vertex. (See Fig. 8–9.)

Figure 8–9

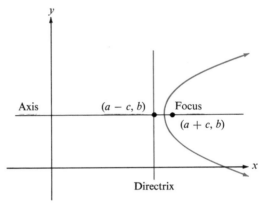

In a similar fashion if we consider a parabola with focus at $(a, b + c)$ and directrix $\{(x, b - c)\}$ we obtain the description

$$\{(x, y) \mid (x - a)^2 = 4c(y - b)\}.$$

In this case the parabola has (a, b) as vertex; the axis is $\{(a, y)\}$.

Example 8–9

Describe the relation $\{(x, y) \mid 4y - x^2 - 4x = 0\}$.

Solution. We write the equation $4y - x^2 - 4x = 0$ as

$$4y = x^2 + 4x$$
$$= (x^2 + 4x + 4) - 4$$
$$= (x + 2)^2 - 4.$$

Then,

$$\{(x, y) \mid 4y - x^2 - 4x = 0\} = \{(x, y) \mid (x + 2)^2 = 4(y + 1)\}.$$

The relation is now easy to recognize as a parabola with vertex at $(-2, -1)$; focus at $(-2, -1 + 1) = (-2, 0)$ (since $c = 1$); directrix: $\{(x, -2)\}$. (See Fig. 8–10 for the graph of this relation.)

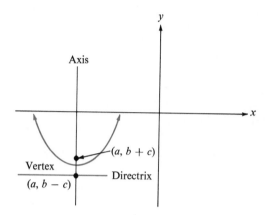

Figure
8–10

Each of the following relations is a parabola. In each case give the vertex, directrix, axis, focus and sketch its graph.

14. $\{(x, y) \mid y^2 = 2x\}$

15. $\{(x, y) \mid x^2 = 16y\}$

16. $\{(x, y) \mid x^2 = 4(y - 1)\}$

17. $\{(x, y) \mid (x - 1)^2 - 8(y + 2) = 0\}$

18. $\{(x, y) \mid y = 7x^2 + 2\}$

19. $\{(x, y) \mid x + 8 - y^2 + 2y = 0\}$

In each of the Exercises 20–23, write a description of a parabola and sketch its graph.

20. Focus: $(0, 1)$; directrix: $\{(x, -1)\}$.

21. Focus: $(-2, 0)$; directrix: $\{(2, y)\}$.

22. Focus: $(2, 0)$; directrix $\{(0, y)\}$.

23. Focus: $(-1, 8)$; directrix: $\{(x, 2)\}$.

24. Use the definition of parabola to write an analytic description of a parabola whose focus is $(0, 0)$ and directrix is $\{(x, y) \mid y = x + 1\}$.

8.3
The Ellipse

The next relation that will be studied in some detail is called the ellipse.

Definition

Let F_1, F_2 be points and let a be a positive number. Then the relation

$$E = \{(x, y) \mid d((x, y), F_1) + d((x, y), F_2) = 2a\}$$

is called an *ellipse*. The points F_1, F_2 are called the *foci* of the ellipse.

It should be noted that in the definition the important aspect is that the sum $d((x, y), F_1) + d((x, y), F_2)$ is constant; it is only for convenience that we take this constant to be $2a$.

If we take $F_1 = (c, 0)$ and $F_2 = (-c, 0)$ for some number c such that $0 \leq c < a$, then the equation

$$d((x, y), F_1) + d((x, y), F_2) = 2a$$

takes the form

$$\sqrt{(x - c)^2 + y^2} + \sqrt{(x + c)^2 + y^2} = 2a.$$

From this last equation we obtain the following sequence of equations:

$$\sqrt{(x - c)^2 + y^2} = 2a - \sqrt{(x + c)^2 + y^2}$$
$$(x - c)^2 + y^2 = 4a^2 - 4a\sqrt{(x + c)^2 + y^2} + (x + c)^2 + y^2$$
$$a\sqrt{(x + c)^2 + y^2} = cx + a^2$$
$$a^2(x + c)^2 + a^2y^2 = c^2x^2 + 2cxa^2 + a^4$$
$$a^2x^2 + a^2c^2 + a^2y^2 = c^2x^2 + a^4$$
$$(a^2 - c^2)x^2 + a^2y^2 = a^4 - a^2c^2 = a^2(a^2 - c^2)$$
$$\frac{x^2}{a^2} + \frac{y^2}{a^2 - c^2} = 1.$$

This shows that in this case

$$E \subseteq \{(x, y) \mid x^2/a^2 + y^2/(a^2 - c^2) = 1\} = S.$$

Conversely, if $(x, y) \in S$, then

$$\frac{x^2}{a^2} + \frac{y^2}{a^2 - c^2} = 1$$

and the steps to obtain the above equations may be reversed to show that $(x, y) \in E$, and hence that $S \subseteq E$. In this case, then, the ellipse is the set

$$\{(x, y) \mid x^2/a^2 + y^2/(a^2 - c^2) = 1\}.$$

See Fig. 8–11 for the graph of this ellipse. It will be clear from Fig. 8–11 why the restriction $0 \leq c < a$ is made; for if $a \leq c$ we have

$$d((x, y), F_1) + d((x, y), F_2) = 2a \leq 2c$$

which indicates that the sum of the lengths of two sides of the triangle $F_1 F_2 X$ is at most equal to the length of the third. Therefore, without the restriction $c < a$ we would have at most one point on the ellipse.

Figure
8–11

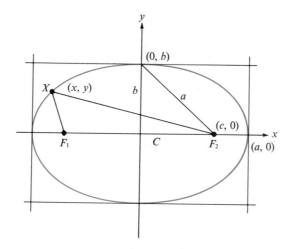

Since $0 \leq c < a$, it follows that $c^2 < a^2$ and $0 < a^2 - c^2$. Now let $b = \sqrt{a^2 - c^2}$. The ellipse above may be expressed as

$$\{(x, y) \mid x^2/a^2 + y^2/b^2 = 1\}.$$

This is said to be the *standard form* of the ellipse.

The points $(a, 0), (-a, 0), (0, b), (0, -b)$ are on the ellipse; the line containing the two foci F_1, F_2 is called the *major axis* and its midpoint is called the *center* of the ellipse. The line perpendicular to the major axis at the center of the ellipse is the *minor axis*.

For the case just discussed with $F_1 = (c, 0), F_2 = (-c, 0)$, the points of the ellipse $(a, 0), (-a, 0)$ are on the major axis; the points of the ellipse $(0, b), (0, -b)$ are on the minor axis. If the foci are taken to be $F_1 = (0, c)$, $F_2 = (0, -c)$, then the roles of the x-axis and y-axis are reversed: the points $(0, a), (0, -a)$ are on the major axis and the points $(b, 0), (-b, 0)$ are on the minor axis. In this case the ellipse is described by

$$\{(x, y) \mid x^2/b^2 + y^2/a^2 = 1\}.$$

Notice that in these two cases the major axis of the ellipse coincides with either the x-axis or the y-axis and that this is determined by whether x^2 or y^2 has the larger coefficient.

If a point (x, y) is on the ellipse $\{(x, y) \mid x^2/a^2 + y^2/b^2 = 1\}$, we have that

$$\frac{x^2}{a^2} + \frac{y^2}{b^2} = 1$$

and hence,

$$x^2/a^2 \leq 1 \qquad \text{and} \qquad y^2/b^2 \leq 1.$$

Therefore,

$$x^2 \leq a^2 \qquad \text{and} \qquad y^2 \leq b^2.$$

Thus,

$$-a \leq x \leq a \qquad \text{and} \qquad -b \leq y \leq b.$$

All the points (x, y) satisfying these two conditions must lie within the rectangle of Fig. 8–12. The graph of the ellipse will be contained with' this rectangle.

Figure 8–12

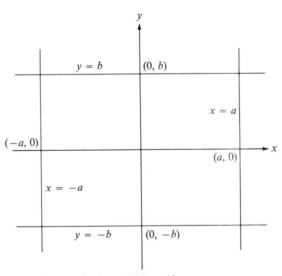

Example 8–10

Graph the ellipse $\{(x, y) \mid x^2 + y^2/4 = 1\}$.

Solution. In this case $b^2 = 1$, $a^2 = 4$; the minor axis is on the x-axis and the major axis is on the y-axis. The graph lies within the rectangle defined by $-1 \leq x \leq 1$, $-2 \leq y \leq 2$. The foci may be determined by solving the equation $b^2 = a^2 - c^2$ or $1 = 4 - c^2$. Then $c = \sqrt{3}$. (See Fig. 8–13.)

Figure
8–13

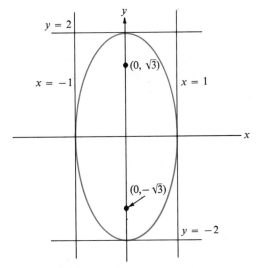

Graph the ellipse $\{(x, y) \mid 2x^2 + 8y^2 - 16 = 0\}$.

Example
8–11

Solution. First, express the ellipse in standard form:

$$\{(x, y) \mid 2x^2 + 8y^2 - 16 = 0\} = \{(x, y) \mid x^2/8 + y^2/2 = 1\}.$$

Then $a^2 = 8$, $b^2 = 2$. The major axis is the x-axis, the minor axis is the y-axis. Since $b^2 = a^2 - c^2$, $c = \sqrt{a^2 - b^2} = \sqrt{6}$ and the foci are $(\sqrt{6},\ 0)$, $(-\sqrt{6},\ 0)$. The graph lies within the rectangle defined by $-2\sqrt{2} \le x \le 2\sqrt{2}$, $-\sqrt{2} \le y \le \sqrt{2}$. (See Fig. 8–14.)

Figure
8–14

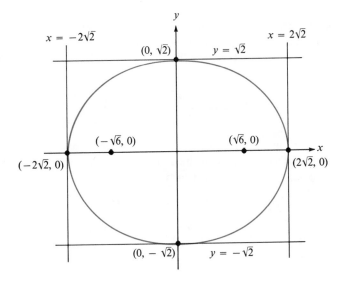

The ellipse $\{(x, y) \mid x^2/a^2 + y^2/b^2 = 1\}$ is a circle if $a = b$ with center at $(0, 0)$ and radius a. Since the ellipse lies within the rectangle $-a \le x \le a$, $-b \le y \le b$, it is evident that the smaller b is, when compared with a, the more the ellipse deviates from a circle. (See Fig. 8–15.) A very convenient way of measuring an ellipse's deviation from a circle is with the number $\sqrt{1 - (b/a)^2}$, which is called the *eccentricity* of the ellipse. Since $b^2 = a^2 - c^2$ — by definition — and since $c < a$, we have $0 < b^2 \le a^2$ and $(b/a)^2 \le 1$. Therefore,

$$0 \le 1 - (b/a)^2 < 1$$
$$0 \le \sqrt{1 - (b/a)^2} < 1.$$

Figure
8–15

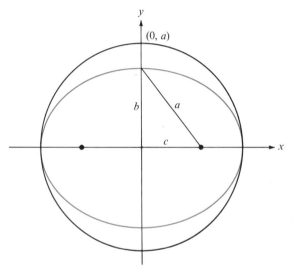

Thus, the eccentricity of an ellipse is a number between 0 and 1. We see from this that *an ellipse has eccentricity 0 if and only if it is a circle;* the closer the eccentricity is to 1, the "flatter" is the ellipse. The eccentricity is also the number c/a since

$$\sqrt{1 - (b/a)^2} = \sqrt{\frac{a^2 - b^2}{a^2}} = \sqrt{\frac{c^2}{a^2}} = \frac{c}{a}.$$

If the foci F_1, F_2 are taken to be $(d + c, e)$, $(d - c, e)$ respectively instead of $(c, 0)$, $(-c, 0)$ as above, then the ellipse — by using arguments similar to those above — may be shown to be

$$\{(x, y) \mid (x - d)^2/a^2 + (y - e)^2/b^2 = 1\}.$$

The center is then (d, e), the major axis is the line $\{(x, e)\}$, and the minor axis is the line $\{(d, y)\}$. (See Fig. 8–16.)

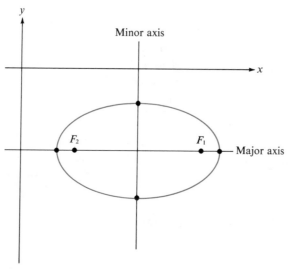

Figure
8–16

As before, the roles of the x-axis and y-axis may be interchanged by taking the foci as $(d, e + c)$, $(d, e - c)$. The ellipse is

$$\{(x, y) \mid (x - d)^2/b^2 + (y - e)^2/a^2 = 1\}.$$

(See Fig. 8–17.)

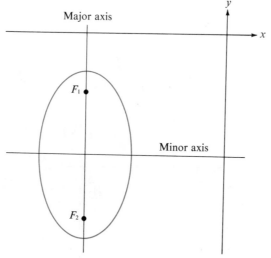

Figure
8–17

Graph the relation

$$\{(x, y) \mid 9x^2 - 18x + 4y^2 + 8y - 23 = 0\}.$$

Example
8–12

Solution. We first write the equation $9x^2 - 18x + 4y^2 + 8y - 23 = 0$ in a more convenient form by completing the squares in x and y:

$$9(x^2 - 2x) + 4(y^2 + 2y) = 23$$
$$9(x^2 - 2x + 1) + 4(y^2 + 2y + 1) = 23 + 9 + 4$$
$$9(x - 1)^2 + 4(y + 1)^2 = 36$$
$$\frac{(x - 1)^2}{4} + \frac{(y + 1)^2}{9} = 1.$$

This is now recognized as an ellipse with center $(1, -1)$. Also $b = 2$, $a = 3$ so that the major axis is parallel to the y-axis and $c = \sqrt{a^2 - b^2} = \sqrt{9 - 4} = \sqrt{5}$. Hence, the foci are $(1, -1 + \sqrt{5})$, $(1, -1 - \sqrt{5})$. In this case

$$\frac{(x - 1)^2}{4} \le 1 \qquad \text{and} \qquad \frac{(y + 1)^2}{9} \le 1$$

or,

$$(x - 1)^2 \le 4 \qquad \text{and} \qquad (y + 1)^2 \le 9.$$

Therefore,

$$-2 \le x - 1 \le 2 \qquad \text{and} \qquad -3 \le y + 1 \le 3.$$

This shows that the graph lies within the rectangle defined by

$$-1 \le x \le 3 \qquad \text{and} \qquad -4 \le y \le 2.$$

See Fig. 8–18 for the graph.

Figure 8–18

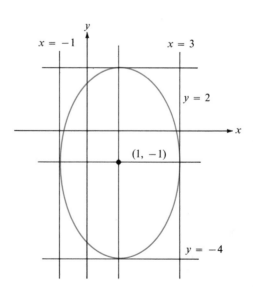

25. Graph each of the following relations.

(a) $\{(x, y) \mid x^2/4 + y^2 = 1\}$

(b) $\{(x, y) \mid x^2 + 8y^2 = 8\}$

(c) $\{(x, y) \mid 4x^2 + 9y^2 = 36\}$

(d) $\{(x, y) \mid x^2 + (y - 1)^2 = 1\}$

(e) $\{(x, y) \mid 6x^2 + (y + 2)^2 = 4\}$

(f) $\{(x, y) \mid 2x^2 - 2y^2 = 0\}$

(g) $\left\{(x, y) \left| \dfrac{(2x - 1)^2}{4} + \dfrac{(y + 2)^2}{16} = 1 \right.\right\}$

(h) $\{(x, y) \mid x^2 + 16x + 4y^2 - 16y + 76 = 0\}$

(i) $\{(x, y) \mid 4x^2 - 4x + 8y^2 + 8y - 1 = 0\}$

26. For each of the relations in Exercise 25 that are ellipses give the eccentricity.

27. Graph each of the following:

(a) $\{(x, y) \mid x^2 + 8(y^2 - 1) \leq 0\}$

(b) $\{(x, y) \mid x^2 + 8(y^2 - 1) \leq 0 \text{ and } x \leq y\}$

(c) $\{(x, y) \mid 4(4x^2 - 1) + (y + 2)^2 \leq 0\}$

(d) $\{(x, y) \mid x^2 - y^2 = 0 \text{ and } x^2 + y^2 \leq 1\}$

(e) $\left\{(x, y) \left| x^2 + y^2 \leq 4 \text{ and } \dfrac{x^2}{4} + y^2 \geq 1 \right.\right\}$

8.4
The Hyperbola

The definition of a hyperbola is very close to that of an ellipse; instead of sums of distances — as with the ellipse — the hyperbola is defined by differences of certain distances.

Let a be a positive number and let F_1, F_2 be points. Then the relation

Definition

$$\{(x, y) \mid |d((x, y), F_1) - d((x, y), F_2)| = 2a\}$$

is called a *hyperbola*. The points F_1, F_2 are called the *foci* of the hyperbola.

As with the ellipse, we may show that for foci $F_1 = (c, 0)$ and $F_2 = (-c, 0)$ the hyperbola is

$$\{(x, y) \mid x^2/a^2 - y^2/(c^2 - a^2) = 1\}.$$

Notice that this description of the hyperbola differs in two ways from the corresponding description of the ellipse: the left-hand side of the equation is a difference instead of a sum and $c^2 - a^2$ occurs rather than $a^2 - c^2$. For the ellipse, c was a number such that $0 \leq c < a$; for the hyperbola c must satisfy the condition $a < c$. This condition $a < c$ is illustrated in Fig. 8–19 where $d_1 < d_2 + 2c$. Hence, $2a = d_1 - d_2 < 2c$ or $a < c$. Therefore, the number $c^2 - a^2$ is positive so that we may define $b = \sqrt{c^2 - a^2}$. The hyperbola may be expressed more simply therefore as

$$\{(x, y) \mid x^2/a^2 - y^2/b^2 = 1\}.$$

Figure 8–19

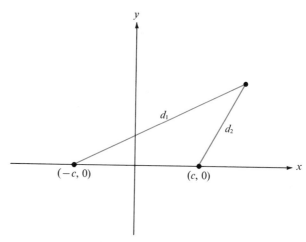

Example 8–13

Graph the relation $\{(x, y) \mid x^2 - y^2 = 1\}$.

Solution. This is a hyperbola in which $a = 1, b = 1$ and thus $1 = \sqrt{c^2 - a^2}$, $c = \sqrt{2}$. The foci are $(\sqrt{2}, 0), (-\sqrt{2}, 0)$. If (x, y) is any point on the graph of this relation, $x^2 - y^2 = 1$ and so

$$y^2 = x^2 - 1$$
$$y = \pm\sqrt{x^2 - 1}.$$

Consider $y = \sqrt{x^2 - 1}$ for $x \geq 1$. Then $y = \sqrt{x^2 - 1} < \sqrt{x^2} = |x| = x$ (since $x > 0$). This implies that the second coordinate on the graph of $y = \sqrt{x^2 - 1}$ is less than the corresponding second coordinate on the graph of $y = x$. It also appears that the larger is x, the closer $\sqrt{x^2 - 1}$ is to $\sqrt{x^2} = x(x > 0)$. (See Fig. 8–20 for the graph of $y = \sqrt{x^2 - 1}, x \geq 1$.) The graph of $\{(x, y) \mid x^2 - y^2 = 1\}$ may now be completed by graphing $y = -\sqrt{x^2 - 1}$ for $x \geq 1$, $y = \sqrt{x^2 - 1}$ for $x \leq -1$, $y = -\sqrt{x^2 - 1}$ for $x \leq -1$. (See Fig. 8–21.)

**Figure
8–20**

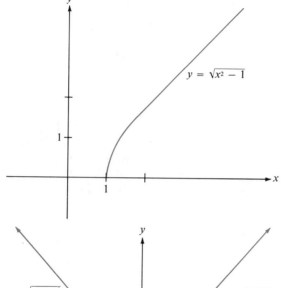

**Figure
8–21**

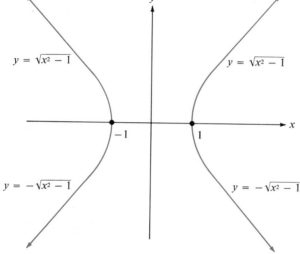

The numbers a, b associated with the hyperbola $\{(x, y) \mid x^2/a^2 - y^2/b^2 = 1\}$ define a rectangle as is the case with an ellipse; the hyperbola is, however, outside the rectangle rather than contained in it. This may be seen by considering a point (x, y) on the hyperbola:

$$\{(x, y) \mid x^2/a^2 - y^2/b^2 = 1\}.$$

Then,

$$y^2/b^2 = x^2/a^2 - 1$$

and hence,

$$y^2 = b^2\left(\frac{x^2 - a^2}{a^2}\right)$$

$$y = \pm(b/a)\sqrt{x^2 - a^2}.$$

This shows at once that we must have $x^2 - a^2 \geq 0$ and

$$a^2 \leq x^2.$$

This implies that $x \geq a$ or $x \leq -a$. Also for $x \geq a$, $\sqrt{x^2 - a^2} < \sqrt{x^2} = |x| = x$. Thus,

$$(b/a)\sqrt{x^2 - a^2} < (b/a)x.$$

Therefore, a point (x, y) on the hyperbola with $x \geq a$ and $y = (b/a)\sqrt{x^2 - a^2}$ lies below the corresponding point on the line $y = (b/a)x$. (See Fig. 8–22.) Similar considerations with $y = -(b/a)\sqrt{x^2 - a^2}$ show that the graph of the hyperbola $\{(x, y) \mid x^2/a^2 - y^2/b^2 = 1\}$ lies between the lines $\{(x, (b/a)x)\}$ and $\{(x, -(b/a)x)\}$ as shown in Fig. 8–22.

Figure 8–22

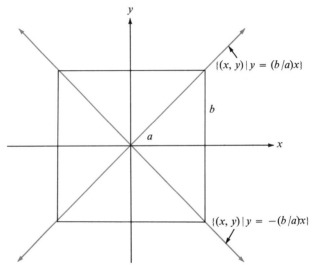

The *major axis* and *minor axis* of a hyperbola are defined in the same way that they are defined for an ellipse.

A more general situation is obtained if we take foci $(d + c, e), (d - c, e)$. Then, the hyperbola is

$$\{(x, y) \mid (x - d)^2/a^2 - (y - e)^2/b^2 = 1\}.$$

Here the major axis is the line $\{(x, e)\}$, the minor axis is $\{(d, y)\}$ and the center is (d, e). The graph lies between the lines $y - e = (b/a)(x - d)$ and $y - e = -(b/a)(x - d)$ as shown in Fig. 8–23. These lines are called the *asymptotes* of the hyperbola.

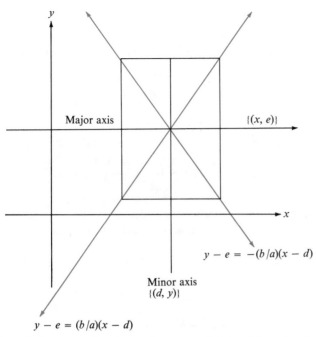

Figure
8–23

$$y - e = (b/a)(x - d)$$

The roles of the x- and y-axes are reversed by taking the foci to be $(d, e + c)$ and $(d, e - c)$. Then, the hyperbola is

$$\{(x, y) \mid (y - e)^2/b^2 - (x - d)^2/a^2 = 1\}.$$

In this situation the major axis is parallel to the y-axis as shown in Fig. 8–24.

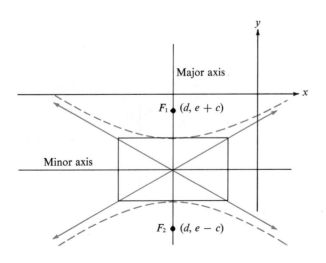

Figure
8–24

Example 8–14

Graph the relation $\{(x, y) \mid 4x^2 - 8x - y^2 - 2y - 1 = 0\}$.

Solution. Completing the squares in x and y in the equation $4x^2 - 8x - y^2 - 2y - 1 = 0$:

$$4(x^2 - 2x) - (y^2 + 2y + 1) = 0$$
$$4(x^2 - 2x + 1) - (y^2 + 2y + 1) = 4$$
$$4(x - 1)^2 - (y + 1)^2 = 4$$
$$(x - 1)^2 - \frac{(y + 1)^2}{4} = 1.$$

Hence, the graph is a hyperbola with center at $(1, -1)$ and major axis parallel to the x-axis. The graph is contained between the lines

$$y + 1 = 2(x - 1) \qquad \text{and} \qquad y + 1 = -2(x - 1)$$

as shown in Fig. 8–25. Since $a = 1$ and $b = 2$, we have

$$c = \sqrt{a^2 + b^2} = \sqrt{1 + 4} = \sqrt{5}.$$

The foci are hence $(1 + \sqrt{5}, -1)$ and $(1 - \sqrt{5}, -1)$.

Figure 8–25

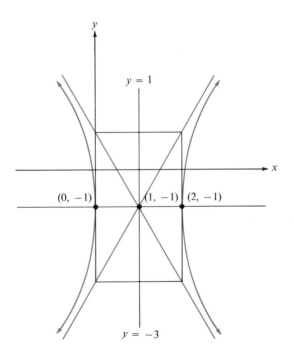

28. Graph each of the following.
    (a) $\{(x, y) \mid x^2/4 - y^2/16 = 1\}$
    (b) $\{(x, y) \mid y^2/16 - x^2/4 = 1\}$
    (c) $\{(x, y) \mid (x + 2)^2/4 - y^2/16 = 1\}$
    (d) $\{(x, y) \mid y^2/16 - (x + 2)^2/4 = 1\}$
    (e) $\{(x, y) \mid (y - 1)^2/9 - (x - 1)^2/2 = 1\}$
    (f) $\{(x, y) \mid (x - 1)^2/2 - (y - 1)^2/9 = 1\}$
    (g) $\{(x, y) \mid (2y + 1)^2 - (2x - 1)^2 = 4\}$
    (h) $\{(x, y) \mid (2y + 1)^2 - (2x - 1)^2 = 1\}$
    (i) $\{(x, y) \mid 12x^2 - 12x - 8y^2 - 21 = 0\}$
    (j) $\{(x, y) \mid 12y^2 - 12y - 8x^2 - 21 = 0\}$

29. In each of the following, determine a hyperbola satisfying the given conditions.
    (a) Foci: $(1, 0), (-1, 0)$; asymptotes: $y = \pm(\frac{1}{7})x$.
    (b) Asymptotes: $y + 1 = \pm(x - 1)$; the point $(4, 1)$ is on the hyperbola.

We consider now two different coordinate systems in the plane. An example of what is meant by this is given in Fig. 8–26 where corresponding axes are taken parallel to each other. In coordinate system II, the axis of first coordinates is called the $\bar{x}$-axis; the axis of second coordinates is the $\bar{y}$-axis. Each of these coordinate systems is said to be a *translation* of the other.

**Figure
8–26**

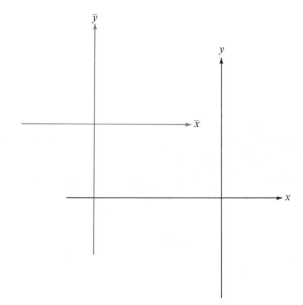

Each point P in the plane now has an ordered pair (x, y) associated with it by system I and also an ordered pair $(\bar{x}, \bar{y})$ associated with it by system II. An obvious question here is whether there is a routine method of obtaining one of these ordered pairs if given the other one. The answer to this is as follows: *if the origin of coordinate system II, has coordinates* h, k *when referred to system I, then the coordinates* (x, y) *and* $(\bar{x}, \bar{y})$ *of a point* P *are related by the equations*

$$\bar{x} = x - h$$
$$\bar{y} = y - k.$$

This is illustrated in Fig. 8–27. With two coordinate systems I and II as described above, we say that the origin of the first system has been translated to the point (h, k) — where h, k refer to the first system.

Figure
8–27

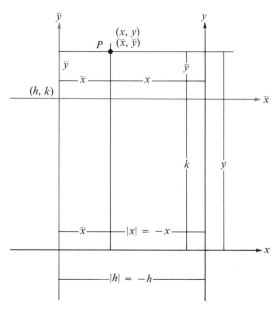

Example
8–15

If a coordinate system is translated to the point $(1, -\sqrt{2})$, then a point P which has coordinates (x, y) in the first system has coordinates $(x - 1, y + \sqrt{2})$ in the second system. Points on the y-axis of the first system have coordinates $(0, c)$; these points in the second system have coordinates $(-1, c + \sqrt{2})$.

Example
8–16

If a coordinate system is translated to the point $(1, -\sqrt{2})$, how is the description of the unit circle with center at the origin affected?

Solution. This relation $\{(x, y) \mid x^2 + y^2 = 1\}$ describes the unit circle with center at the origin. Let P be a point on this circle and have coordinates (a, b) in System I. Then after translating to $(1, -\sqrt{2})$, P has new coordinates $(\bar{a}, \bar{b})$ in System II such that $\bar{a} = a - 1, \bar{b} = b + \sqrt{2}$. Hence, $a = \bar{a} + 1$ and $b = \bar{b} - \sqrt{2}$. Since P is on the circle, $a^2 + b^2 = 1$ and substitution gives

$$(\bar{a} + 1)^2 + (\bar{b} - \sqrt{2})^2 = 1.$$

This unit circle therefore is described in System II as

$$\{(\bar{x}, \bar{y}) \mid (\bar{x} + 1)^2 + (\bar{y} - \sqrt{2})^2 = 1\}.$$

(See Fig. 8–28.)

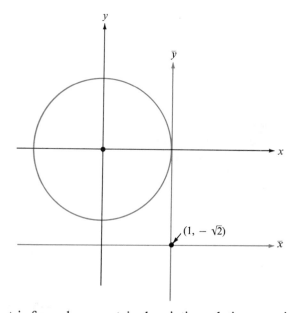

**Figure
8–28**

If a geometric figure has a certain description relative to a given coordinate system, it may be that the description can be replaced by one much simpler by making a translation. For instance, the hyperbola H may be under consideration and it may be described as

$$H = \{(x, y) \mid (x - e)^2/a^2 - (y - d)/b^2 = 1\}$$

in the given coordinate system. Translation to the point (e, d) produces the simpler description:

$$\{(\bar{x}, \bar{y}) \mid \bar{x}^2/a^2 - \bar{y}^2/b^2 = 1\}.$$

Consider now two coordinate systems — I and II — both having the same origin but such that the x-axis and $\bar{x}$-axis make an angle φ as in Fig. 8–29. System II is said to be a *rotation* of System I. As before, a

Figure 8–29

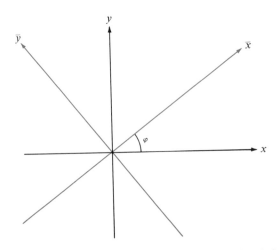

point P has coordinates (x, y) from I and coordinates $(\bar{x}, \bar{y})$ from II. How are these coordinates related? In Fig. 8–30, we let r be the distance that P is from the origin of the two systems and α is the angle as shown. Then $x = r \cos \alpha$, $y = r \sin \alpha$, $\bar{x} = r \cos(\alpha - \varphi)$, $\bar{y} = r \sin(\alpha - \varphi)$. Hence,

$$\bar{x} = r[\cos(\alpha - \varphi)]$$
$$= r[\cos \alpha \cos \varphi + \sin \alpha \sin \varphi]$$
$$= (r \cos \alpha) \cos \varphi + (r \sin \alpha) \sin \varphi$$
$$= x \cos \varphi + y \sin \varphi$$

and

$$\bar{y} = r \sin(\alpha - \varphi)$$
$$= r[\sin \alpha \cos \varphi - \cos \alpha \sin \varphi]$$
$$= (r \sin \alpha) \cos \varphi - (r \cos \alpha) \sin \varphi$$
$$= y \cos \varphi - x \sin \varphi.$$

This gives the equations:

(8-6)
$$\bar{x} = x \cos \varphi + y \sin \varphi$$
$$\bar{y} = y \cos \varphi - x \sin \varphi.$$

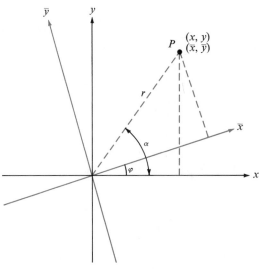

**Figure
8–30**

These equations may be used to compute the coordinates $(\bar{x}, \bar{y})$ if we are given the coordinates (x, y). From Fig. 8–31 we have — by computing $r \cos (\alpha + \varphi)$ and $r \sin (\alpha + \varphi)$ — the equations

$$x = \bar{x} \cos \varphi - \bar{y} \sin \varphi \qquad \textbf{(8-7)}$$

$$y = \bar{y} \cos \varphi + \bar{x} \sin \varphi.$$

The Eq. **(8-7)** may be used to compute the coordinates (x, y) if we are given $(\bar{x}, \bar{y})$.

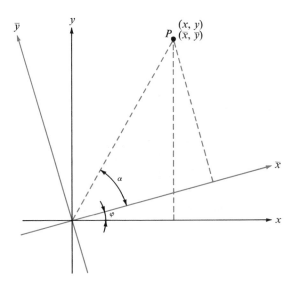

**Figure
8–31**

**Example
8–17**

Suppose that a coordinate system is given and the line bisecting the first and third quadrants is under consideration. When referred to the given coordinate system, the line is described by $\{(x, y) \mid x = y\} = \{(x, x)\}$. Suppose now that the given system is rotated by $\pi/4$ radians. What is the description of the line when referred to the new system?

Solution. The old and new coordinates of every point are related by Eq. **(8-7)**, where in this case $\varphi = \pi/4$:

$$x = \bar{x} \cos(\pi/4) - \bar{y} \sin(\pi/4)$$
$$y = \bar{y} \cos(\pi/4) + \bar{x} \sin(\pi/4)$$

or

$$x = \tfrac{1}{2}\sqrt{2}\,\bar{x} - \tfrac{1}{2}\sqrt{2}\,\bar{y}$$
$$y = \tfrac{1}{2}\sqrt{2}\,\bar{y} + \tfrac{1}{2}\sqrt{2}\,\bar{x}.$$

For a point on the line in question, $x = y$ and therefore,

$$\tfrac{1}{2}\sqrt{2}\,\bar{x} - \tfrac{1}{2}\sqrt{2}\,\bar{y} = \tfrac{1}{2}\sqrt{2}\,\bar{y} + \tfrac{1}{2}\sqrt{2}\,\bar{x}.$$

This simplifies to

$$\bar{y} = 0.$$

This implies that when referred to the new coordinate system every point $(\bar{x}, \bar{y})$ of the given line is such that $\bar{y} = 0$; *i.e.*, the given line is now described as the $\bar{x}$-axis or

$$\{(\bar{x}, \bar{y}) \mid \bar{y} = 0\} = \{(\bar{x}, 0)\}.$$

(See Fig. 8–32.)

**Figure
8–32**

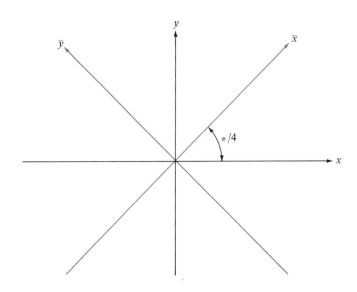

Let us consider the graph of the relation $\{(x, y) \mid xy = 1\}$. If a rotation of the coordinate system is made, the graph of $\{(x, y) \mid xy = 1\}$ will not change; this description of the graph will change, however. If the rotation is through $\pi/4$ radians as above, we have

Example 8–18

$$x = (\tfrac{1}{2})\sqrt{2}\bar{x} - (\tfrac{1}{2})\sqrt{2}\bar{y}$$
$$y = (\tfrac{1}{2})\sqrt{2}\bar{y} + (\tfrac{1}{2})\sqrt{2}\bar{x}.$$

Then,

$$xy = \tfrac{1}{2}(\bar{x})^2 - \tfrac{1}{2}(\bar{y})^2$$

and if $xy = 1$,

$$(\bar{x})^2 - (\bar{y})^2 = 2.$$

The original graph is described in the new system as

$$\{(\bar{x}, \bar{y}) \mid (\bar{x})^2 - (\bar{y})^2 = 2\}.$$

We recognize this as the hyperbola

$$\{(\bar{x}, \bar{y}) \mid \bar{x}^2/(\sqrt{2})^2 - \bar{y}^2/(\sqrt{2})^2 = 1\}$$

in standard form with $a = b = \sqrt{2}$.

A relation of the form

$$\{(x, y) \mid Ax^2 + By^2 + Cxy + Dx + Ey + F = 0\}$$

is called a *quadratic* relation where A, B, C, D, E, F are numbers and not all of A, B, C are zero. The circle, parabola, ellipse and hyperbola are all graphs of quadratic relations. If the quadratic relation is such that $C = 0$, then we may determine its graph by completing the square in x and y as was done in Example 8–14. If $C \neq 0$, a rotation of axes may be made in such a way that the new quadratic relation has no xy-term. Then the graph may be determined by completing the squares.

Graph the relation $\{(x, y) \mid x^2 + 3xy + y^2 = 1\}$.

Example 8–19

Solution. If a rotation of axes is made through φ radians, then the old and new coordinates are related by

$$x = \bar{x} \cos \varphi - \bar{y} \sin \varphi$$
$$y = \bar{y} \cos \varphi + \bar{x} \sin \varphi.$$

If the coordinates (x, y) of a point satisfy the condition $x^2 + 3xy + y^2 = 1$, then substitution from above gives

$$(\bar{x} \cos \varphi - \bar{y} \sin \varphi)^2 + 3(\bar{x} \cos \varphi - \bar{y} \sin \varphi)(\bar{y} \cos \varphi + \bar{x} \sin \varphi)$$
$$+ (\bar{y} \cos \varphi + \bar{x} \sin \varphi)^2 = 1.$$

This last equation may be written as follows:

$$(\bar{x})^2 \cos^2 \varphi - 2\bar{x}\bar{y} \cos \varphi \sin \varphi + (\bar{y})^2 \sin^2 \varphi$$
$$+ 3(\bar{x}\bar{y} \cos^2 \varphi - \bar{x}\bar{y} \sin^2 \varphi + (\bar{x})^2 \sin \varphi \cos \varphi - (\bar{y})^2 \sin \varphi \cos \varphi)$$
$$+ (\bar{y})^2 \cos^2 \varphi + 2\bar{x}\bar{y} \sin \varphi \cos \varphi + (\bar{x})^2 \sin^2 \varphi = 1$$

or

$$(1 + 3) \sin \varphi \cos \varphi)(\bar{x})^2 + (1 - 3 \sin \varphi \cos \varphi)(\bar{y})^2$$
$$+ 2(\cos^2 \varphi - \sin^2 \varphi)\bar{x}\bar{y} = 1$$

or

$$(1 + \tfrac{3}{2} \sin 2\varphi)(\bar{x})^2 + (1 - \tfrac{3}{2} \sin 2\varphi)(\bar{y})^2 + (2 \cos 2\varphi)\bar{x}\bar{y} = 1.$$

From this it is easy to see that a very convenient choice for φ is $\pi/4$; for then $(2 \cos 2\varphi)\bar{x}\bar{y} = (2 \cos \pi/2)\bar{x}\bar{y} = 0$. Making this choice for φ, the above equation is

$$(\tfrac{5}{2})(\bar{x})^2 - (\tfrac{1}{2})(\bar{y})^2 = 1.$$

We conclude that the graph of the original relation is the graph of

$$\{(\bar{x}, \bar{y}) \mid (\bar{x})^2/(\tfrac{2}{5}) - (\bar{y})^2/2 = 1\}.$$

This graph is easily seen to be a hyperbola with $a = \sqrt{\tfrac{2}{5}}$, $b = \sqrt{2}$ and $c = \sqrt{\tfrac{2}{5} + 2} = 2\sqrt{\tfrac{3}{5}}$. When referred to the $\bar{x}, \bar{y}$ system the center of this hyperbola is at $(0, 0)$, the foci are $(2\sqrt{\tfrac{3}{5}}, 0), (-2\sqrt{\tfrac{3}{5}}, 0)$. The asymptotes are the lines

$$\{(\bar{x}, \bar{y}) \mid \bar{y} = \pm(\sqrt{5})\bar{x}\}.$$

If we wish a description of these points and lines in terms of the original x, y-coordinate system, we must use the equations relating the x, y-system to the $\bar{x}, \bar{y}$-system. We use the equations (from **(8-7)**)

$$x = \bar{x} \cos \pi/4 - \bar{y} \sin \pi/4$$
$$y = \bar{y} \cos \pi/4 + \bar{x} \sin \pi/4$$

or

$$x = \tfrac{1}{2}\sqrt{2}\bar{x} - \tfrac{1}{2}\sqrt{2}\bar{y}$$
$$y = \tfrac{1}{2}\sqrt{2}\bar{y} + \tfrac{1}{2}\sqrt{2}\bar{x}.$$

Thus, in the x, y-system, the point $(0, 0)$ has coordinates

$$x = \tfrac{1}{2} 2 \cdot 0 - \tfrac{1}{2} 2 \cdot 0 = 0$$
$$y = \tfrac{1}{2} 2 \cdot 0 + \tfrac{1}{2} 2 \cdot 0 = 0.$$

The foci $(2\sqrt{\tfrac{3}{5}}, 0), (-2\sqrt{\tfrac{3}{5}}, 0)$ in the x, y-system have coordinates

$$x = (\tfrac{1}{2}\sqrt{2})(2\sqrt{\tfrac{3}{5}}) - \tfrac{1}{2}\sqrt{2} \cdot 0 = \sqrt{\tfrac{6}{5}}$$
$$y = \tfrac{1}{2}\sqrt{2} \cdot 0 + \tfrac{1}{2}\sqrt{2}(2\sqrt{\tfrac{3}{5}}) = \sqrt{\tfrac{6}{5}}$$

and

$$x = (\tfrac{1}{2}\sqrt{2})(-2\sqrt{\tfrac{3}{5}}) - \tfrac{1}{2}\sqrt{2} \cdot 0 = -\sqrt{\tfrac{6}{5}}$$
$$y = (\tfrac{1}{2}\sqrt{2}) \cdot 0 + \tfrac{1}{2}\sqrt{2}(-2\sqrt{\tfrac{3}{5}}) = -\sqrt{\tfrac{6}{5}}.$$

We use the equations (from **(8-6)**)

$$\bar{x} = (\tfrac{1}{2})\sqrt{2}x + (\tfrac{1}{2})\sqrt{2}y$$
$$\bar{y} = (\tfrac{1}{2})\sqrt{2}y - (\tfrac{1}{2})\sqrt{2}x$$

to obtain a description of the asymptotes in the x, y-system. Since the asymptotes are described by

$$\{(\bar{x}, \bar{y}) \mid \bar{y} = (\pm\sqrt{5})\bar{x}\}$$

in the $\bar{x}, \bar{y}$-system, substitution into the equation $\bar{y} = (\pm\sqrt{5})\bar{x}$ gives

$$\tfrac{1}{2}\sqrt{2}y - \tfrac{1}{2}\sqrt{x} = (\pm\sqrt{5})[\tfrac{1}{2}\sqrt{2}x + \tfrac{1}{2}\sqrt{2}y]$$

which simplifies to

$$y = \left(\frac{1 + \sqrt{5}}{1 - \sqrt{5}}\right)x \quad (\text{for} + \sqrt{5})$$

and

$$y = \left(\frac{1 - \sqrt{5}}{1 + \sqrt{5}}\right)x \quad (\text{for} - \sqrt{5}).$$

Hence, the asymptotes in the x, y-system are

$$\{(x, y) \mid y = ax \quad \text{or} \quad y = a^{-1}x\}$$

where $a = (1 + \sqrt{5})/(1 - \sqrt{5})$. (See Fig. 8–33.)

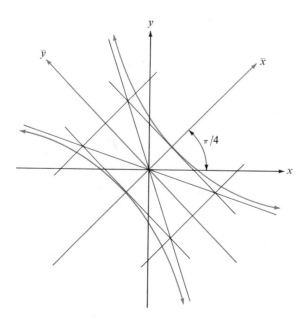

**Figure
8–33**

30. A coordinate system is translated to the point $(\frac{1}{2}, -\frac{3}{5})$. Each of the following gives the coordinates of points referred to the original system; compute the coordinates of these points in the new system.

(a) $(0, 0)$ (b) $(1, 0)$ (c) $(0, 5)$
(d) $(1, 1)$ (e) $(\sqrt{2}, \frac{1}{2})$ (f) $(x + \frac{1}{2}, -\frac{3}{5})$

31. Give a description of the graph of each of the following after a translation to the point $(\frac{1}{2}, -\frac{3}{5})$ has been made.
(a) $\{(x, y) \mid x = y\}$
(b) $\{(x, y) \mid x + \frac{1}{2} = y - \frac{3}{5}\}$
(c) $\{(x, y) \mid x - \frac{1}{2} = y + \frac{3}{5}\}$
(d) $\{(x, y) \mid (y + \frac{3}{5})^2 = 4(x - \frac{1}{2})\}$
(e) $\{(x, y) \mid (y - \frac{3}{5})^2 = 4(x + \frac{1}{2})\}$
(f) $\{(x, y) \mid x^2 + 2x + 3y = 1\}$

32. Describe the graph of each of the following after a rotation of $\pi/6$ radians has been made.
(a) $\{(x, y) \mid y = (1/\sqrt{3})x\}$
(b) $\{(x, y) \mid x(x - y) = 1\}$
(c) $\{(x, y) \mid y^2 - 2xy + x = 7\}$
(d) $\{(x, y) \mid x^2 + y^2 = 1\}$

33. Graph each of the following (if possible):
(a) $\{(x, y) \mid x^2 + y^2 = 0\}$
(b) $\{(x, y) \mid 2x^2 + y^2 + 1 = 0\}$
(c) $\{(x, y) \mid x^2 + 2xy + y^2 = 1\}$
(d) $\{(x, y) \mid x^2 - y^2 = 0\}$
(e) $\{(x, y) \mid 2x^2 - 2xy + 2y^2 = 1\}$
(f) $\{(x, y) \mid x^2 + \sqrt{3}xy = 1\}$
(g) $\{(x, y) \mid 2x^2 - 3xy + y^2 - 2x - 3y = 0\}$

34. If the graphs of any of the relations in 32 above are parabolas, ellipses or hyperbolas, find the foci, center and asymptotes (for hyperbolas).

For any real number t consider the point $(\cos t, \sin t)$ in the plane. Since $\cos^2 t + \sin^2 t = 1$, this point lies on the unit circle $\{(x, y) \mid x^2 + y^2 = 1\}$. We have already seen that if $P(x, y)$ is any point (not the origin) in the plane and r is the distance between the origin and P, then

$$x = r \cos t$$

$$y = r \sin t$$

where t is a measure of an angle made by the x-axis and the line on P and the origin. (See Fig. 8–34.) If (x, y) is a point of the unit circle with center at the origin, then $r = 1$ and

$$x = \cos t$$
$$y = \sin t.$$

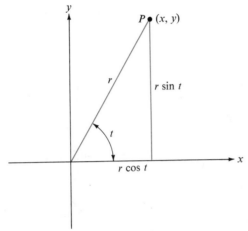

Figure
8–34

This shows that

$$\{(x, y) \mid x^2 + y^2 = 1\} = \{(\cos t, \sin t) \mid t \text{ is a real number}\}.$$

It is common practice to express this fact by saying that the equations

$$x = \cos t$$
$$y = \sin t$$

are *parametric* equations, or a *parametric representation*, for the unit circle, center at the origin, and that t is a *parameter*.

If u, s are functions, the set

$$\{(u(t), s(t)) \mid t \in \mathcal{D}(u) \cap \mathcal{D}(s)\}$$

determines a graph in the plane. We say that the graph is given *parametrically* by

$$x = u(t)$$
$$y = s(t)$$

with parameter t. It is often convenient to think of t as representing time and to imagine the pairs $(u(t), s(t))$ traversing the points of the curve as t takes on values in $\mathcal{D}(u) \cap \mathcal{D}(s)$. (See Fig. 8–35.)

295

**Figure
8–35**

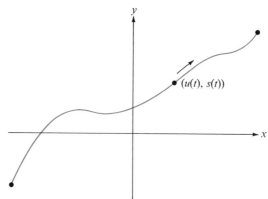

$(u(t), s(t))$

**Example
8–20**

Graph the curve given by the parametric equations

$$x = \cos t$$
$$y = \sin t \quad \text{for} \quad t \in [0, \pi].$$

Solution. All the points $(\cos t, \sin t)$ are on the unit circle with center at the origin. However, not all the points of the unit circle are given by $(\cos t, \sin t)$ for some $t \in [0, \pi]$. In fact,

$$\mathfrak{R}(\cos t) = [-1, 1] \quad \text{and} \quad \mathfrak{R}(\sin t) = [0, 1]$$

for $t \in [0, \pi]$. Hence, the second coordinate of the pairs $(\cos t, \sin t)$ is never negative and the graph is the top half of the unit circle. (See Fig. 8–36.)

**Figure
8–36**

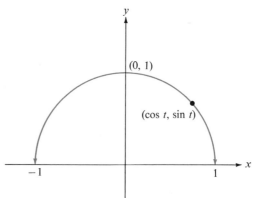

$(0, 1)$

$(\cos t, \sin t)$

-1

1

**Example
8–21**

Graph the curve given by the parametric equations

$$x = t^2$$
$$y = t^4$$

where t is any real number.

Solution. It is clear that $x \geq 0$, $y \geq 0$ and that $y = x^2$. Hence, the graph is given by

$$\{(x, y) \mid y = x^2 \text{ and } x \geq 0\}.$$

This has a graph that is part of a parabola as shown in Fig. 8–37.

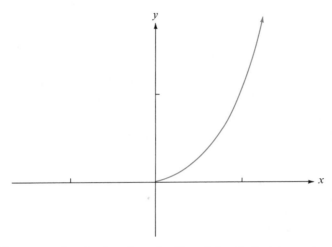

Figure
8–37

Consider two points in the plane (a, b) and (c, d) that are not on a line parallel to the y-axis. Then $a \neq c$ and the slope of the line on these points is

$$m = (d - b)/(c - a).$$

The line on these two points is then the graph of

$$\{(x, y) \mid y - b = m(x - a)\}.$$

(See Exercise 5.)

Parametric equations for this line are

$$x = a + t(c - a)$$
$$y = b + t(d - b) \quad \text{for} \quad t \in \mathbf{R}.$$

Notice that for $t = 0$ the equations give (a, b) and for $t = 1$, (c, d). For $t \in (0, 1)$, the parametric equations represent points between (a, b) and (c, d). To see this, suppose that $c - a$ and $d - b$ are positive. Then if $0 < t < 1$ we have

$$a < a + t(c - a) < c$$

and

$$b < b + t(d - b) < d.$$

(See Fig. 8–38.)

**Figure
8–38**

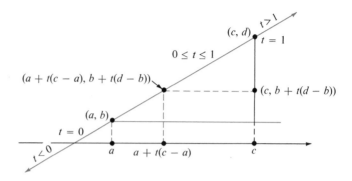

Exercises

35. Graph the following parametric equations.
   (a) $x = 2 \cos t$
        $y = 2 \sin t, t \in [0, \pi]$
   (b) $x = 1 + \cos t$
        $y = 1 - \sin t, t \in [0, \pi]$
   (c) $x = \sin t$
        $y = \cos 2t, t \in (0, \pi/2)$
   (d) $x = \tan t$
        $y = \sec t, t \in (-\pi/2, \pi/2)$
   (e) $x = 1 - t$
        $y = t + 1, t \in \mathbf{R}$
   (f) $x = 2 \cos t$
        $y = 3 \sin t, t \in [0, \pi]$
   (g) $x = a \cos t$
        $y = b \sin t, a, b > 0, t \in [0, \pi]$
   (h) $x = 2^t$
        $y = 2^t(2^t + 1), t \geq 0$
   (i) $x = 3^t$
        $y = 3^{-t}, t \geq 0.$
   (j) $x = \frac{1}{2}t$
        $y = \sqrt{1 - t^2}, t \in [-1, 1]$

36. (a) Show that the graph of the parametric equations

$$x = a + t(c - a)$$
$$y = b + t(d - b), t \in \mathbf{R}$$

   is the line on the points (a, b) and (c, d) if $(a, b) \neq (c, d)$.
   (b) It was shown above that the equations in (a) represent points of
       the line segment between (a, b), (c, d) for $0 < t < 1$ in case
       $c - a > 0$ and $d - b > 0$. Enumerate the other cases and show
       that this result holds in each.

(c) Show that every line is the graph of some parametric equations

$$x = a + t \cos \varphi$$
$$y = b + t \sin \varphi, \, t \in \mathbf{R}$$

for some number φ.

37. Show that the graph of the following parametric equations is an ellipse:

$$x = c + a \cos t$$
$$y = d + b \sin t, \, t \in [0, 2\pi]$$

where a, b, c, d are real numbers and $a > 0$, $b > 0$.

There are other coordinate systems besides the usual rectangular co-ordinate systems which we have considered up to this point. One such system is called the system of *polar coordinates*. The object of the system is to associate pairs of numbers with points in the plane, but in a way which is different from the rectangular system. Choose a point O in the plane and a ray M on O as shown in Fig. 8–39. The ray M is called the *pole*. If P is any point in the plane different from the origin, O, then the position of P is uniquely determined by the distance $r = |OP|$ and the angle θ which $\overline{OP}$ makes with $\overline{OM}$ (see Fig. 8–39). In this manner the pair (r, θ) is associated with P. The pair (r, θ) associated with P is not unique. For example, the pairs $(r, \theta + 360°)$, $(r, \theta + 720°)$ are also associated with P. The angle θ can be taken in either the counterclockwise (positive) or clockwise (negative) direction.

Figure
8–39

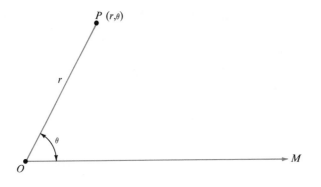

It is convenient to have a meaning for (r, θ) when $r \leq 0$ as well as for $r > 0$. For $r = 0$ (r, θ) means O for all θ and if $r < 0$, we locate the appropriate point by rotating the segment of length $|r|$ through an angle of $\theta + 180°$ (see Fig. 8–40).

Figure 8–40

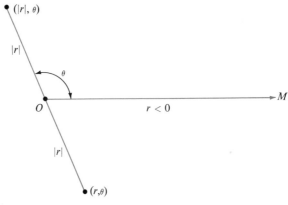

Example 8–22

Plot the points having polar coordinates $(1, 30°)$, $(2, 90°)$, $(2, 120°)$, $(-1, 30°)$, $(1, 210°)$, $(2, -150°)$, $(-1, -45°)$.

Solution. These points are plotted in Fig. 8–41. Notice that $(1, 210°)$ and $(-1, 30°)$ are the same point.

Figure 8–41

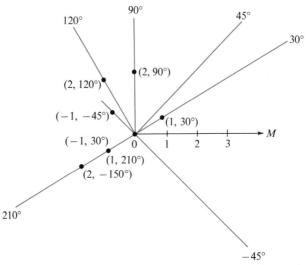

We shall now consider some simple equations involving polar coordinates and their graphs.

Find the graph of $r = a$, where $a > 0$.

Example
8–23

Solution. The problem is to describe the set $\{(r, \theta) \mid r = a, \theta \in R\}$. Since the radius r has constant length a, as θ takes on all real values a circle is generated. Hence, the graph of $r = a$ is a circle (see Fig. 8–42).

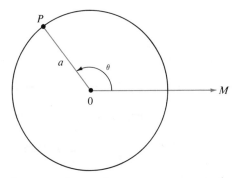

Figure
8–42

Find the graph of $\theta = a$, where a is any non-zero number.

Example
8–24

Solution. Here we wish to plot all the points of $\{(r, \theta) \mid \theta = a, r \in R\}$. In this case θ does not change while r takes all real values. We therefore get the line through the origin shown in Fig. 8–43.

Figure
8–43

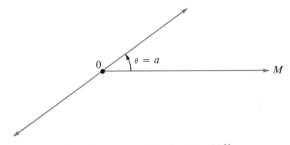

Find the graph of $r = \cos \theta$ for $\theta \in [0, 90°]$.

Example
8–25

Solution. In this case $r \in [0, 1]$. We may use Table I to compute r for some convenient values of θ:

θ	0°	15°	30°	45°	60°	75°	90°
r	1	0.97	0.87	0.71	0.5	0.26	0

These points are plotted as in Fig. 8–44 and the graph is completed by appropriately connecting these points.

**Figure
8–44**

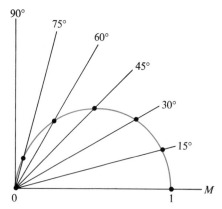

If a rectangular coordinate system is taken so that the x-axis coincides with the pole of a polar coordinate system and the origins are the same, it is an easy matter to describe the relation which exists between the two. As is indicated in Fig. 8–45, if $r > 0$, we have

(8-8) $$x = r \cdot \cos \theta \quad \text{and} \quad y = r \cdot \sin (\theta)$$

**Figure
8–45**

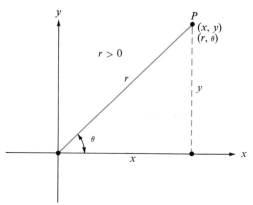

If $r < 0$, as in Fig. 8–46,

$$-x = |r| \cdot \cos \theta \quad \text{and} \quad -y = |r| \cdot \sin \theta$$

or since $r < 0$,

$$-x = -r \cdot \cos \theta \quad \text{and} \quad -y = -r \cdot \sin \theta.$$

Since these last equations are the same as Eq. **(8-8)**, it is evident that Eq. **(8-8)** holds in all cases.

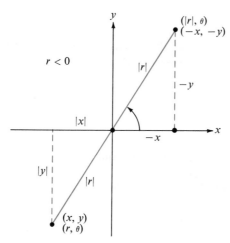

Figure
8–46

Express the polar equation $r = \cos \theta + \sin \theta$ as an equation in rectangular coordinates.

Example
8–26

Solution. We use Eq. **(8-8)**. If $r = \cos \theta + \sin \theta$, then $r^2 = r \cos \theta + r \sin \theta$ or $r^2 = x + y$. Also from Eq. **(8-8)** $x^2 + y^2 = r^2 \cos^2 \theta + r^2 \sin^2 \theta = r^2 (\cos^2 \theta + \sin^2 \theta) = r^2$. Then $x^2 + y^2 = x + y$.

Express the equation $y = x^2$ in polar coordinates.

Example
8–27

Solution. From Eq. **(8-8)** $y = r \sin \theta$ and $x^2 = r^2 \cos^2 \theta$ so that $r \sin \theta = r^2 \cos^2 \theta$.

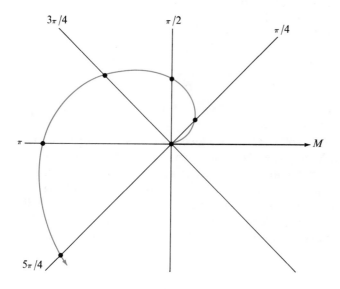

Figure
8–47

Example 8–28

Plot the graph of the equation $r = \theta$, for $\theta \geq 0$.

Solution. In a case such as this θ must be given in radians since r must measure lengths. As θ increases so does r. Hence, the graph is a spiral as given in Fig. 8–47.

Exercises

38. Sketch the graph of each polar equation below.
    (a) $r = \sin \theta$
    (b) $r = 3 \sin \theta$
    (c) $2r = \sin \theta$
    (d) $r = 2 \cos \theta + 1$
    (e) $r = \sin (2\theta)$
    (f) $r = 2 \cos (3\theta)$
    (g) $r^2 = 2 \cos (2\theta)$
    (h) $r = \cos \theta - \sin \theta$
    (i) $r = 4(\sin \theta - \cos \theta)$
    (j) $r\theta = 1$

39. Express each of the following polar equations in rectangular coordinates.

    (a) $r = 2 \cos \theta$ (b) $3r = 2 \sin \theta$
    (c) $r = \sin (2\theta)$ (d) $r^2 = 2 \cos (2\theta)$

40. Express each of the following equations in rectangular coordinates in polar coordinates.
    (a) $x + 2y + 1 = 0$
    (b) $x^2 + y^2 = 1$
    (c) $2x^2 + 2x + y^2 - y = 1$
    (d) $x^2 - y^2 = 4$
    (e) $x^2 + 2y^2 = 4$

8.8
Solid Analytic Geometry

The basis for solid analytic geometry is the association with each point in 3-dimensional space (also called 3-space) an ordered triple (x, y, z), where x, y, z are real numbers. In order to accomplish this association, let three mutually perpendicular lines be given as in Fig. 8–48. Let each of the lines, called the x-axis, y-axis, z-axis respectively, have a coordinate system with the positive directions indicated in Fig. 8–48. The three lines of Fig. 8–48 determine three planes: the xy-plane, the xz-plane, the yz-plane.

Figure
8–48

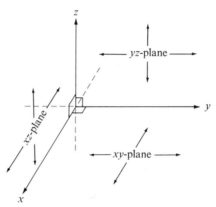

If P is any point in 3-space, there is a plane on P which is parallel to the yz-plane; this plane intersects the x-axis in a unique point having some number associated with it. Therefore, we obtain in this fashion a unique number, a say, on the x-axis associated with P. In a similar way we obtain a number b on the y-axis and a number c on the z-axis. Hence, with P is associated a unique triple (a, b, c) of real numbers (see Fig. 8–49). By re-

Figure
8–49

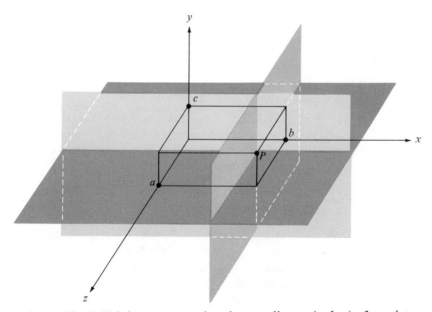

ferring to Fig. 8–50 it is easy to see that the coordinates (a, b, c) of a point P indicate how far the point is from the xy-, xz-, yz-planes. Point P is a distance of $|a|$ from the yz-plane, $|b|$ from the xz-plane and $|c|$ from the xy-plane.

**Figure
8–50**

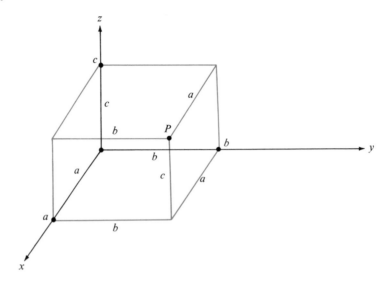

**Example
8–29**

If the point P has coordinates $(1, -5, -3)$, how far is P from the xy-, xz-, yz-planes?

Solution. P is 1 unit from the yz-plane, 5 units from the xz-plane and 3 units from the xy-plane.

**Example
8–30**

Describe the set $S = \{(x, y, z) \mid y = 3\}$.

Solution. If $(x, y, z) \in S$, then $y = 3$ and this indicates that the point is 3 units from the xz-plane. Conversely, every point which is 3 units from the xz-plane and on the same side of the xz-plane as the positive y-axis is in S. But the plane which intersects the y-axis at 3 and is parallel to the xz-plane contains all these points and only these points. Hence, S is the plane parallel to the xz-plane and intersects the y-axis at 3 (see Fig. 8–51).

A rectangular "box" is shown in Fig. 8–52. Each face of this solid figure is a rectangle. Suppose that the dimensions of the box are p, q, r as shown in Fig. 8–52. By the Pythagorean Theorem applied to the right $\triangle ABC$ $|AC|^2 = r^2 + q^2$. Applying the same theorem to right $\triangle ACD$ gives $|AC|^2 + p^2 = |AD|^2$ or $r^2 + q^2 + p^2 = |AD|^2$. Then

$$|AD| = \sqrt{p^2 + q^2 + r^2}.$$

This shows that a *rectangular box having dimensions* p, q, r *has a diagonal of length* $\sqrt{p^2 + q^2 + r^2}$.

Figure
8–51

z

xz-plane

1 2 3

y

x

Figure
8–52

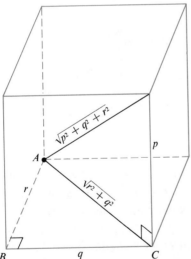

$\sqrt{p^2 + q^2 + r^2}$

p

A

r

$\sqrt{r^2 + q^2}$

B q C

Let $P = (p_1, p_2, p_3)$ and $Q = (q_1, q_2, q_3)$ be points in 3-space. Then the distance, $|PQ|$, between P and Q is given by the formula

$$|PQ| = \sqrt{(p_1 - q_1)^2 + (p_2 - q_2)^2 + (p_3 - q_3)^2}.$$

Proof. As indicated in Fig. 8–53 we may view P, Q as endpoints of a diagonal of a rectangular box and this box has dimensions as indicated. Then by the above result

$$|PQ| = \sqrt{|p_1 - q_1|^2 + |p_2 - q_2|^2 + |p_3 - q_3|^2}$$
$$= \sqrt{(p_1 - q_1)^2 + (p_2 - q_2)^2 + (p_3 - q_3)^2}.$$

Figure
8–53

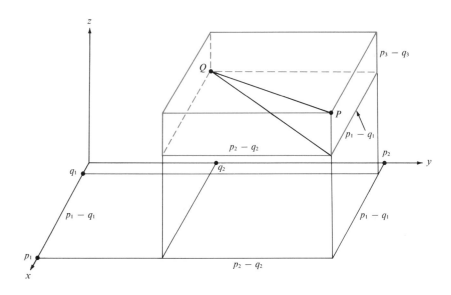

**Example
8–31**

Compute the distance between the points $(-2, 0, 1)$ and $(1, 1, 4)$.

Solution. By the formula just developed the distance is

$$\sqrt{(-2 - 1)^2 + (0 - 1)^2 + (1 - 4)^2}$$
$$= \sqrt{(-3)^2 + (-1)^2 + (-3)^2}$$
$$= \sqrt{19}.$$

Let P, Q be two distinct points in 3-space. Associated with the line segment $\overline{PQ}$ are two *directed line segments* $\overrightarrow{PQ}$ and $\overrightarrow{QP}$. The directed line segment $\overrightarrow{PQ}$ has *initial* point P and *terminal* point Q. We say that $\overrightarrow{PQ}$ is *directed from P to Q*. If M is a line on P, Q, we may assign a direction to M by using P and Q, *i.e.*, M either has the same direction as $\overrightarrow{PQ}$ or the direction of $\overrightarrow{QP}$. Then we say that M is a *directed line*.

Suppose M is a line in which a direction, $\overrightarrow{PQ}$, has been given, where P, Q are points on M (see Fig. 8–54). At P we may construct three line segments $\overrightarrow{PX}$, $\overrightarrow{PY}$, $\overrightarrow{PZ}$ such that $\overrightarrow{PX}$ is parallel to and has the same direction as the x-axis, $\overrightarrow{PY}$ is parallel to and has the same direction as the y-axis, $\overrightarrow{PZ}$ is parallel to and has the same direction as the z-axis. The angles

$$\alpha = \angle QPX, \quad \beta = \angle QPY, \quad \gamma = \angle QPZ$$

are called the direction angles for the line M. It follows immediately that *two lines are parallel and have the same direction if and only if they have the same direction angles*. Each direction angle has a radian measure which lies in the interval $[0, \pi]$.

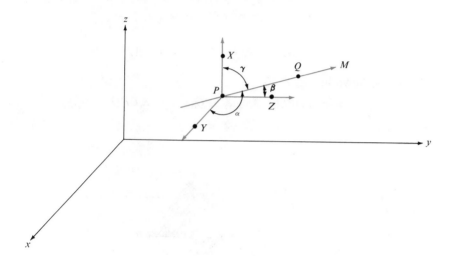

**Figure
8–54**

Let right $\triangle ABC$ be given as in Fig. 8–55. It will be recalled from Chapter 5 that $\cos(\measuredangle A) = b/c$ and that

$$b = c \cdot \cos(\measuredangle A).$$

This result is needed in the proof of the next theorem.

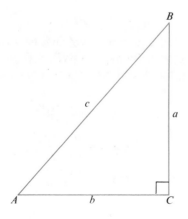

**Figure
8–55**

Let P, Q be two distinct points in space such that $P = (p_1, p_2, p_3)$ and $Q = (q_1, q_2, q_3)$. Let M be a line on P, Q with direction $\overrightarrow{PQ}$ and direction angles α, β, γ. Then

$$q_1 - p_1 = |PQ| \cdot \cos \alpha$$
$$q_2 - p_2 = |PQ| \cdot \cos \beta$$
$$q_3 - p_3 = |PQ| \cdot \cos \gamma$$

and

$$\cos^2 (\alpha) + \cos^2 (\beta) + \cos^2 (\gamma) = 1.$$

Proof. As shown in Fig. 8–56

$$\alpha = \measuredangle APQ, \quad \beta = \measuredangle BPQ, \quad \gamma = \measuredangle CPQ.$$

**Figure
8–56**

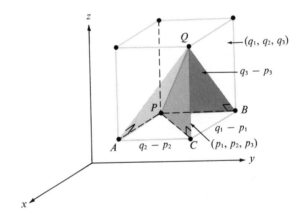

Consider the right $\triangle APQ$. Since $|AP| = q_1 - p_1$ and $\measuredangle APQ = \alpha$ we have

$$q_1 - p_1 = |PQ| \cdot \cos \alpha.$$

To obtain the other two equations involving the p's and q's consider right $\triangle BPQ$ and right $\triangle CPQ$. By hypothesis $P \neq Q$ and $|PQ| \neq 0$. Then the equations involving the directions cosines ($\cos \alpha$, $\cos \beta$, $\cos \gamma$) may be written

$$\cos \alpha = \frac{q_1 - p_1}{|PQ|}$$

$$\cos \beta = \frac{q_2 - p_2}{|PQ|}$$

$$\cos \gamma = \frac{q_3 - p_3}{|PQ|}.$$

Hence,

$$\cos^2(\alpha) + \cos^2(\beta) + \cos^2(\gamma) = \frac{(q_1 - p_1)^2}{|PQ|^2} + \frac{(q_2 - p_2)^2}{|PQ|^2} + \frac{(q_3 - p_3)^2}{|PQ|^2}$$

$$= \frac{(q_1 - p_1)^2 + (q_2 - p_2)^2 + (q_3 - p_3)^2}{|PQ|^2}$$

$$= 1 \quad \text{by Theorem 31.}$$

If $P = (-2, 0, 1)$, $Q = (1, 1, 4)$, compute the direction cosines of the directed line on $\overrightarrow{PQ}$.

Example
8–32

Solution. Use Theorem 32. $|PQ| = \sqrt{19}$ from Example 8–31. Then if α, β, γ are the direction angles

$$\cos \alpha = (1 - (-2))/\sqrt{19} = 3/\sqrt{19}$$
$$\cos \beta = (1 - 0)/\sqrt{19} = 1/\sqrt{19}$$
$$\cos \gamma = (4 - 1)/\sqrt{19} = 3/\sqrt{19}.$$

41. Give the distance that each point below is from the xy-, xz- and yz-planes.

Exercises

(a) $(0, 1, 0)$ (b) $(2, 1, -1)$ (c) $(-12, 8, 3)$
(d) $(4, -7, \frac{1}{2})$ (e) $(-4\frac{1}{2}, 10, 8)$ (f) $(1, 2, 2)$
(g) (a, a, b) (h) (a, b, a) (i) (b, a, a)

42. Describe in geometrical language each of the following sets of points in 3-space and make a sketch.

(a) $S = \{(x, y, z) \mid z = 10\}$
(b) $S = \{(x, y, z) \mid z = -5\}$
(c) $S = \{(x, y, z) \mid y = 0\}$
(d) $S = \{(x, y, z) \mid x = 4\}$
(e) $S = \{(x, y, z) \mid z = 0 \text{ and } x = y\}$
(f) $S = \{(x, y, z) \mid x = 0 \text{ and } y = z\}$
(g) $S = \{(x, y, z) \mid y = 0 \text{ and } x = z\}$
(h) $S = \{(x, y, z) \mid z = 0 \text{ and } y = 2x\}$
(i) $S = \{(x, y, z) \mid x = 0 \text{ and } z = 3y\}$
(j) $S = \{(x, y, z) \mid y = 0, x \geq 0, z \geq 0 \text{ and } x^2 + z^2 = 1\}$
(k) $S = \{(x, y, z) \mid z = 0 \text{ and } x^2 + y^2 = 1\}$
(l) $S = \{(x, y, z) \mid x = y\}$
(m) $S = \{(x, y, z) \mid x = z\}$
(n) $S = \{(x, y, z) \mid y = z\}$

43. Compute the distance between each pair of the following points.
    (a) $P = (0, 0, 0)$, $Q = (1, 0, 0)$
    (b) $P = (0, 0, 0)$, $Q = (1, 1, 0)$
    (c) $P = (0, 0, 0)$, $Q = (1, 0, 1)$
    (d) $P = (0, 0, 0)$, $Q = (1, 1, 1)$
    (e) $P = (0, 0, 0)$, $Q = (-1, -1, 1)$
    (f) $P = (0, 1, 0)$, $Q = (2, -1, 1)$
    (g) $P = (-1, -1, -1)$, $Q = (2, 0, 1)$
    (h) $P = (20, 19, 18)$, $Q = (15, 18, 18)$
    (i) $P = (1, -\frac{1}{2}, 2)$, $Q = (\frac{1}{2}, 0, 1)$
    (j) $P = (5, -11, 4\frac{1}{2})$, $Q = (-1, -9, \frac{1}{2})$

44. For each pair of points P, Q given in exercise 43 above compute the direction cosines of the directed line on $\overrightarrow{PQ}$.

45. Let $P = (p_1, p_2, p_3)$, $Q = (q_1, q_2, q_3)$ be two points in 3-space and let $M = (a, b, c)$ where

$$a = (p_1 + q_1)/2, \ b = (p_2 + q_2)/2, \ c = (p_3 + q_3)/2.$$

Compute $|PM|$ and $|QM|$. What conclusion do you draw from this?

46. For each pair of points of exercise 43 above compute the midpoint of the segment $|PQ|$ (see Exercise 45).

8.9

Lines in 3-Space

Let α, β, γ be the direction angles of the directed line M. Then we say that r, s, t are direction numbers for M if and only if there is some positive number d such that

$$r = d \cdot \cos \alpha$$
$$s = d \cdot \cos \beta$$
$$t = d \cdot \cos \gamma.$$

Theorem 33

If $P = (p_1, p_2, p_3)$, $Q = (q_1, q_2, q_3)$ are any two distinct points on a line M and M has direction $\overrightarrow{PQ}$, then $q_1 - p_1$, $q_2 - p_2$, $q_3 - p_3$ are direction numbers for M.

Proof. If α, β, γ are the direction angles of $\overrightarrow{PQ}$, then

$$q_1 - p_1 = |PQ| \cdot \cos \alpha$$
$$q_2 - p_2 = |PQ| \cdot \cos \beta$$
$$q_3 - p_3 = |PQ| \cdot \cos \gamma$$

by Theorem 32. It then follows by definition that $q_1 - p_1, q_2 - p_2, q_3 - p_3$ are direction numbers for M since $|PQ| > 0$.

Let r, s, t be direction numbers for a directed line M and let $\Delta = \sqrt{r^2 + s^2 + t^2}$. Then if α, β, γ are the direction angles of M,

$$\Delta \cdot \cos \alpha = r$$
$$\Delta \cdot \cos \beta = s$$
$$\Delta \cdot \cos \gamma = t.$$

Proof. By definition of direction numbers, there is a positive number d such that

$$r = d \cdot \cos \alpha$$
$$s = d \cdot \cos \beta$$
$$t = d \cdot \cos \gamma.$$

Then

$$\begin{aligned} r^2 + s^2 + t^2 &= d^2 \cos^2 (\alpha) + d^2 \cos^2 (\beta) + d^2 \cos^2 (\gamma) \\ &= d^2[\cos^2 (\alpha) + \cos^2 (\beta) + \cos^2 (\gamma)] \\ &= d^2 \cdot 1 \quad \text{by Theorem 32} \\ &= d^2. \end{aligned}$$

Therefore, since $d > 0$,

$$d = \sqrt{r^2 + s^2 + t^2} = \Delta.$$

If α, β, γ are direction angles for the directed line M and $\cos \alpha = \frac{1}{2}$, $\cos \beta = \frac{1}{2}$, $\cos \gamma = 1/\sqrt{2}$, find two different sets of direction numbers for M.

Example 8–33

Solution. By definition of direction numbers, $d \cdot \cos \alpha$, $d \cos \beta$, $d \cdot \cos \gamma$ are direction numbers where d is any positive number. Hence, for $d = 1$

$$d \cdot \cos \alpha = 1 \cdot \tfrac{1}{2} = \tfrac{1}{2}$$
$$d \cdot \cos \beta = 1 \cdot \tfrac{1}{2} = \tfrac{1}{2}$$
$$d \cdot \cos \gamma = 1 \cdot 1/\sqrt{2} = 1/\sqrt{2}.$$

For $d = 2$

$$d \cdot \cos \alpha = 2 \cdot \tfrac{1}{2} = 1$$
$$d \cdot \cos \beta = 2 \cdot \tfrac{1}{2} = 1$$
$$d \cdot \cos \gamma = 2 \cdot 1/\sqrt{2} = 2/\sqrt{2}.$$

Then $\tfrac{1}{2}, \tfrac{1}{2}, 1/\sqrt{2}$ is one set of direction numbers and $1, 1, 2/\sqrt{2}$ is another.

Example 8–34

If $P = (1, 8, -4)$ and $Q = (-2, 7, 1)$, find direction numbers for a line on P, Q having direction $\overrightarrow{PQ}$. Also find direction numbers for a line on P, Q having direction $\overrightarrow{QP}$.

Solution. Use Theorem 33. A line on P, Q with direction $\overrightarrow{PQ}$ has direction numbers

$$-2 - 1 = -3$$
$$7 - 8 = -1$$
$$1 - (-4) = 5.$$

A line on P, Q having direction $\overrightarrow{QP}$ has direction numbers

$$1 - (-2) = 3$$
$$8 - 7 = 1$$
$$-4 - 1 = -5.$$

Example 8–35

If r, s, t are direction numbers for a directed line M and a is any positive number, show that ar, as, at are also direction numbers for M.

Solution. By definition of direction numbers there is a positive number d such that

$$r = d \cdot \cos \alpha$$
$$s = d \cdot \cos \beta$$
$$t = d \cdot \cos \gamma$$

where α, β, γ are the direction angles for M. Then

$$ar = ad \cdot \cos \alpha$$
$$as = ad \cdot \cos \beta$$
$$at = ad \cdot \cos \gamma$$

and since $ad > 0$ (both a and d are positive), ar, as, and at are direction numbers for M by definition of direction numbers.

Theorem 35

Let $P = (a, b, c)$ be a point on the directed line M and let k, q, n be direction numbers for M. Then

$$M = \{(x, y, z) \mid x = a + kw, \, y = b + qw, \, z = c + nw, \, w \in R\}.$$

Proof. Let

$$S = \{(x, y, z) \mid x = a + kw, \, y = b + qw, \, z = c + nw, \, w \in R\}.$$

We will show that $M \subseteq S$ and $S \subseteq M$. Suppose that a point Q is on M. If $Q = P$, then $Q = (a, b, c)$ and

$$a = a + k \cdot 0$$
$$b = b + q \cdot 0$$
$$c = c + n \cdot 0.$$

This shows that $Q \in S$ in this case. If $Q \neq P$, let $Q = (x, y, z)$. Since P, Q are both on M, M has the direction of $\overrightarrow{PQ}$ or $\overrightarrow{QP}$. If M has the direction of $\overrightarrow{PQ}$, then $x - a$, $y - b$, $z - c$ are direction numbers for M. Since k, q, n are also direction numbers for M we may write

$$\Delta \cdot \cos \alpha = k$$
$$\Delta \cdot \cos \beta = q$$
$$\Delta \cdot \cos \gamma = n$$

by Theorem 34, where α, β, γ are direction angles and $\Delta = \sqrt{k^2 + q^2 + n^2}$. Then

$$x - a = d \cdot \cos \alpha = d \cdot (k/\Delta) = k(d/\Delta)$$
$$y - b = d \cdot \cos \beta = d \cdot (q/\Delta) = q(d/\Delta)$$
$$z - c = d \cdot \cos \gamma = d \cdot (n/\Delta) = n(d/\Delta)$$

for some $d > 0$. Therefore,

$$x = a + kw$$
$$y = b + qw$$
$$z = c + nw$$

where $w = d/\Delta$. This implies that $Q = (x, y, z) \in S$. If M has the direction $\overrightarrow{QP}$, then $a - x$, $b - y$, $c - z$ are direction numbers and

$$a - x = k(d/\Delta)$$
$$b - y = q(d/\Delta)$$
$$c - z = n(d/\Delta)$$

or

$$x = a + k(-d/\Delta) = a + kw$$
$$y = b + q(-d/\Delta) = b + qw$$
$$z = c + n(-d/\Delta) = c + nw$$

where $w = -d/\Delta$. Again it follows that $Q = (x, y, z) \in S$. This shows that $M \subseteq S$. Now for some $w \in R$ let

$$x = a + kw$$
$$y = b + qw$$
$$z = c + nw$$

i.e., let $Q = (x, y, z)$ be a point in S. If $w = 0$, then $Q = (x, y, z) = (a, b, c) = P$ and Q is on M. Using Theorem 34 again $k = \Delta \cos \alpha$, $q = \Delta \cos \beta$, $n = \Delta \cos \gamma$. Then

$$x - a = w \cdot \Delta \cos \alpha$$
$$y - b = w \cdot \Delta \cos \beta$$
$$z - c = w \cdot \Delta \cos \gamma.$$

If $w > 0$, $w\Delta > 0$ and $x - a, y - b, z - c$ are direction numbers for the line having direction angles α, β, γ; *i.e.*, for the line M. Hence, Q is on M. If $w < 0$, $a - x, b - y, c - z$ are direction numbers for the line having direction angles α, β, γ. Again Q is on M. We conclude that $S \subseteq M$ and from the first part of the proof $S = M$.

When a line in space is described by the equations

$$x = a + kw, \; y = b + qw, \; z = c + nw$$

we say that w is a *parameter* and that the line is given by parametric equations (compare this with p. 297). In case the three direction numbers k, q, n are all non-zero, we may solve the parametric equations for the parameter w to obtain:

$$w = (x - a)/k, \quad w = (y - b)/q, \quad w = (z - c)/n.$$

Then the three equations

(8-9) $$\frac{x - a}{k} = \frac{y - b}{q} = \frac{z - c}{n}$$

are another description of the line on point $P = (a, b, c)$ and having direction numbers k, q, n.

Suppose that $P = (p_1, p_2, p_3)$ and $Q = (q_1, q_2, q_3)$ are two distinct points. Then $q_1 - p_1, q_2 - p_2, q_3 - p_3$ are direction numbers for the line M on P, Q having direction $\overrightarrow{PQ}$. Then parametric equations for M may, therefore, be written as

(8-10)
$$x = p_1 + (q_1 - p_1)w,$$
$$y = p_2 + (q_2 - p_2)w,$$
$$z = p_3 + (q_3 - p_3)w$$

or

(8-11)
$$x = (1 - w)p_1 + q_1 w$$
$$y = (1 - w)p_2 + q_2 w$$
$$z = (1 - w)p_3 + q_3 w.$$

Write parametric equations for the line on the points $P = (1, 5, 3)$, $Q = (-1, 0, 1)$ and directed from P to Q.

Example
8–36

Solution. Direction numbers for the line are

$$-1 - 1 = -2$$
$$0 - 5 = -5$$
$$1 - 3 = -2.$$

Then parametric equations for the line are

$$x = 1 + (-2)w = 1 - 2w$$
$$y = 5 + (-5)w = 5 - 5w$$
$$z = 3 + (-2)w = 3 - 2w \qquad \text{for } w \in R.$$

In Theorem 35 k, q, n are arbitrary direction numbers for the line M. By Example 8–35 $k' = k/\Delta, q' = q/\Delta, n' = n/\Delta$ are also direction numbers for M where $\Delta = \sqrt{k^2 + q^2 + n^2}$. The parametric equations for M may thus be written as

$$x = a + k'w, \quad y = b + q'w, \quad z = c + n'w.$$

Then for $P = (a, b, c)$ and $Q = (x, y, z)$ we have by Theorem 31

$$\begin{aligned}
|PQ|^2 &= (x - a)^2 + (y - b)^2 + (z - c)^2 \\
&= (k'w)^2 + (q'w)^2 + (n'w)^2 \\
&= [(k')^2 + (q')^2 + (n')^2]w^2 \\
&= (k^2/\Delta^2 + q^2/\Delta^2 + n^2/\Delta^2)w^2 \\
&= \left(\frac{k^2 + q^2 + n^2}{\Delta^2}\right)w^2 \\
&= 1 \cdot w^2 \\
&= w^2.
\end{aligned}$$

Therefore, $|PQ| = |w|$. This shows that we may choose parametric equations for any line so that the parameter w in the parametric equations

$$x = a + kw, \quad y = b + qw, \quad z = c + nw$$

is either the distance between $P = (a, b, c)$ and $Q = (x, y, z)$ or is the negative of this distance.

Let $P = (p_1, p_2, p_3)$, $Q = (q_1, q_2, q_3)$ be two distinct points and let M be a line on P and Q with direction $\overrightarrow{PQ}$. If $X = (x, y, z)$ is a point on M given by the parametric equations

$$x = p_1 + (q_1 - p_1)w$$
$$y = p_2 + (q_2 - p_2)w$$
$$z = p_3 + (q_3 - p_3)w$$

for $w \in R$, then

    (*a*) if $0 \leq w \leq 1$, X lies on the line segment $\overline{PQ}$;
    (*b*) if $w > 1$, X is on the opposite side of Q from P;
    (*c*) if $w < 0$, X is on the opposite side of P from Q.

(See Fig. 8–57.)

Figure 8–57

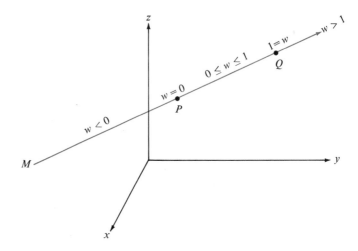

Proof. By Theorem 31

$$|PX| = \sqrt{(x - p_1)^2 + (y - p_2)^2 + (z - p_3)^2}$$
$$= \sqrt{(q_1 - p_1)^2 w^2 + (q_2 - p_2)^2 w^2 + (q_3 + p_3)^2 w^2}$$
$$= \sqrt{w^2}\sqrt{(q_1 - p_1)^2 + (q_2 - p_2)^2 + (q_3 - p_3)^2}$$
$$= |w| \cdot |PQ|.$$

If $w = 0$, $X = P$ and if $w = 1$, $X = Q$. For $0 < w < 1$, we have

$$|PX| = |w| \cdot |PQ| = w\,|PQ| < |PQ|$$

and the equations for M show that $\overrightarrow{PQ}$ and $\overrightarrow{PX}$ have the same direction numbers which implies that X is between P and Q proving (*a*). If $w > 1$, $|PX| = w\,|PQ| > |PQ|$. Also, the equations for the line M show that $\overrightarrow{PQ}$ and $\overrightarrow{PX}$ have the same direction numbers. Hence, X is on the opposite side of Q from P. This proves (*b*). For (*c*), if $w < 0$, the parametric equations show that $\overrightarrow{PQ}$ and $\overrightarrow{PX}$ have direction numbers which differ in sign only. Therefore, X is on the opposite side of P from Q.

Let $P = (p_1, p_2, p_3)$, $Q = (q_1, q_2, q_3)$ be two distinct points in 3-space, and suppose that $|OP| = 1$, $|OQ| = 1$ (see Fig. 8–58). If α, β, γ are direction angles for $\overrightarrow{OP}$ and δ, ϵ, η are direction angles for $\overrightarrow{OQ}$, then by Theorem 32

$$p_1 = |OP| \cos \alpha = \cos \alpha$$
$$p_2 = |OP| \cos \beta = \cos \beta$$
$$p_3 = |OP| \cos \gamma = \cos \gamma$$
$$q_1 = |OQ| \cos \delta = \cos \delta$$
$$q_2 = |OQ| \cos \epsilon = \cos \epsilon$$
$$q_3 = |OQ| \cos \eta = \cos \eta.$$

Figure 8–58

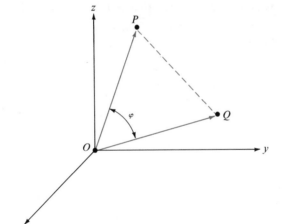

Now consider the angle φ between $\overrightarrow{OP}$ and $\overrightarrow{OQ}$. Apply the Law of Cosines to $\triangle OPQ$:

$$|PQ|^2 = |OP|^2 + |OQ|^2 - 2\,|OP||OQ| \cos \varphi$$
$$= 1 + 1 - 2 \cos \varphi$$
$$= 2(1 - \cos \varphi).$$

But the quantity $|PQ|$ may also be calculated using the distance formula:

$$|PQ|^2 = (p_1 - q_1)^2 + (p_2 - q_2)^2 + (p_3 - q_3)^2$$
$$= p_1{}^2 + p_2{}^2 + p_3{}^2 + q_1{}^2 + q_2{}^2 + q_3{}^2 - 2(p_1 q_1 + p_2 q_2 + p_3 q_3).$$

Now, because of the equations above

$$p_1{}^2 + p_2{}^2 + p_3{}^2 = \cos^2 (\alpha) + \cos^2 (\beta) + \cos^2 (\gamma) = 1$$

and similarly $q_1{}^2 + q_2{}^2 + q_3{}^2 = 1$. Then we have

$$|PQ|^2 = 1 + 1 - 2(p_1q_1 + p_2q_2 + p_3q_3).$$

Using the two expressions for $|PQ|^2$ gives

$$\cos \varphi = p_1q_1 + p_2q_2 + p_3q_3.$$

Theorem 37

Let M, N be two distinct, intersecting, directed lines which have direction cosines m_1, m_2, m_3 and n_1, n_2, n_3 respectively. If φ is the angle between the positive directions of M, N, then

$$\cos \varphi = m_1n_1 + m_2n_2 + m_3n_3.$$

Proof. Choose points P, Q such that $|OP| = 1$, $|OQ| = 1$, $\overrightarrow{OP}$ has the same direction angles as M and $\overrightarrow{OQ}$ has the same direction angles as N (see Fig. 8–59). Then the direction cosines of $\overrightarrow{OP}$ are the same as those of M and the direction cosines of $\overrightarrow{OQ}$ are the same as those of N. The result now follows by the argument of the preceding paragraph.

Figure 8–59

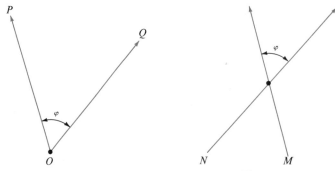

Example 8–37

If directed line M has direction cosines $\frac{1}{2}$, $\frac{1}{2}$, $1/\sqrt{2}$ and directed line N has direction cosines 0, 1, 0, find the angle φ between M and N.

Solution. By Theorem 37

$$\cos \varphi = \tfrac{1}{2}\cdot 0 + \tfrac{1}{2}\cdot 1 + 0\cdot 1/\sqrt{2}$$
$$= \tfrac{1}{2}.$$

Then $\varphi \doteq 1.0472$ radians or $60°$ by Table I.

Example 8–38

If $P = (1, 1, 1)$ and $Q = (1/\sqrt{2},\ 0,\ -1/\sqrt{2})$, find the angle φ between $\overrightarrow{OP}$ and $\overrightarrow{OQ}$.

Solution. By Theorem 32 if α, β, γ are direction angles for $\overrightarrow{OP}$

$$1 = |OP| \cos \alpha = \sqrt{3} \cdot \cos \alpha$$
$$1 = |OP| \cos \beta = \sqrt{3} \cdot \cos \beta$$
$$1 = |OP| \cos \gamma = \sqrt{3} \cdot \cos \gamma$$

so that $\cos \alpha = \cos \beta = \cos \gamma = 1/\sqrt{3}$. Also, if δ, ϵ, η are direction angles for $\overrightarrow{OQ}$

$$1/\sqrt{2} = |OQ| \cos \delta = 1 \cdot \cos \delta$$
$$0 = |OQ| \cos \epsilon = 1 \cdot \cos \epsilon$$
$$-1/\sqrt{2} = |OQ| \cos \eta = 1 \cdot \cos \eta.$$

Then

$$\cos \delta = 1/\sqrt{2}$$
$$\cos \epsilon = 0$$
$$\cos \eta = -1/\sqrt{2}.$$

By Theorem 37

$$\cos \varphi = (1/\sqrt{3})(1/\sqrt{2}) + (1/\sqrt{3}) \cdot 0 + (1/\sqrt{3})(-1/\sqrt{2}) = 0.$$

It follows that $\varphi = \pi$ so that $\overrightarrow{OP}$ and $\overrightarrow{OQ}$ are perpendicular.

47. For each pair of points P, Q below find direction numbers, direction cosines and direction angles to the nearest degree for $\overrightarrow{PQ}$.

Exercises

(a) $P = (-2, 1, -1)$, $Q = (1, 0, 1)$. *Solution.* Direction numbers are $1 - (-2) = 3$, $0 - 1 = -1$, $1 - (-1) = 2$. By Theorem 32

$$3 = |PQ| \cos \alpha$$
$$-1 = |PQ| \cos \beta$$
$$2 = |PQ| \cos \gamma$$

where α, β, γ are direction angles for $\overrightarrow{PQ}$.

$$|PQ| = \sqrt{3^2 + (-1)^2 + 2^2} = \sqrt{14}$$

and the direction cosines are

$$\cos \alpha = 3/\sqrt{14}$$
$$\cos \beta = -1/\sqrt{14}$$
$$\cos \gamma = 2/\sqrt{14}.$$

Using $\sqrt{14} \doteq 3.742$

$$\cos \alpha \doteq 3/(3.742)$$
$$\cos \beta \doteq -1/(3.742)$$
$$\cos \gamma \doteq 2/(3.742).$$

If we use common logarithms for these computations we find that

$$3/(3.742) \doteq 0.8017$$
$$-1/(3.742) \doteq -0.2672$$
$$2/(3.742) \doteq 0.5345.$$

These numbers are approximately the direction cosines for $\overrightarrow{PQ}$. To obtain the direction angles α, β, γ the equations

$$\cos \alpha = 0.8017$$
$$\cos \beta = -0.2672$$
$$\cos \gamma = 0.5345$$

are solved. To the nearest degree $\alpha = 37°$, $\beta = 105°$, $\gamma = 68°$.

(b) $P = (0, 0, 0)$, $Q = (1, 1, 1)$
(c) $P = (0, 1, 0)$, $Q = (1, 1, 1)$
(d) $P = (1, 1, 0)$, $Q = (1, 1, 1)$
(e) $P = (-1, -1, 1)$, $Q = (1, 1, 1)$
(f) $P = (0, 1, 2)$, $Q = (2, -1, 0)$
(g) $P = (-1, 1, -1)$, $Q = (\frac{1}{2}, -1, 2)$

48. Write parametric equations for the directed line on point P and having the given direction numbers in each case below.
    (a) $P = (0, 0, 0)$, direction numbers 1, 1, 1
    (b) $P = (3, -5, 2)$, direction numbers 2, 0, -1
    (c) $P = (-1, 0, 1)$, direction numbers -3, 12, 8
    (d) $P = (\frac{1}{2}, \frac{3}{4}, 11)$, direction numbers -1, 4, -1
    (e) $P = (22, -7, 2)$, direction numbers 0, 3, $-\frac{1}{2}$
    (f) $P = (\pi, \sqrt{5}, -1)$, direction numbers $2\sqrt{2}$, -4, 5

49. For each P, Q of Exercise 47, write parametric equations for the directed line on $\overrightarrow{PQ}$. Then write non-parametric equations for the lines.

50. For each pair of points P, Q of Exercise 47, except (b), find the angle between $\overrightarrow{OP}$ and $\overrightarrow{OQ}$.

51. In each case below find a value for the unknown so that the three given points lie on a line.

(a) $(1, -1, 0), (4, 1, -1), (p, -3, 1)$
(b) $(0, 2, -1), (-2, 1, 0), (-4, 0, r)$
(c) $(5, 10, 4), (4, 15, 1), (0, q, -11)$
(d) $(25, 11, -17), (12, 18, -21), (-1, q, -25)$
(e) $(-3, -8, 2), (-21, 1, -9), (p, 82, -108)$.

52. Let $P = (p_1, p_2, p_3)$, $Q = (q_1, q_2, q_3)$ be two distinct points both different from the origin. Show that the line on P, Q is on the origin if and only if there is a number a such that $q_1 = ap_1$, $q_2 = ap_2$, $q_3 = ap_3$.

53. If two distinct, intersecting, directed lines have direction numbers h, m, n and j, k, i respectively,
    (a) show that there is a positive number a such that

    $$a \cdot \cos \varphi = hj + mk + ni,$$

    where φ is the angle between the lines, and

    (b) show that the two lines are perpendicular if and only if

    $$hj + mk + ni = 0.$$

54. For each P, Q of Exercise 47 write parametric equations for some line on P and perpendicular to $\overrightarrow{PQ}$.

If π is a plane in 3-space, P is any point in π, and Q is a point not in π, such that $\overrightarrow{QP}$ is perpendicular to π, then π consists of all those points and only those points X such that $\overrightarrow{QP}$ is perpendicular to $\overrightarrow{PX}$. Using this characterization we may write an equation for π. The directed line segment $\overrightarrow{QP}$ is called a *normal* for π.

Let P be a point in a plane π and $Q = (a, b, c)$ a point not in π. Suppose that $\overrightarrow{QP}$ is a normal for π with direction cosines m, n, k. Then $X = (x, y, z)$ is a point of π if and only if

$$|PQ| = (x - a)m + (y - b)n + (z - c)k.$$

Proof. Let u, v, w be direction cosines for $\overrightarrow{QX}$. Then by Theorem 32

$$x - a = du$$
$$y - b = dv$$
$$z - c = dw$$

where $d = |QX|$. The point X is in π if and only if $\overrightarrow{QP}$ is perpendicular to $\overrightarrow{PX}$ (see Fig. 8–60). If φ is the angle between $\overrightarrow{QP}$ and $\overrightarrow{QX}$, $\overrightarrow{QP}$ is perpendicular to $\overrightarrow{PX}$ if and only if $|PQ| = d \cdot \cos \varphi$. By Theorem 37

$$\cos \varphi = mu + nv + kw$$

so that

$$\begin{aligned} |PQ| &= d \cdot (mu + nv + kw) \\ &= (du)m + (dv)n + (dw)k \\ &= (x - a)m + (y - b)n + (z - c)k. \end{aligned}$$

Figure
8–60

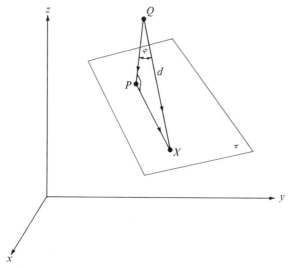

As a consequence of this theorem we may say that if π is a plane, then

$$\pi = \{(x, y, z) \mid (x - a)m + (y - b)n + (z - c)k = |PQ|\}$$

where P is in π, $\overrightarrow{QP}$ is a normal to π with direction cosines m, n, k and $Q = (a, b, c)$. In case π does not contain the origin, we may choose $Q = (0, 0, 0) = O$ and our equation simplifies to

$$mx + ny + kz = |PO|.$$

The equation

$$(x - a)m + (y - b)n + (z - c)k = |PQ|$$

may be expressed as

$$mx + ny + kz + (-|PQ| - am - bn - ck) = 0$$

or

$$mx + ny + kz + g = 0$$

where $g = -|PQ| - am - bn - ck$. This shows, of course, that the points S of π have coordinates (x, y, z) which satisfy the equation

$$mx + ny + kz + g = 0$$

or that *the coordinates of every point in a plane satisfy a linear equation.*

The points $X = (x, y, z)$ of a plane π satisfy a linear equation

$$mx + ny + kz + g = 0$$

where m, n, k are direction numbers for any normal to π.

**Theorem
39**

We could now use similar techniques to show that the set of all points $X = (x, y, z)$ which satisfy a linear equation

$$mx + ny + kz + g = 0$$

where not all of m, n, k are zero, is a plane π. A different procedure will be used, however; one which will expose another property of planes.

Let C be a set of points in 3-space. C is said to be a *convex set* if and only if C has the following property:

Definition

for all points P, Q in C, $\overline{PQ} \subseteq C$.

3-space is a convex set; every line and every line segment is also convex.

**Example
8–39**

Euclid defined a plane to be a convex set which is not all of 3-space and which contains at least 3 non-collinear points, *i.e.* 3 points not on a line. Using this definition let us show that the set π of all points $X = (x, y, z)$ which satisfy the linear equation

$$mx + ny + kz + g = 0$$

is a plane (where not all m, n, k are zero). If $g \neq 0$, then $m \cdot 0 + n \cdot 0 + k \cdot 0 + g \neq 0$ and $(0, 0, 0) \notin \pi$. If $g = 0$, then at least one of m, n, k is not zero, say $m \neq 0$. Then $m \cdot 1 + n \cdot 0 + k \cdot 0 \neq 0$ so that $(1, 0, 0) \notin \pi$. In either case π is not all of 3-space. Now suppose that $P = (p_1, p_2, p_3)$ and $Q = (q_1, q_2, q_3)$ are two points of π. By definition of π

$$mp_1 + np_2 + kp_3 + g = 0 \qquad \text{and} \qquad mq_1 + nq_2 + kq_3 + g = 0.$$

We may write parametric equations for the line on $\overrightarrow{PQ}$ as

$$x = p_1 + (q_1 - p_1)w$$
$$y = p_2 + (q_2 - p_2)w$$
$$z = p_3 + (q_3 - p_3)w.$$

Then

$$
\begin{aligned}
mx + ny + kz + g &= m(p_1 + (q_1 - p_1)w) + n(p_2 + (q_2 - p_2)w) \\
&\quad + k(p_3 + (q_3 - p_3)w) + g \\
&= (mp_1 + np_2 + kp_3 + g) \\
&\quad + [m(q_1 - p_1) + n(q_2 - p_2) + k(q_3 - p_3)]w \\
&= 0 + [mq_1 + nq_2 + kq_3 - (mp_1 + np_2 + kp_3)]w \\
&= (-g - (-g))w \\
&= 0 \cdot w = 0.
\end{aligned}
$$

It follows that the line on $\overrightarrow{PQ}$ is contained in π and that π is a convex set. To finish the proof that π is a plane it must be shown that π contains three non-collinear points. This is left as an exercise (see Exercise 64).

Example 8–40

Find parametric equations for the line of intersection of the two planes given by the equations

$$2x - y + z = 0$$
$$-x + 3y - 2z + 4 = 0.$$

Solution. First, we find two points on the intersection of the two planes. Multiplying the second equation by 2 and adding it to the first gives

$$5y - 3z + 8 = 0.$$

In a similar way we obtain the equations

$$5x + z + 4 = 0 \quad \text{and} \quad 3x + y + 4 = 0.$$

One solution to the first equation is $y = -1$, $z = 1$. Using this value for z in the second equation gives $x = -1$. Now the point $(-1, -1, 1)$ is seen to lie on both planes. In the third equation if $x = 1$, $y = -7$ and from the second equation this implies that $z = -9$. Thus a second point on both planes is $(1, -7, -9)$. Using these points we may write parametric equations for the line of intersection of the planes:

$$x = -1 + (1 + 1)w = -1 + 2w$$
$$y = -1 + (-7 + 1)w = -1 - 6w$$
$$z = 1 + (-9 - 1)w = 1 - 10w.$$

55. (*a*) Write an equation for the *xy*-plane.
    (*b*) Write an equation for the *xz*-plane.
    (*c*) Write an equation for the *yz*-plane.
    (*d*) Write an equation for the plane on the point $(1, 0, 0)$ and parallel to the *yz*-plane.
    (*e*) Write an equation for the plane on the point $(0, 1, 0)$ and parallel to the *zx*-plane.
    (*f*) Write an equation for the plane on the point $(0, 0, 1)$ and parallel to the *xy*-plane.
    (*g*) Write an equation for the plane on the point $(-2, 6, 5)$ and parallel to the *xy*-plane.
    (*h*) Write an equation for the plane on the point $(-2, 6, 5)$ and parallel to the *xz*-plane.
    (*i*) Write an equation for the plane on the point $(-2, 6, 5)$ and parallel to the *yz*-plane.

56. In each case below you are given a point *P* and direction numbers. Write an equation for a plane on the point and having the given direction numbers.
    (*a*) $P = (0, 0, 0)$, direction numbers: $1, 1, 1$.
    (*b*) $P = (0, 0, 0)$, direction numbers: $2, -1, 4$.
    (*c*) $P = (1, 2, 3)$, direction numbers: $1, 1, 1$
    (*d*) $P = (0, 1, 5)$, direction numbers: $3, -5, 1$
    (*e*) $P = (-1, 0, 7)$, direction numbers: $-1, 1, -2$
    (*f*) $P = (2, -1, 0)$, direction numbers: $2, 3, -5$

57. (*a*) Write an equation for a plane on the points $(0, -2, 3)$, $(0, 1, 0)$ and parallel to the *x*-axis.
    (*b*) Write an equation for a plane on the points $(1, 0, -3)$, $(4, 0, 6)$ and parallel to the *y*-axis.
    (*c*) Write an equation for the plane on the two points $(1, 1, 0)$, $(2, -5, 0)$ and parallel to the *z*-axis.

58. In each case below write an equation for the plane on the three given points.
    (*a*) $(0, 0, 0)$, $(1, 1, 1)$, $(-1, 1, 1)$
    (*b*) $(0, 0, 0)$, $(1, 1, 1)$, $(1, -1, 1)$
    (*c*) $(0, 0, 0)$, $(1, 1, 1)$, $(1, 1, -1)$
    (*d*) $(0, 1, 0)$, $(1, 0, 0)$, $(0, 0, 1)$
    (*e*) $(1, 0, 0)$, $(0, 2, 0)$, $(0, 0, 3)$
    (*f*) $(1, 0, 0)$, $(-1, 0, 0)$, $(0, 0, 1)$
    (*g*) $(1, 0, 3)$, $(-1, 0, 3)$, $(0, -1, 0)$
    (*h*) $(3, 0, 0)$, $(0, -2, 0)$, $(-4, 1, 7)$

59. In each case below you are to write an equation for the plane which intersects the xy-plane in the given line and contains the given point.
    (a) $3x - 2y + 1 = 0$, $(1, 1, 1)$
    (b) $x - y - 2 = 0$, $(2, -1, 3)$
    (c) $2x - y + 2 = 0$, $(-2, 1, 1)$
    (d) $-x + 2y - 1 = 0$, $(-1, 2, 1)$
    (e) $x + y + 1 = 0$, $(1, 1, 1)$
    (f) $-y + x - 1 = 0$, $(-1, 1, 2)$

60. (a) Show that a plane parallel to the plane given by the equation $ax + by + cz + d = 0$ has equation $ax + by + cz + e = 0$ for some number e.
    (b) Find an equation for the plane on the point $(-3, 2, -1)$ and parallel to the plane having equation $4x - 3y + 2z + 1 = 0$.
    (c) Find an equation for the plane on the point $(1, -1, 1)$ and parallel to the plane having equation $-x + 2y - z + 2 = 0$.
    (d) Find an equation for the plane on the point $(5, -2, 3)$ and parallel to the plane having equation $2x + 3y + 4z - 1 = 0$.

61. (a) Show that a plane π on the point (a, b, c) and having a normal with direction numbers m, n, k has an equation

$$m(x - a) + n(y - b) + k(z - c) = 0.$$

    (b) Show that the line with equations

$$\frac{x - a}{m} = \frac{y - b}{n} = \frac{z - c}{k}$$

    is a normal to the plane π above.

62. Show that if two planes π_1 and π_2 have equations

$$m_1 x + n_1 y + k_1 z + g_1 = 0$$

and

$$m_2 x + n_2 y + k_2 z + g_2 = 0$$

then π_1 and π_2 are perpendicular if and only if

$$m_1 m_2 + n_1 n_2 + k_1 k_2 = 0.$$

63. Show that any point $X = (x, y, z)$ on the line connecting two distinct points $(0, b, c)$ and $(0, d, e)$ has $x = 0$.

64. Show that the solution set of the equation $mx + ny + kz + g = 0$ contains 3 non-collinear points, where not all of m, n, k are zero. This may be accomplished by the following cases:

Case 1 (at least 2 of m, n, k are not zero, say $k \neq 0$ and $n \neq 0$.)

  Subcase 1 ($g \neq 0$). In this case show that $(0,\ 0,\ -g/k)$, $(1,\ 0,\ (-g-m)/k)$, $(0,\ -g/n,\ 0)$ are distinct, in the solution set and are not collinear.

  Subcase 2 ($g = 0$). In this case show that $(0,\ k,\ -n)$, $(0, 2/n, -2/k)$, $(1,\ 1,\ -(m+n))$ are distinct, in the solution set and are not collinear.

Case 2 (only one of m, n, k is not zero, say $k \neq 0$). In this case the equation is $kz + g = 0$. Show that the points $(0, 0, t), (0, 1, t), (1, 0, t)$ are distinct, in the solution set and are not collinear, where $t = -g/k$.

We wish to consider here the problem of graphs of equations in x, y, z and of *functions of two variables*. By a function of two variables we mean a function whose domain is a subset of $R \times R$ and whose range is contained in R. For example we may be given a function such as $f(x, y) = x$. By the graph of this function we mean $\{(x, y, z) \mid z = f(x, y) = x\}$. From the previous section we know that this is the graph of a plane. For any function f of two variables the graph is $\{(x, y, z) \mid z = f(x, y)\}$.

Describe the graph of the function $f(x, y) = \sqrt{1 - x^2 - y^2}$.

Example
8–41

Solution. By definition the graph is $\{(x, y, z) \mid z = \sqrt{1 - x^2 - y^2}\}$. But if $z = \sqrt{1 - x^2 - y^2}$, then $x^2 + y^2 + z^2 = 1$. Also

$$x^2 + y^2 + z^2 = (x - 0)^2 + (y - 0)^2 + (z - 0)^2$$

and this shows that we are considering points (x, y, z) the square of whose distance from the origin is 1, *i.e.* points which are 1 unit from the origin. Since we must have $f(x, y) > 0$ our graph is a hemispherical surface.

When we are considering the graph of a function of two variables, it is natural to use knowledge of the graphs of functions of one variable whenever possible. For instance, consider

$$f(x, y) = x^2 - y.$$

To determine the graph we must consider all (x, y, z) such that $z = x^2 - y$. If $z = 0$, our points (x, y, z) are in the xy-plane and the equation is $0 = x^2 - y$ or $y = x^2$. But the graph of $y = x^2$ is a parabola. Hence, in the xy-plane the graph is a parabola (see Fig. 8–61). If $z = k$, where k is

Figure
8–61

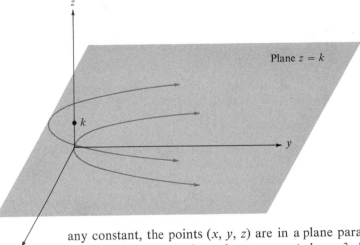

any constant, the points (x, y, z) are in a plane parallel to the xy-plane and the equation is $k = x^2 - y$ or $y + k = x^2$. Thus in any plane parallel to the xy-plane the graph is a parabola. For $y = k$ the points (x, y, z) are in a plane parallel to the xz-plane; the equation is $z = x^2 - k$ or $z + k = x^2$ and again the graph in planes parallel to the xz-plane is a parabola (see Fig. 8–62). Finally, if $x = k$, the points (x, y, z) are in a

Figure
8–62

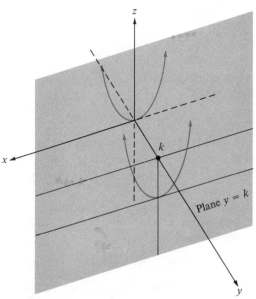

plane parallel to the yz-plane and the equation is $z + y = k^2$ which has a line as graph; the graph in planes parallel to the yz-plane is a line. Thus we have the surface indicated in Fig. 8–63.

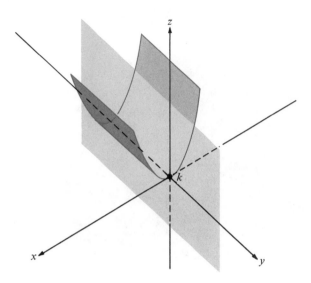

Figure
8–63

The method we have just used for the graph of $f(x, y) = x^2 - y$ is known as *taking slices*.

Suppose that we wish to determine the graph of the equation

$$x^2 + y^2 + z^2 = 1$$

i.e. we wish to determine $\{(x, y, z) \mid x^2 + y^2 + z^2 = 1\}$. If we let $z = k$, we take a slice parallel to the xy-plane and obtain the equation

$$x^2 + y^2 = 1 - k^2.$$

This shows that $1 - k^2 \geq 0$ or $|k| \leq 1$ so that a slice taken for $k > 1$ or $k < -1$ misses the graph. But for $1 - k^2 \geq 0$ the graph is a circle. We obtain the same results in the other planes and see that the graph is a spherical surface as shown in Fig. 8–64.

If the above equation had been

$$x^2 + y^2 + 2z^2 = 1$$

slices parallel to the xy-plane would still be circles. But slices parallel to the xz-plane and yz-plane are ellipses. For example if $y = k$, the equation is

$$x^2 + 2z^2 = 1 - k^2$$

and the graph is an ellipse. This is an example of an *ellipsoid*.

Some Graphs in 3-Space

Figure
8-64

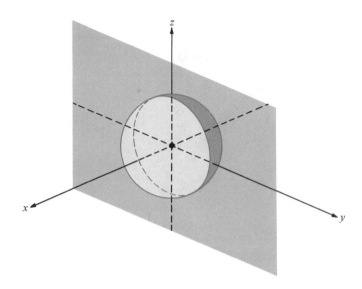

65. Discuss the graph of each equation below in 3-space.

(a) $x^2 + y^2 = 1$ (b) $x^2 - y^2 = 1$
(c) $y = z^2$ (d) $2x^2 + z^2 = 1$
(e) $2x^2 - z^2 = 1$ (f) $x^2 - y^2 = 0$

66. Discuss the graph of each function given below.

(a) $f(x, y) = 1$ (b) $f(x, y) = -4$
(c) $f(x, y) = x$ (d) $f(x, y) = -2y + x$
(e) $f(x, y) = x^2 + y$ (f) $f(x, y) = x - 2y^2$
(g) $f(x, y) = x^2 + y^2$ (h) $f(x, y) = x^2 - y^2$
(i) $f(x, y) = |x|$ (j) $f(x, y) = |x| + |y|$
(k) $f(x, y) = \cos(x)$ (l) $f(x, y) = [y]$

67. Sketch the graph of $4z = x^2 + 4y^2$.

68. Sketch the graph of $x^2/4 + y^2 + z^2/2 = 1$.

69. Sketch the graph of $x^2/4 - y^2 + z^2/2 = 1$.

70. Sketch the graph of $x^2/4 - y^2 - z^2/2 = 1$.

71. Sketch the graph of $x^2/4 - y^2 - z^2/2 = 0$.

COMPLEX NUMBERS

This chapter is devoted to a brief introduction to the study of the complex number system. The complex numbers are, in a certain sense, an enlargement of the real number system. An expansion of the set of real numbers is desirable for several reasons; one of the most important reasons is that in the system of complex numbers we can solve equations of the type $x^2 + 1 = 0$ which have no real number solution.

By the *complex number system* **C** we mean the set **R** $\times$ **R** (the set of all ordered pairs of real numbers) with an addition and multiplication defined as follows:

Addition of Complex Numbers

$$(a, b) + (c, d) = (a + c, b + d)$$

for all ordered pairs (a, b), (c, d).

Multiplication of Complex Numbers

$$(a, b) \cdot (c, d) = (ac - bd, ad + bc)$$

for all ordered pairs (a, b), (c, d).

It should be noted that in the equation

$$(a, b) + (c, d) = (a + c, b + d),$$

defining addition of complex numbers, the "$+$" on the left is being *defined* and the "$+$" on the right denotes ordinary real number addition. Thus the symbol "$+$" is being used in two different ways here. Similar remarks are in order for the multiplication definition. It also should be observed that since complex numbers are ordered pairs of real numbers, we have the usual condition governing equality of ordered pairs:

$$(a, b) = (c, d) \text{ if, and only if, } a = c \text{ and } b = d.$$

Definitions

The complex numbers satisfy many of the properties of real numbers. For example, the definitions of addition and multiplication show that the sum and product of complex numbers are each unique complex numbers.

Theorem 40

The complex numbers **C** have the following properties:

1. The sum, $\alpha + \beta$, and product, $\alpha\beta$, of any complex numbers α, β are unique complex numbers.

2. For any two complex numbers α, β

$$\alpha + \beta = \beta + \alpha \quad \text{and} \quad \alpha\beta = \beta\alpha.$$

3. If α, β, γ are complex numbers, then

$$\alpha + (\beta + \gamma) = (\alpha + \beta) + \gamma \quad \text{and} \quad \alpha(\beta\gamma) = (\alpha\beta)\gamma.$$

4. If α, β, γ are complex numbers, then

$$\alpha(\beta + \gamma) = \alpha\beta + \alpha\gamma.$$

5. If α is any complex number, then
   (a) $\alpha + (0, 0) = \alpha$
   (b) $\alpha \cdot (1, 0) = \alpha$.

6. (a) For each complex number α, there is a complex number $-\alpha$ such that

$$\alpha + (-\alpha) = (0, 0).$$

   (b) For each complex number $\alpha \neq (0, 0)$, there is a complex number α^{-1} such that

$$\alpha \cdot \alpha^{-1} = (1, 0).$$

Proof: We have already noted that the sum and product of complex numbers is each a unique complex number. The proof of 4 and 6 will be given and the others will be left to the student.

To prove 4, let $\alpha = (a, b)$, $\beta = (c, d)$ and $\gamma = (e, f)$ for real numbers a, b, c, d, e, f. Then, using the definitions of addition and multiplication,

$$
\begin{aligned}
\alpha(\beta + \gamma) &= (a, b) \cdot ((c, d) + (e, f)) \\
&= (a, b)(c + e, d + f) \\
&= (a(c + e) - b(d + f), a(d + f) + b(c + e)) \\
\alpha\beta &= (a, b)(c, d) = (ac - bd, ad + bc) \\
\alpha\gamma &= (a, b)(e, f) = (ae - bf, af + be) \\
\alpha\beta + \alpha\gamma &= (ac - bd, ad + bc) + (ae - bf, af + be) \\
&= (ac - bd + ae - bf, ad + bc + af + be).
\end{aligned}
$$

Now, by the condition for equality of complex numbers, $\alpha(\beta + \gamma) = \alpha\beta + \alpha\gamma$ if and only if

$$a(c + e) - b(d + f) = ac - bd + ae - bf$$

and

$$a(d + f) + b(c + e) = ad + bc + af + be.$$

Since these last two equations are seen to be true, we may conclude that

$$\alpha(\beta + \gamma) = \alpha\beta + \alpha\gamma.$$

If $\alpha = (a, b)$, then by definition of addition

$$\alpha + (-a, -b) = (a, b) + (-a, -b)$$
$$= (a - a, b - b).$$

Then, since $a - a = 0$ and $b - b = 0$, we have $\alpha + (-a, -b) = (0, 0)$ by complex number equality. Therefore, we take $-\alpha$ to *mean* the complex number $(-a, -b)$. This proves 6 (*a*). For 6 (*b*), the condition that $\alpha \neq (0, 0)$ implies that if $\alpha = (a, b)$, $a \neq 0$ or $b \neq 0$. Then $a^2 + b^2 \neq 0$. Define α^{-1} to be the complex number

$$\left(a/(a^2 + b^2), -b/(a^2 + b^2)\right).$$

By definition of multiplication,

$$\alpha \cdot \alpha^{-1} = (a, b)\left(\frac{a}{a^2 + b^2}, \frac{-b}{a^2 + b^2}\right)$$
$$= \left(\frac{a^2}{a^2 + b^2} + \frac{b^2}{a^2 + b^2}, \frac{-ab}{a^2 + b^2} + \frac{ab}{a^2 + b^2}\right)$$
$$= \left(\frac{a^2 + b^2}{a^2 + b^2}, 0\right)$$
$$= (1, 0).$$

The last two lines above follow from the previous lines because of the condition for equality. This completes the proof of 6 (*b*).

Consider the set of all complex numbers $(a, 0)$. These numbers add and multiply as follows:

$$(a, 0) + (b, 0) = (a + b, 0 + 0)$$
$$= (a + b, 0)$$
$$(a, 0) \cdot (b, 0) = (ab - 0 \cdot 0, a \cdot 0 + 0 \cdot b)$$
$$= (ab, 0).$$

This shows that the additive and multiplicative properties of these numbers depend only on the real numbers a, b and for this reason the complex numbers $(a, 0)$ are identified with the corresponding real numbers a. It is in this sense that the complex numbers are an expansion of the real number system.

Theorem 41

The equation

$$x^2 + 1 = 0$$

has a solution in the complex number system.

Proof: By the principle, stated above, for identifying real numbers a with their corresponding complex numbers $(a, 0)$, the equation in question is

$$x^2 + (1, 0) = (0, 0).$$

The complex number $(0, 1)$ is a solution:

$$\begin{aligned}
(0, 1)^2 + (1, 0) &= (0, 1) \cdot (0, 1) + (1, 0) \\
&= (0 \cdot 0 - 1 \cdot 1, 0 \cdot 1 + 1 \cdot 0) + (1, 0) \\
&= (0 - 1, 0 + 0) + (1, 0) \\
&= (-1, 0) + (1, 0) \\
&= (0, 0).
\end{aligned}$$

The complex number $(0, 1)$, shown above to be a solution of the equation

$$x^2 + 1 = 0,$$

is usually denoted by i. Using this notation, every complex number (a, b) may be expressed

$$a + bi$$

as the following computation shows:

$$\begin{aligned}
(a, b) &= (a + 0, 0 + b) \\
&= (a, 0) + (0, b) \\
&= (a, 0) + (b \cdot 0 - 0 \cdot 1, b \cdot 1 + 0 \cdot 0) \\
&= (a, 0) + (b, 0) \cdot (0, 1) \\
&= a + bi.
\end{aligned}$$

Example 9–1

Solve the equation

$$(1, 3)\alpha = (7, 1).$$

Solution. Since $(1, 3) \neq (0, 0)$, compute

$$(1, 3)^{-1} = \left(1/(1^2 + 3^2), -3/(1^2 + 3^2)\right) = (\tfrac{1}{10}, \tfrac{-3}{10}).$$

Then,

$$(\tfrac{1}{10}, \tfrac{-3}{10})(1, 3)\alpha = (\tfrac{1}{10}, \tfrac{-3}{10})(7, 1)$$

or

$$\alpha = (\tfrac{7}{10} + \tfrac{3}{10}, \tfrac{1}{10} - \tfrac{21}{10})$$
$$\alpha = (1, -2).$$

To check this solution,

$$(1, 3)(1, -2) = (1 + 6, -2 + 3)$$
$$= (7, 1).$$

Solve the equation $x^2 + 8 = 0$.

Example 9–2

Solution. We seek a complex number x such that $x^2 = -8$. Since

$$i^2 = (0, 1)(0, 1)$$
$$= (-1, 0) = -1$$

and $(2\sqrt{2})^2 = 8$, it seems reasonable that $2i\sqrt{2}$ might be a solution:

$$(2i\sqrt{2})^2 + 8 = (2i\sqrt{2})(2i\sqrt{2}) + 8$$
$$= i^2(2\sqrt{2})^2 + 8$$
$$= -1 \cdot 8 + 8$$
$$= 0.$$

Hence, $2i\sqrt{2}$ is a solution and $-2i\sqrt{2}$ also may be shown to be a solution.

Express the complex number $\dfrac{2 + 3i}{1 + i}$ in the form $a + bi$.

Example 9–3

Solution. The number $\dfrac{2 + 3i}{1 + i}$ means $(2 + 3i) \cdot (1 + i)^{-1}$. Hence,

$$\frac{2 + 3i}{1 + i} = (2 + 3i) \cdot (1 + i)^{-1}$$
$$= (2, 3) \cdot (1, 1)^{-1}$$
$$= (2, 3) \cdot (\tfrac{1}{2}, -\tfrac{1}{2})$$
$$= (1 + \tfrac{3}{2}, -1 + \tfrac{3}{2})$$
$$= (\tfrac{5}{2}, \tfrac{1}{2})$$
$$= (\tfrac{5}{2}) + (\tfrac{1}{2})i.$$

Exercises

1. Compute the following sums and products.
   (a) $(1, 5) + (\frac{1}{2}, 4)$
   (b) $(\frac{7}{8}, -9) + (-\frac{1}{2}, 2)$
   (c) $(1, 2) \cdot (3, 2)$
   (d) $(0, 5) \cdot (\sqrt{2}, \frac{1}{8})$
   (e) $(1 + i) + (-3 - 5i)$
   (f) $(2 + 6i) + \frac{1}{2}$
   (g) $3(2 + 7i) - 2(1 - i)$
   (h) $(\frac{1}{8} - i) \cdot (-1 + i)$
   (i) $(\sqrt{2} + i) \cdot (\sqrt{2} - i)$

2. Compute the multiplicative inverse of each of the following; *i.e.*, compute α^{-1}.
   (a) $\alpha = (1, 0)$ (b) $\alpha = (0, 1)$
   (c) $\alpha = 1 - i$ (d) $\alpha = \frac{1}{2} + (\frac{3}{4})i$
   (e) $\alpha = 2i$ (f) $\alpha = 4$

3. Express each of the following in the form $a + bi$.
   (a) $(1, 0)$ (b) $(7, 2)$
   (c) $4 \cdot (\frac{1}{2}, 1)$ (d) $\dfrac{1 - i}{1 + i}$
   (e) $\dfrac{\frac{1}{2} + i}{i}$ (f) $\dfrac{-9 + 3i}{7(2 - 9i)}$

4. Solve each of the following equations:
   (a) $(2 + i)x + 5 = i$ (b) $ix - 7i = 1$
   (c) $x^2 + 2 = 0$ (d) $x^2 + i = 0$

5. Prove parts 2, 3, 5 of Theorem 40.

6. Prove each of the following by mathematical induction.
   (a) $i^{4n} = 1$, for all positive integers n.
   (b) $i^{4n+1} = i$, for all positive integers n.
   (c) $i^{4n+2} = -1$, for all positive integers n.

7. Prove: $a(c, d) = (ac, ad)$ or $a(c + di) = ac + (ad)i$ and that
   $[a(c, d)][b(e, f)] = ab(c, d)(e, f)$ [*Hint:* $a = (a, 0)$.]

9.2
Polar Representation

Complex numbers are ordered pairs of real numbers and may therefore be represented as points in the plane. If the point $P(a, b)$ is a distance r from the origin, we have the equations

$$a = r \cos \varphi$$
$$b = r \sin \varphi$$

where φ measures the angle shown in Fig. 9–1. Then

$$a + bi = (a, b)$$
$$= (r \cos \varphi, r \sin \varphi)$$
$$= r(\cos \varphi, \sin \varphi) \quad \text{(See Exercise 7.)}$$
$$= r(\cos \varphi + i \sin \varphi).$$

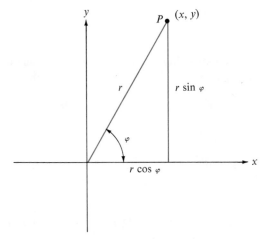

Figure
9–1

Hence, every complex number $a + bi$ may be expressed as

$$r(\cos \varphi + i \sin \varphi)$$

where $r = \sqrt{a^2 + b^2}$ and φ is a suitably chosen number. This is known as the *polar form* of the complex number.

Represent the complex number $1 + i$ in polar form.

Example
9–4

Solution. The point $P(1, 1)$ in the plane is on the line bisecting the first and third quadrants. Thus, we may take $\pi/4$ for the angle measure. Also $P(1, 1)$ is $\sqrt{1^2 + 1^2} = \sqrt{2}$ distance from the origin. Then,

$$1 + i = \sqrt{2}(\cos \pi/4 + i \sin \pi/4)$$

so that $\sqrt{2}(\cos \pi/4 + i \sin \pi/4)$ is the polar form of $1 + i$.

Show that every complex number of the form $\cos \varphi + i \sin \varphi$ represents a point on the unit circle, center at the origin.

Example
9–5

Solution. The complex number $\cos \varphi + i \sin \varphi$ is the ordered pair $(\cos \varphi, \sin \varphi)$; and since $\cos^2 \varphi + \sin^2 \varphi = 1$, it is a point of the unit circle, center at the origin: $\{(x, y) \mid x^2 + y^2 = 1\}$.

The polar form of a complex number is convenient especially for multiplying. Suppose that the polar form of two complex numbers α, β is $r_1(\cos \varphi_1 + i \sin \varphi_1)$ and $r_2(\cos \varphi_2 + i \sin \varphi_2)$, respectively. Then,

$$\alpha\beta = [r_1(\cos \varphi_1 + i \sin \varphi_1)][r_2(\cos \varphi_2 + i \sin \varphi_2)]$$
$$= r_1r_2(\cos \varphi_1, \sin \varphi_1)\cdot(\cos \varphi_2, \sin \varphi_2)$$
$$= r_1r_2(\cos \varphi_1 \cos \varphi_2 - \sin \varphi_1 \sin \varphi_2, \cos \varphi_1 \sin \varphi_2 + \sin \varphi_1, \cos \varphi_2)$$
$$= r_1r_2[\cos (\varphi_1 + \varphi_2), \sin (\varphi_1 + \varphi_2)]$$
$$= r_1r_2[\cos (\varphi_1 + \varphi_2) + i \sin (\varphi_1 + \varphi_2)].$$

This shows — roughly speaking — that complex numbers in polar form may be multiplied by multiplying their distances from the origin and adding their angles.

Example 9–6

Multiply the complex numbers

$$\sqrt{2}(\cos \pi/4 + i \sin \pi/4) \qquad \text{and} \qquad 2(\cos \pi/3 + i \sin \pi/3).$$

Solution.

$$[\sqrt{2}(\cos \pi/4 + i \sin \pi/4)]\cdot[2(\cos \pi/3 + i \sin \pi/3)]$$
$$= 2\sqrt{2}[\cos (\pi/4 + \pi/3) + i \sin (\pi/4 + \pi/3)]$$
$$= 2\sqrt{2}(\cos 7\pi/12 + i \sin 7\pi/12).$$

Example 9–7

$$[\sqrt{2}(\cos \pi/4 + i \sin \pi/4)]^2$$
$$= \sqrt{2}\cdot\sqrt{2}(\cos (\pi/4 + \pi/4) + i \sin (\pi/4 + \pi/4))$$
$$= 2(\cos (\pi/2) + i \sin (\pi/2)).$$

Hence, since $1 + i = \sqrt{2}(\cos \pi/4 + i \sin \pi/4)$,

$$(1 + i)^2 = 2i.$$

Theorem 42 (De Moivre's Theorem)

If n is a natural number, then

$$[r(\cos \varphi + i \sin \varphi)]^n = r^n(\cos n\varphi + i \sin n\varphi).$$

Proof: Apply the induction principle to the statement $P(n)$. $P(1)$ is

$$[r(\cos \varphi + i \sin \varphi)]^1 = r^1(\cos 1\cdot\varphi + i \sin 1\cdot\varphi)$$

which is seen to be true. $P(k)$ and $P(k + 1)$ are respectively

$$[r(\cos \varphi + i \sin \varphi)]^k = r^k(\cos k\varphi + i \sin k\varphi)$$

and

$$[r(\cos \varphi + i \sin \varphi)]^{k+1} = r^{k+1}[\cos (k + 1)\varphi + i \sin (k + 1)\varphi].$$

If $P(k)$ is true, we have

$$
\begin{aligned}
[r(\cos \varphi + i \sin \varphi)]^{k+1} &= [r(\cos \varphi + i \sin \varphi)]^k \cdot [r(\cos \varphi + i \sin \varphi)] \\
&= [r^k(\cos k\varphi + i \sin k\varphi)] \cdot [r(\cos \varphi + i \sin \varphi)] \\
&= r^{k+1}[\cos (k\varphi + \varphi) + i \sin (k\varphi + \varphi)] \\
&= r^{k+1}[\cos (k + 1)\varphi + i \sin (k + 1)\varphi].
\end{aligned}
$$

This shows that if $P(k)$ is true, then so is $P(k + 1)$ and proves the theorem.

Compute $(1 + i)^{64}$.

Example
9–8

Solution. $1 + i = \sqrt{2}(\cos \pi/4 + i \sin \pi/4)$ and therefore,

$$
\begin{aligned}
(1 + i)^{64} &= [\sqrt{2}(\cos \pi/4 + i \sin \pi/4)]^{64} \\
&= (\sqrt{2})^{64}(\cos 64 \cdot \pi/4 + i \sin 64 \cdot \pi/4) \\
&= 2^{32}(\cos 16\pi + i \sin 16\pi) \\
&= 2^{32}(1 + i \cdot 0) = 2^{32}.
\end{aligned}
$$

Find a complex number $a + bi$ such that $(a + bi)^3 = 1 + i$.

Example
9–9

Solution. Suppose that $a + bi = r(\cos \varphi + i \sin \varphi)$ and that $(a + bi)^3 = 1 + i$. Then,

$$[r(\cos \varphi + i \sin \varphi)]^3 = 1 + i = \sqrt{2}(\cos \pi/4 + i \sin \pi/4)$$

or

$$r^3(\cos 3\varphi + i \sin 3\varphi) = \sqrt{2}(\cos \pi/4 + i \sin \pi/4)$$

or

$$(r^3 \cos 3\varphi, r^3 \sin 3\varphi) = (\sqrt{2} \cos \pi/4, \sqrt{2} \sin \pi/4).$$

Hence,

$$r^3 \cos 3\varphi = \sqrt{2} \cos \pi/4 = \sqrt{2}(\tfrac{1}{2}\sqrt{2}) = 1$$

and

$$r^3 \sin 3\varphi = \sqrt{2} \sin \pi/4 = \sqrt{2}(\tfrac{1}{2}\sqrt{2}) = 1.$$

Therefore, $r^3(\cos 3\varphi - \sin 3\varphi) = 0$; and since r^3 cannot be zero, we must have

$$\cos 3\varphi = \sin 3\varphi.$$

Then

$$\sin^2 3\varphi + \cos^2 3\varphi = 1$$

and

$$2 \sin^2 3\varphi = 1$$
$$\sin^2 3\varphi = \tfrac{1}{2}.$$

Thus, $\sin 3\varphi = \pm\tfrac{1}{2}\sqrt{2}$, but $\sin 3\varphi$ cannot be negative ($r^3 \sin 3\varphi = 1$ and $r^3 > 0$) so

$$\sin 3\varphi = \tfrac{1}{2}\sqrt{2} \qquad \text{and} \qquad \cos 3\varphi = \tfrac{1}{2}\sqrt{2}.$$

One solution of these equations is obtained if $3\varphi = \pi/4$ or $\varphi = \pi/12$. If we take this number for φ and substitute into

$$r^3 \sin 3\varphi = 1,$$

we obtain

$$r^3 = \sqrt{2}$$

and

$$r = (2^{1/2})^{1/3} = 2^{1/6}.$$

We conclude from this that the complex number

$$2^{1/6}\left(\cos \frac{\pi}{12} + i \sin \frac{\pi}{12}\right)$$

may be a solution to the problem. That it is indeed a solution follows from the computations below:

$$\left[2^{1/6}\left(\cos \frac{\pi}{12} + i \sin \frac{\pi}{12}\right)\right]^3 = (2^{1/6})^3\left(\cos 3 \cdot \frac{\pi}{12} + i \sin 3 \cdot \frac{\pi}{12}\right)$$
$$= 2^{1/2}\left(\cos \frac{\pi}{4} + i \sin \frac{\pi}{4}\right)$$
$$= \sqrt{2}(\tfrac{1}{2}\sqrt{2} + i(\tfrac{1}{2}\sqrt{2}))$$
$$= 1 + i.$$

Theorem 43

Let α be a non-zero complex number and let n be a natural number. Then there are exactly n complex numbers $\beta_0, \beta_1, \beta_2, \beta_3, \ldots, \beta_{n-1}$ such that for each $k = 0, 1, 2, 3, \ldots, n - 1$,

$$(\beta_k)^n = \alpha.$$

Proof: (The word "exactly" is used here to mean that there are n of these numbers and only n.) Suppose that in polar form $\alpha = r(\cos \varphi + i \sin \varphi)$; and suppose that $\gamma = s(\cos \theta + i \sin \theta)$ is a complex number such that $\gamma^n = \alpha$. Thus, by De Moivre's Theorem, we have

$$s^n(\cos n\theta + i \sin n\theta) = r(\cos \varphi + i \sin \varphi)$$

or

$$s^n \cos n\theta + is^n \sin n\theta = r \cos \varphi + ir \sin \varphi.$$

It follows from this that

$$s^n \cos n\theta = r \cos \varphi \qquad \text{and} \qquad s^n \sin n\theta = r \sin \varphi \qquad \textbf{(9-1)}$$

and hence,

$$\begin{aligned}
r^2 &= r^2(\sin^2 \varphi + \cos^2 \varphi) \\
&= r^2 \sin^2 \varphi + r^2 \cos^2 \varphi \\
&= s^{2n} \sin^2 n\theta + s^{2n} \cos^2 n\theta \\
&= s^{2n}(\sin^2 n\theta + \cos^2 n\theta) \\
&= s^{2n}.
\end{aligned}$$

Since both r and s must be positive, we conclude that $r = s^n$ or $s = r^{1/n}$. Using this result in Eq. **(9-1)** leads to

$$\cos n\theta = \cos \varphi \qquad \text{and} \qquad \sin n\theta = \sin \varphi. \qquad \textbf{(9-2)}$$

From Section 5.7 $\cos n\theta = \cos \varphi$ if, and only if, $n\theta = \varphi + 2k\pi$ or $n\theta = -\varphi + 2k\pi$ for some integer k. Also, $\sin n\theta = \sin \varphi$ if and only if $n\theta = \varphi + 2m\pi$ or $n\theta = -\varphi + (2m + 1)\pi$ for some integer m. Therefore, since both equations in **(9-2)** must hold, there is some integer k such that $n\theta = \varphi + 2k\pi$ or $\theta = \dfrac{1}{n} (\varphi + 2k\pi)$. We thus have shown that if $\gamma = s(\cos \theta + i \sin \theta)$ is a complex number such that $\gamma^n = \alpha$, then there is some integer k such that

$$s = r^{1/n} \qquad \text{and} \qquad \theta = (1/n)(\varphi + 2k\pi).$$

Now define, for each integer k,

$$\beta_k = r^{1/n}\left[\cos \left(\frac{\varphi + 2k\pi}{n}\right) + i \sin \left(\frac{\varphi + 2k\pi}{n}\right)\right].$$

By De Moivre's Theorem,

$$\begin{aligned}
(\beta_k)^n &= r[\cos (\varphi + 2k\pi) + i \sin (\varphi + 2k\pi)] \\
&= r[\cos \varphi + i \sin \varphi] \\
&= \alpha.
\end{aligned}$$

This shows that all the numbers β_k have the required property; *i.e.*, that $(\beta_k)^n = \alpha$. These numbers are, however, not all distinct. In fact, for *every* integer k, β_k is one of the numbers $\beta_0, \beta_1, \beta_2, \ldots, \beta_{n-1}$. For if k is any integer, we may divide k by n and obtain a quotient and remainder with the remainder less than n. That is, we may find integers q and j such that

$$k = nq + j \qquad \text{and} \qquad 0 \le j < n.$$

Then,

$$\cos\left(\frac{\varphi + 2k\pi}{n}\right) = \cos\left(\frac{\varphi + 2(nq + j)\pi}{n}\right)$$

$$= \cos\left(\frac{\varphi + 2j\pi}{n} + 2q\pi\right)$$

$$= \cos\left(\frac{\varphi + 2j\pi}{n}\right);$$

and similarly,

$$\sin\left(\frac{\varphi + 2k\pi}{n}\right) = \sin\left(\frac{\varphi + 2j\pi}{n}\right).$$

From this,

$$\beta_k = r^{1/n}\left[\cos\left(\frac{\varphi + 2k\pi}{n}\right) + i \sin\left(\frac{\varphi + 2k\pi}{n}\right)\right]$$

$$= r^{1/n}\left[\cos\left(\frac{\varphi + 2j\pi}{n}\right) + i \sin\left(\frac{\varphi + 2j\pi}{n}\right)\right]$$

$$= \beta_j$$

and j is one of the integers $0, 1, 2, 3, \ldots, n - 1$. Now suppose that $\beta_h = \beta_j$ for $0 \le h < j \le n - 1$. Then by definition of β_h and β_j,

$$\cos\left(\frac{\varphi + 2h\pi}{n}\right) = \cos\left(\frac{\varphi + 2j\pi}{n}\right)$$

$$\sin\left(\frac{\varphi + 2h\pi}{n}\right) = \sin\left(\frac{\varphi + 2j\pi}{n}\right).$$

As in the earlier part of the proof, we conclude that

$$\frac{\varphi + 2h\pi}{n} = \frac{\varphi + 2j\pi}{n} + 2m\pi$$

for some integer m. Then,

$$\frac{2h\pi}{n} = \frac{2j\pi}{n} + 2m\pi$$

or

$$h = j + 2mn;$$

and consequently, $j - h = -2mn$. This implies that $j - h$, which is a positive number, is a multiple of n. But since

$$0 < j - h \leq n - 1,$$

we must have $m = 0$ and $j = h$. However, this is a contradiction since we assumed that $h < j$. Thus the numbers

$$\beta_0, \beta_1, \beta_2, \beta_3, \ldots, \beta_{n-1}$$

are exactly n in number and this completes the proof.

The numbers $\beta_0, \beta_1, \beta_2, \ldots, \beta_{n-1}$ are called the *nth roots* of the complex number α.

Find the 4th roots of $1 + i$.

Example
9–10

Solution. In polar form $1 + i$ is $\sqrt{2}(\cos \pi/4 + i \sin \pi/4)$. The 4th roots are, according to Theorem 43, $\beta_0, \beta_1, \beta_2, \beta_3$ where

$$\beta_k = (\sqrt{2})^{1/4}\left[\cos\left(\frac{\pi/4 + 2k\pi}{4}\right) + i \sin\left(\frac{\pi/4 + 2k\pi}{4}\right)\right]$$

$$= 2^{1/8}[\cos(\pi/16 + \tfrac{1}{2}k\pi) + i \sin(\pi/16 + \tfrac{1}{2}k\pi)].$$

Therefore,

$$\beta_0 = 2^{1/8}(\cos \pi/16 + i \sin \pi/16)$$
$$\beta_1 = 2^{1/8}[\cos(\pi/16 + \pi/2) + i \sin(\pi/16 + \pi/2)]$$
$$\beta_2 = 2^{1/8}[\cos(\pi/16 + \pi) + i \sin(\pi/16 + \pi)]$$
$$\beta_3 = 2^{1/8}[\cos(\pi/16 + 3\pi/2) + i \sin(\pi/16 + 3\pi/2)].$$

Find the 5th roots of 1.

Example
9–11

Solution. In polar form $1 = 1 \cdot (\cos 0 + i \sin 0)$ and so

$$\beta_k = 1^{1/5}\left[\cos\left(\frac{0 + 2k\pi}{5}\right) + i \sin\left(\frac{0 + 2k\pi}{5}\right)\right]$$

for $k = 0, 1, 2, 3, 4$:

$$\beta_0 = \cos 0 + i \sin 0 = 1$$
$$\beta_1 = \cos(2\pi/5) + i \sin(2\pi/5)$$
$$\beta_2 = \cos(4\pi/5) + i \sin(4\pi/5)$$
$$\beta_3 = \cos(6\pi/5) + i \sin(6\pi/5)$$
$$\beta_4 = \cos(8\pi/5) + i \sin(8\pi/5).$$

8. Express each of the following complex numbers in polar form.

    (a) $(\frac{1}{2}) + (\frac{1}{2})i$ (b) $(\frac{1}{2}) - (\frac{1}{2})i$
    (c) $-7 + 7i$ (d) 2
    (e) $2i$ (f) 0
    (g) $-2 - 2\sqrt{3}i$ (h) $1 + \sqrt{3}i$
    (i) $\sqrt{3} - i$

9. Use De Moivre's Theorem to compute the following:

    (a) i^{14} (b) $[(\frac{1}{2}) - (\frac{1}{2})i]^6$
    (c) $(-2 - 2\sqrt{3}i)^5$ (d) $(1 + \sqrt{3}i)^8$
    (e) $(\sqrt{3} - i)^6$

10. Find the square root of each of the following:

    (a) -1 (b) $-\sqrt{2}$
    (c) -7 (d) -4
    (e) d, where d is a real number such that $d < 0$

11. Let $Q(x) = ax^2 + bx + c$ be a quadratic function and suppose that a, b, c are real numbers such that $b^2 - 4ac < 0$. It has been shown in Chapter 4 that Q has no zeros which are real numbers. Show that Q does have zeros which are complex numbers.

12. Solve the following equations:

    (a) $x^2 + x + 1 = 0$ (b) $x^3 - 1 = 0$
    (c) $x^2 + 2x + 2 = 0$ (d) $x^2 - x + 1 = 0$
    (e) $2x^2 - x + 1 = 0$ (f) $x^4 + 1 = 0$
    (g) $y^6 + 2y^3 + 2 = 0$ (h) $z^6 - 1 = 0$

13. (a) Compute the 3rd roots of 1.
    (b) Compute the 3rd roots of -1.
    (c) Compute the 2nd roots of $1 - i$.
    (d) Compute the 5th roots of $\cos(\pi/8) + i \sin(\pi/8)$.
    (e) Compute the 7th roots of $5[\cos(\pi/3) + i \sin(\pi/3)]$.

APPENDIX

ANSWERS
TO SELECTED
PROBLEMS

1. (*a*) True   (*b*) False   (*c*) True   (*d*) True (*e*) True   (*f*) True   (*g*) False
(*h*) False

2. (*a*) {9}   (*c*) {1, 5}   (*d*) {1, 2, 3, 4, 6, 12}   (*f*) {0, 1}

3. (*a*) True   (*b*) False   (*c*) True   (*d*) True   (*e*) False

4. (*a*) 3   (*b*) An infinite number.   (*c*) $A \subseteq B$ is true.

5. (*a*) $V \subseteq W$ is true.   (*b*) (i) False   (ii) True   (iii) True

6. (*a*) (i) True   (ii) True   (iii) True   (iv) True   (*b*) (i) True   (ii) True
(iii) False   (iv) True   (*c*) (i) True   (ii) False   (iii) False   (iv) False

8. (*a*) 4   (*b*) 8   (*c*) 2^4, 2^5

12. (*a*) {7, 8, 11, 13, 4, 2, 17, 9}   (*b*) {7, 8, 11, 13, 4, 1, 5, 19}
(*c*) {2, 4, 8, 17, 9, 1, 7, 5, 11, 19}   (*d*) {8, 4}   (*e*) {7, 11}   (*f*) $\varnothing$
(*g*) {7, 11, 13}   (*h*) {2, 17, 9}   (*i*) {8, 13, 4}   (*j*) {1, 5, 19}   (*k*) {2, 4, 8, 17, 9}
(*l*) {1, 7, 5, 11, 19}

13. (*g*) {6}   (*h*) {6, 12}   (*i*) {7, 8, 9, 10, . . .}   (*j*) {3, 4, 7, 8, 9, 10, . . .}
(*k*) {1, 2, 3, 4}

14. (*a*) $X = \varnothing, X = \{1\}, X = \{2\}$, or $X = \{1, 2\}$ (*c*) $X \subseteq Y$
(*e*) $X = \{2, 3\}, X = \{2, 3, 4\}, X = \{2, 3, 5\}$, or $X = \{2, 3, 4, 5\}$
(*f*) $1 \in M$ and $2 \notin M$ (*h*) $\{1, 2\} \subseteq X$ and $7 \notin X$
(*i*) $Y = \{3, 5, 9\}, Y = \{3, 5, 9, 1\}, Y = \{3, 5, 9, 2\}$ or $Y = \{3, 5, 9, 1, 2\}$

22. (*b*) $A' \cap B \cap C$ (*c*) $A \cap B$ (*d*) $A \cap C'$ (*e*) $A' \cap (B \cup C)$
(*f*) $A \cap B' \cap C$

26. (*a*) Not a function   (*b*) A function   (*c*) A function   (*d*) Not a function
(*e*) A function   (*f*) A function   (*g*) Not a function

27. (*a*) $g(1) = 4$, $g(2) = 4$, $g(5) = 6$ (*b*) $h(0) = b$, $h(0.5) = 7$, $h(9) = 0.7$
(*c*) $i(x) = 2$ (*d*) $k(y) = y + 1$

29. (*a*) All numbers except 0

30. (*a*) All numbers except 1

31. (*a*) All non-negative numbers

32. (*a*) The set of all numbers except 2 and −1

33. $A(r) = \pi r^2$

34. $A(h) = \frac{1}{2}h^2$

35. $V(w) = w^3$

36. $A(w) = 2w^2$

37. (*a*) $(h + g)(x) = 3x + 5$, $(g \cdot h)(x) = 10x + 2x^2$

40. (*a*) $h \circ g(x) = h(g(x)) = 2(5 + x) = 10 + 2x$

41. (*a*) $g \circ h(x) = g(h(x)) = 5 + (2x)$

42. (*a*) $h(x) = \sqrt{x}$, $g(x) = 7x + 5$ (*b*) $h(x) = x + 1$, $g(x) = x^3$
(*c*) $h(x) = x + 2x^2 + 7x^3$, $g(x) = x^3$

44. (*a*) $\frac{1}{2}x$ (*b*) h (*c*) $x - 5$ (*d*) h (*e*) $(1/35)(28 - x)$ (*f*) h (*g*) $\frac{4}{x} - 2, x \neq 0$

Chapter
2

1. (*a*) $1/4$ (*b*) $43/36$ (*c*) $16/21$ (*d*) $6/5$ (*e*) $-1/2$ (*f*) $5/2$ (*g*) $1/100$
(*h*) $112/100000$ (*i*) $5625/10000$

2. (*a*) 2.1 (*b*) $1.\overline{6}$ (*c*) 9 (*d*) $0.\overline{14}$ (*e*) 4.8260869565217391304347
(*f*) $-5.\overline{137}$ (*g*) 0.00228310502 (*h*) 115.5 (*i*) $-68.\overline{72}$

3. (*a*) $4/33$ (*b*) $3509/999$ (*c*) $6600/1111$ (*d*) $-1331/990$
(*e*) $3616103/499950$ (*f*) $-54137919/2499750$ (*i*) 1 (*j*) $277/100$ (*k*) 350

4. (*a*) Irrational   (*b*) Irrational   (*c*) Irrational

7. (*a*) $-a < 0$ (*b*) $a < 0$ (*c*) $-a > 0$ (*d*) $-x = -5$ (*e*) $-y = 2$
(*f*) $x - 2 > 0$ (*g*) $1 - w > 0$ (*h*) $u + 1 > 4$ (*i*) $2u < 6$ (*j*) $|k| = k$
(*k*) $|k| = -k$ (*l*) $|-k| = -k$ (*m*) $|-k| = k$ (*n*) $|\frac{1}{2} - w| = \frac{1}{2} - w$
(*o*) $|xy| = -xy$

8. (*a*) $x \geq -7/32$ (*c*) $x > 17/3$ (*d*) $x \leq -16$ (*f*) $x \leq 51$

9. (*a*) $\{x \mid x < -1\}$ (*b*) $\{x \mid -1 < x < 1\}$ (*c*) $\emptyset$ (*d*) $\{x \mid 2/3 \leq x \leq 8/3\}$
(*e*) $\{x \mid 1 \leq x < 7\}$ (*f*) $\emptyset$ (*g*) $\{x \mid 1 \leq x < 6\}$

10. (*a*) $x = 0$ (*b*) $x = 1$ (*c*) $\{-10, 10\}$ (*d*) $\{-1, 3/7\}$ (*e*) No solution
(*f*) $\{-1/4, 1/4\}$

11. (*a*) $x = -1$ (*b*) $x = -17$ (*c*) $x = -14$ (*d*) $x = 2$

12. (*a*) $x = -\frac{1}{2}$ (*b*) $\{-10, 10/3\}$ (*c*) $\{4, 2/3\}$ (*d*) $\{x \mid x \leq -1\}$

13. (*a*) $\{x \mid -6 < x < 2\}$ (*b*) $\{x \mid -11/4 \leq x \leq 13/4\}$
(*c*) $\{x \mid 0.1 \leq x \leq 0.3\}$ (*d*) $\{x \mid x < -2\} \cup \{x \mid x > 0\}$
(*e*) $\{x \mid x < -2.5\} \cup \{x \mid x > -0.5\}$ (*f*) $\{x \mid x \leq 0.65\} \cup \{x \mid x \geq 0.75\}$

14. (a) $\{x \mid x < 1.5\}$ (b) $\{x \mid x \leq 0\}$ (c) $\{x \mid -\frac{1}{2} < x\}$

20. (a) If $x \leq -\sqrt{2}$, then x is a lower bound. If $y \geq 0.78$, then y is an upper bound.   (b) If $x \leq 0$, then x is a lower bound. No upper bound exists.
(c) If $x \leq 0$, then x is a lower bound. No upper bound exists.
(d) If $x \leq 0$, then x is a lower bound. No upper bound exists.
(e) If $x \leq 0$, then x is a lower bound. If $x \geq 1$, then x is an upper bound.
(f) If $x \leq 0$, then x is a lower bound. If $x \geq 1$, then x is an upper bound.

21. (a) $-\sqrt{2}$ is the greatest lower bound; 0.78 is the least upper bound.
(b) 0 is the greatest lower bound.   (c) 0 is the greatest lower bound.
(d) 0 is the greatest lower bound.
(e) 0 is the greatest lower bound; 1 is the least upper bound.
(f) 0 is the greatest lower bound; 1 is the least upper bound.

26. (a) $\sqrt[3]{2}$ is the least upper bound of the set $\{y \mid y$ is a positive rational number and $y^3 < 2\}$.
(c) $3^{3/5}$ is the least upper bound of the set $\{y \mid y$ is a positive rational number and $y^5 < 3^3\}$.

10. (a) 1   (b) $\sqrt{130}$ (c) $(5/4)\sqrt{17}$ (d) $\sqrt{2\sqrt{2} + (273/64)}$

11. (a) $\sqrt{(a-1)^2 + (1-1)^2} = |a-1|$ (b) $\sqrt{a^2 - 4a + 5}$
(c) $\sqrt{(a-2)^2 + (2-b)^2}$ (d) $\sqrt{2(b-a)}$

12. (a) $\{(x, y) \mid x^2 + y^2 = 1\}$ (c) $\{(x, y) \mid (x+1)^2 + y^2 = 1\}$
(e) $\{(x, y) \mid (x-3)^2 + (y+7)^2 = 4/9\}$
(g) $\{(x, y) \mid (x - \sqrt{2})^2 + (y+4)^2 = 1/4\}$

15. (a) Circle, center at (0, 0), radius 1   (b) Not a circle   (c) Not a circle
(d) Circle, center at (1, 1), radius 2   (e) Not a circle
(f) Circle, center at $(-8, -5)$, radius $\sqrt[4]{2}$ (g) Circle, center at $(-\frac{1}{2}, 1)$, radius 1

25. (a) $\{(x, y) \mid x \geq 1\}$ (b) $\{(x, y) \mid y - x = 1\}$
(c) $\{(x, y) \mid x = y\} \cup \{(x, y) \mid x = -y\}$ (d) $\{(x, y) \mid y = |x|\}$
(e) $\{(x, y) \mid x = y$ and $x \geq 0\} \cup \{(x, y) \mid x = -y$ and $x \geq 0\}$
(f) $\{(x, y) \mid |x| \leq 1$ and $|y| \leq 1\}$

28. (a) $f(a) = a^2 + 1$ (b) $f(x) = x^2 - 5$ for $x \in [-6, -5]$ (c) $f(c) = \sqrt{c}$
(d) $f(c) = \sqrt{8 - c}$

32. (a) $[\frac{1}{2}, 1)$ (b) $[-1, 0)$ (c) $[10, 15)$ (d) $\{x \mid x < 0\}$ (e) Z (f) $x = 0$

1. (a) 2   (c) 1   (e) 3   (g) k

2. (a)Not a polynomial   (b) 2   (c)Not a polynomial   (d)Polynomial, no degree
(e) 9   (f) Not a polynomial   (g) Not a polynomial

3. (a) $[-1, 0)$ (c) $[2, 3)$ (e) $\{3\}$ (g) $\{0\}$ (i) no zeros

7. (a) $n = -2$ (b) $p = 10$

8. (a) $m = 0$ (b) $n = 2$ or $n = -2$

9. (a) $x = 4, y = -4; xy = -16$ (b) $x = 1, y = 2; xy = 2$

(c) maximum area is 100

10. (a) No   (b) $\{k \mid |k| \leq 10\}$ (c) No   (d) $\{k \mid |k| \leq 5\}$

11. (a) 2   (c) $2, -5$ (e) $1, 11$ (g) $-3/2, 4$ (i) $-3/2, \frac{1}{2}$

12. (a) $1 + \sqrt{2}, 1 - \sqrt{2}$ (c) $(1/4)(1 \pm \sqrt{17})$ (e) $(\frac{1}{2})(-5 \pm \sqrt{5})$
(g) $(\frac{1}{2})(2 \pm \sqrt{6})$ (i) $(1/18)(-1 \pm \sqrt{37})$

13. (a) Discriminant $= -3$, no real zeros
(c) discriminant $= 49/4$, two real zeros (rational)

14. (b) (i) Two zeros   (ii) Two zeros   (iii) None   (iv) None   (v) Two zeros
(d) (i) One zero   (ii) None   (iii) Two zeros   (iv) None   (v) Two zeros

15. $k = 3$

16. Discriminant $= [2(s - t)]^2$

17. $t = -\frac{1}{2}$

18. (a) sum: 1; product: 1/4   (c) sum: $-\sqrt{2}$; product: 1/4
(e) sum: -3 product: 1   (g) sum: 4 product $\frac{1}{2}\sqrt{2}$

19. (a) $x^2 + x - 2$ (c) $x^2 - 4x + 1$ (e) $x^2 - 2\sqrt{5}x + 19/4$

20. (b) $4(x - \frac{1}{2})^2 = (2x - 1)^2$ (d) $(2x - \sqrt{2} - 1)(2x - \sqrt{2} + 1)$
(f) $(2x + 3 + \sqrt{5})(4x + 6 - 2\sqrt{5})$
(h) $(\sqrt{2}x - 2\sqrt{2} + \sqrt{8 - \sqrt{2}})(\sqrt{2}x - 2\sqrt{2} - \sqrt{8 - \sqrt{2}})$

21. Disciminant non-negative for $k \geq 4, k \leq 1$

24. $|m| \leq 2$

25. (a) $\{y \mid |y| \geq 2\}$ (c) $\{y \mid y \geq 1\}$ (e) $\{y \mid y$ is a real number$\}$
(g) $\{y \mid y \geq 1/4\}$

26. (a) $(x - \sqrt{2})^2(x + \sqrt{2})$ (c) $(x - \sqrt{11})(x + \sqrt{11})(x - 1)(x + 1)$
(e) $(x - 2)(x + 2)(x^2 + 3/2)$
(g) $(x - \sqrt{(\frac{1}{2})(5 + \sqrt{39})})(x + \sqrt{(\frac{1}{2})(5 + \sqrt{39})})(x^2 - (\frac{1}{2})(5 - \sqrt{39}))$
(i) $(x - 1)(x^2 + x + 1)(x + 1)(x^2 - x + 1)$

29. $f(1) = 5, f(-1) = 3, f(\frac{1}{2}) = -3/4, f(-\frac{1}{2}) = -7/4, f(-2) = 113$

30. $f(1) = 4, f(-1) = -2, f(\frac{1}{2}) = 7/4, f(-2) = -92, f(2) = 32$

32. $f(3) = 596$

33. $f(a) = 0$

34. $m = 3$

35. $k = -22, n = -24$

36. (a) quotient is $x - 2$; remainder is 3
(c) quotient is $x^4 - 3x^3 + 5x^2 - 18x + 54$; remainder is -170
(e) quotient is $5x^2 + 7x + 21$; remainder is 41

40. $1/x - 1/(x + 1)$

42. $\dfrac{11}{8x + 24} + \dfrac{5}{8x - 8}$

44. $\dfrac{-1}{3x + 3} + \dfrac{x + 4}{3x^2 - 3x + 3}$

46. $\dfrac{1}{x + 1} - \dfrac{1}{(x + 1)^2}$

48. $\dfrac{1}{x + 1} - \dfrac{2}{(x + 1)^2} + \dfrac{1}{(x + 1)^3}$

50. (a) (i) $d = 9$ (ii) Two zeros: 0, 3 (iii) $\{x \mid 0 \le x \le 3\}$
(iv) $\{x \mid x \le 0\} \cup \{x \mid x \ge 3\}$
(c) (i) $d = 40$ (ii) Two zeros: $\pm\sqrt{10}/2$
(iii) $\{x \mid x \le -\sqrt{10}/2\} \cup \{x \mid x \ge \sqrt{10}/2\}$ (iv) $\{x \mid \sqrt{10}/2 \le x \le \sqrt{10}/2\}$
(e) (i) $d = 9$ (ii) Two zeros: $-\frac{1}{2}, 1$ (iii) $\{x \mid x \le -\frac{1}{2}\} \cup \{x \mid x \ge 1\}$
(iv) $\{x \mid -\frac{1}{2} \le x \le 1\}$
(g) (i) $d = -3$ (ii) No zeros (iii) All real numbers (iv) $\varnothing$
(i) (i) $d = 25$ (ii) Two zeros: $-1, 2/3$ (iii) $\{x \mid -1 \le x \le 2/3\}$
(iv) $\{x \mid x \le -1\} \cup \{x \mid x \ge 2/3\}$
(k) (i) $d = 2$ (ii) Two zeros: $-2\sqrt{6} \pm 4$
(iii) $\{x \mid x \le -2\sqrt{6} - 4\} \cup \{x \mid x \ge -2\sqrt{6} + 4\}$
(iv) $\{x \mid -2\sqrt{6} - 4 \le x \le -2\sqrt{6} + 4\}$

51. (a) $\{x \mid x \le (\frac{1}{2})(4 - \sqrt{14})\} \cup \{x \mid x \ge (\frac{1}{2})(4 + \sqrt{14})\}$ (c) $\varnothing$
(e) $\{x \mid 0 \le x \le 1\}$

(g) $\{x \mid x \le a\} \cup \{x \mid x \ge b\}$, where $a = \dfrac{10 + 11\sqrt{2} - \sqrt{386 + 264\sqrt{2}}}{2 + 2\sqrt{2}}$ and
$b = \dfrac{10 + 11\sqrt{2} + \sqrt{386 + 264\sqrt{2}}}{2 + 2\sqrt{2}}$

53. (a) $\{x \mid -1 < x < 4\}$ (c) $\{x \mid x < -1/3\} \cup \{x \mid 8/5 < x < 17/7\}$
(e) $\{x \mid \sqrt{2}/2 < |x| < \sqrt{5/2}\}$

54. (a) $g(x) > 0$; $\{x \mid x < -2\} \cup \{x \mid x > 1\}$
$g(x) < 0$: $\{x \mid -2 < x < 1\}$
$g(x) \ge 0$: $\{x \mid x < -2\} \cup \{x \mid x \ge 1\}$
$g(x) \le 0$: $\{x \mid -2 < x \le 1\}$

(c) $g(x) > 0$: $\{x \mid -3 < x < -1/3\}$
  $g(x) < 0$: $\{x \mid x < -3\} \cup \{x \mid -1/3 < x < \frac{1}{2}\}$
  $g(x) \geq 0$: $\{x \mid -3 < x < -1/3\} \cup \{x \mid x \geq \frac{1}{2}\}$
  $g(x) \leq 0$: $\{x \mid x < -3\} \cup \{x - 1/3 < x \leq \frac{1}{2}\}$

55. (a) $P(-1) = -3 < 0 < 21 = P(3)$ (c) $P(0) = -2 < 0 < 1 = P(1)$
(e) $P(-1) = -4 < 0 < 2 = P(1)$ (g) $P(-11) = 71773 > 0 > -1 = P(-10)$
(i) $P(3) = -1 < 0 < 80282 = P(6)$

56. (a) upper: 2, lower: -2 for example   (c) upper 2, lower -2, for example
(e) upper 2, lower -2, for example   (g) upper 1, lower -11, for example
(i) upper 6, lower -6, for example

57. (a) $P(x) = (6x + 5)(x - 2) + 12$
(b) $R(x) = 6x^2 + 7x + 2 = (6x + 13)(x - 1) + 15$
(c) $P(-1) = 15, P(0) = 2$, no conclusion   (d) $P(0) = 2, P(1) = 1$, no conclusion
(e) $P(1) = 1, P(2) = 12$, no conclusion
(f) The discriminant is 1. There are two real zeros.   (g) The zeros are $1/2, 2/3$.
(h) $P(0) = 2, P(7/12) = -1/24$ (i) $P(7/12) = -1/24, P(1) = 1$

61. (b) 3 iis a factor of $x + y$ and a is a factor of $x + y$ (c) Either $x = 1$ or $x = 2$
(d) $x = 1$ (e) $x = 1$ or $x = -1$ (f) a, b are relatively prime
(g) 2 is a factor of u; 3 is a factor of w
(h) If w and z are relatively prime, then $w = 3$ or $w = 1$ and $z = 1$.

62. (a) 1, 3, 9, 27   (c) 1, 5, 25, 125   (e) 1, 2, 3, 4, 6, 12
(g) 1, 2, 3, 4, 6, 9, 12, 18, 36   (i) 1, 2, 3, 4, 5, 6, 10, 12, 15, 20, 30, 60

64. (a) 1, 2   (c) 1, 2, 3, 4, 6   (e) 1, 2, 7, 14

65. (a) ± 1 (c) ± 1 (e) $\pm 1/2, \pm 1$ (g) $\pm 2/3, \pm 1/3, \pm 2, \pm 1$
(i) $\pm 1/10, \pm 1/5, \pm 2/5, \pm 4/5, \pm 8/5, \pm 1/2, \pm 1, \pm 2, \pm 4, \pm 8$

67. (b) -2 (c) -2 (d) $1/2$ (e) $1/2, -3/2, -1/3$ (f) $1/2$ (g) $\pm 1/2$

68. (a) $P(2) = -1$; no integral zeros   (b) $P(2) = -2$; no integral zeros
(c) $P(-1) = 21$; no integral zeros   (d) $P(1) = -33$; no integral zeros
(e) $P(3) = 67$; no integral zeros

69. (a) No rational zeros   (c) No rational zeros   (e) $1/2, -5/2, -2/3$
(f) $1/2, -1/6, -1, -7$

2. (a) $\sin (\angle A) = 3/5$, $\tan (\angle A) = 3/4$, $\sec (\angle A) = 5/4$
(c) $\cos (\angle A) = 4/5$, $\cot (\angle A) = 4/3$, $\csc (\angle A) = 5/3$

3. (a) $c = \sqrt{10}$ (b) $\sin (\angle A) = 1/\sqrt{10}$, $\tan (\angle A) = 1/3$, $\sec (\angle A) = \sqrt{10}/3$
(e) $\sin (\angle B) = 3/\sqrt{10}$, $\tan (\angle B) = 1/3$, $\sec (\angle B) = \sqrt{10}$

4. (a) 1.8807   (c) 1.0355   (e) 0.9636   (g) 0.9997   (i) 1.0000

5. (*a*) 26.5   (*c*) 18.5   (*e*) 46.5   (*g*) 60.5   (*i*) 1.0

6. (*a*) 78°   (*d*) $m \measuredangle A = 17.5°$, $a = 0.3153$, $c = 1.0485$

7. (*a*) (1, 0)   (*b*) (1, 0)   (*c*) $\left(\dfrac{\sqrt{3}}{2}, \dfrac{1}{2}\right)$ (*d*) $\left(\dfrac{\sqrt{3}}{2}, -\dfrac{1}{2}\right)$ (*e*) $(\frac{1}{2}\sqrt{2}, \frac{1}{2}\sqrt{2})$

(*f*) $(\frac{1}{2}, -\frac{1}{2}\sqrt{3})$ (*g*) (0, 1)   (*h*) (0, −1)   (*i*) (0, −1)   (*j*) $(-\frac{1}{2}, \frac{1}{2}\sqrt{3})$
(*k*) $(-\frac{1}{2}\sqrt{3}, \frac{1}{2})$ (*l*) (−1, 0)   (*m*) $(-\frac{1}{2}\sqrt{2}, -\frac{1}{2}\sqrt{2})$ (*n*) $(-\frac{1}{2}\sqrt{3}, -\frac{1}{2})$
(*o*) $(-\frac{1}{2}, -\frac{1}{2}\sqrt{3})$ (*p*) $(\frac{1}{2}, -\frac{1}{2}\sqrt{3})$ (*q*) $(\frac{1}{2}\sqrt{3}, -\frac{1}{2})$ (*r*) (0, 1)   (*s*) $(-\frac{1}{2}, -\frac{1}{2}\sqrt{3})$
(*t*) $(-\frac{1}{2}\sqrt{3}, -\frac{1}{2})$ (*u*) (−1, 0)   (*v*) $(-\frac{1}{2}\sqrt{2}, \frac{1}{2}\sqrt{2})$ (*w*) $(-\frac{1}{2}\sqrt{3}, \frac{1}{2})$ (*x*) $(\frac{1}{2}, \frac{1}{2}\sqrt{3})$

10.

$F(t)$ is in quadrant	cos (*t*) is	sin (*t*) is
(*a*) II	negative	positive
(*b*) I	positive	positive
(*c*) IV	positive	negative
(*d*) III	negative	negative
(*e*) II	negative	positive

11. (*a*) IV   (*b*) II   (*c*) I   (*d*) III   (*e*) II

12. (*a*) 0   (*b*) 1/2   (*c*) −1/2   (*d*) $-\sqrt{3}/2$ (*e*) $\sqrt{2}/2$ (*f*) −1

14. (*a*) $-\sqrt{3}/2$ (*c*) $\sqrt{2}/4 - \sqrt{6}/4$ (*e*) −1   (*g*) $\sqrt{3}/2$

15. (*a*) 3.4874   (*b*) −0.9613   (*c*) 0.2756   (*d*) 0.5299

16. (*a*) 0.0116   (*b*) −0.0116   (*c*) 0.9999   (*d*) −0.0116   (*e*) 0.0058

17. (*a*) −0.9911   (*b*) 0.1334   (*c*) 7.4295   (*d*) 0.2644   (*e*) −0.9644   (*f*) 0.66
(*g*) 0.75

32. (*a*) 360°   (*c*) 22.5° or 22°30′

33. (*a*) $\pi/45$ (*c*) $13\pi/720$

34. (*a*) 10.5°   (*c*) 4°   (*e*) 48.5°

35. (*a*) 0.0262   (*c*) 0.3403   (*e*) 1.2392

36. (*a*) 0.1299   (*c*) 1.4   (*e*) 1.4465

37. (*a*) 45°   (*c*) 53.5°   (*e*) 29°

38.

reference number	sin (*t*)	cos (*t*)	tan (*t*)
(*a*) 0.0611	−0.0610	−0.9981	0.0612
(*c*) 1.3003	0.9636	0.2672	3.6059
(*e*) 1.5184	−0.9986	0.0523	−19.0811

39.

reference number	sin (*t*)	cos (*t*)	tan (*t*)
(*a*) 85°	0.9962	−0.0872	−11.4301
(*c*) 23°	−0.3907	0.9205	−0.4245
(*e*) 10.5°	−0.1822	0.9833	−0.1853

40. (*a*) 0.7860   (*c*) 0.5089   (*e*) −0.8052   (*g*) −0.5951

41. (a) 19.5°, 0.3403   (d) 56°31′, 0.9864   (f) 67°23′, 1.1761

42. (a) $\sqrt{3}/4$ (c) 50   (e) 1.1816

43. (a) $m \measuredangle C = 91°$, $b = 0.8746$, $c = 0.9998$ (c) No solution
(e) $m \measuredangle C = 96.5°$, $b = 18.0$, $a = 11.0$

44. (a) $m \measuredangle B = 76.0° = m \measuredangle C$, $a = 0.4839$
(c) $b = 1.9$, $m \measuredangle A = 128.5°$, $m \measuredangle C = 6.5°$
(e) $c = 4.4779$, $m \measuredangle A = 28.5°$, $m \measuredangle B = 88°$

45. (a) $m \measuredangle A = 29°$, $m \measuredangle B = 75.5°$, $m \measuredangle C = 75.5°$ (c) No solution
(f) $m \measuredangle A = 3.5°$, $m \measuredangle B = 132°$, $m \measuredangle C = 44.5°$

46. (a) $m \measuredangle B = 30°$, $m \measuredangle C = 120°$, $c = \sqrt{3}$ (b) No solution

47. $2/\sqrt{3}$ or approximately 1.16 miles

49. 34.562′

50. 1.85 miles

51.

	amplitude	period	two zeros	graph lies between
(a)	3	π	$-1/2$, $(1/2)(\pi - 1)$	$y = 3$, $y = -3$
(b)	5	$2\pi/5$	$(\pi + 2)/10$, $(3\pi + 2)/10$	$y = 5$, $y = -5$
(c)	2/3	$2\pi/3$	$(14 - \pi)/6$, $(14 - 3\pi)/6$	$y = 2/3$, $y = -2/3$
(d)	8	2	$7/\pi$, $(\pi + 7)/\pi$	$y = 8$, $y = -8$

52. (a) Amplitude $\frac{1}{3}$; period 2π (b) Amplitude 1; period 2
(c) Amplitude 1; period $2\pi/7$ (d) Amplitude 2; period 2π
(e) Amplitude 1; period π (f) Amplitude 2; period 2

54. No.

57. (a) $x = (3/2 + 2n)\pi$ or $x = (-\frac{1}{2} + 2n)\pi$ for some integer n
(c) $w = -(4n + 1)(\pi/6)$ or $w = (4n + 1)(\pi/10)$ for some integer n
(e) $v = 2n\pi$ or $v = (2/5)n\pi$ for some integer n
(g) $t = 2(s + n)$ or $t = 2(n - s) + 1$ for some integer n

61. (a) $\{\pi/6 + 2n\pi \mid n$ is an integer$\} \cup \{-\pi/6 + k\pi \mid k$ is an odd integer$\}$
(b) $\{k\pi + 1 \mid k$ is an integer$\}$ (c) $\{\pi/2 + 2n\pi \mid n$ is an integer$\}$
(d) $\{2\pi/3 + 2n\pi \mid n$ is an integer$\} \cup \{-2\pi/3 + 2n\pi \mid n$ is an integer$\}$
(e) $\{k\pi \mid k$ is an integer$\} \cup \{2\pi/3 + 2k\pi \mid k$ is an integer$\} \cup \{-2\pi/3 - 2k\pi \mid k$ is an integer$\}$ (f) $\{\pi/2 + m\pi \mid m$ is an integer$\}$

63. (a) $\{3(4n + 1)\pi/8 \mid n$ is an integer$\} \cup \{-3(4n + 1)\pi/4 \mid n$ is an integer$\}$
(b) $\{(4n + 1)\pi/2 \mid n$ is an integer$\} \cup \{-(4n + 1)\pi/6 \mid n$ is an integer$\}$
(c) $\{-(4n + 1)\pi/6 \mid n$ is an integer$\} \cup \{-(4n + 1)\pi/2 \mid n$ is an integer$\}$
(d) $\{t \mid t = n\pi/6$ or $t = (2n + 1)\pi/2$ where n is an integer$\}$

64. (*a*) No solution.
(*b*) If *s* is any number such that sin $s = \frac{1}{3}\sqrt{3}$, then the solution set is
$\{2s + 4n\pi \mid n$ is an integer$\}$ $\cup$ $\{-2s + 2(2n + 1)\pi \mid n$ is an integer$\}$ $\cup$
$\{-2s + 4n\pi \mid n$ is an integer$\}$ $\cup$ $\{2s + 2(2n + 1)\pi \mid n$ is an integer$\}$.
(*c*) $\{\pi/2 + n\pi \mid n$ is an integer$\}$
(*d*) $\{2n\pi \mid n$ is an integer$\}$ $\cup$ $\{\pi/2 + 2n\pi \mid n$ is an integer$\}$

65. (*a*) 0.0873 (*b*) 0.8029 (*c*) 0.5934 (*d*) 0.2705 (*e*) 1.5446 (*f*) 1.2479
(*g*) 0.0175 (*h*) 1.0559

66. (*b*) $\pi - 0.3316 \doteq 2.8100$ (*c*) -1.4574 (*d*) -0.9861 (*e*) 2.0333
(*f*) -1.4923

67. (*a*) -0.1745 (*b*) 2.8100 (*c*) -1.4574 (*d*) -0.9861 (*e*) 2.0333
(*f*) -1.4923

68. (*a*) 0.1810, -0.1810 (*c*) 0.9179, 2.2237 (*e*) 1.7440, 1.3976

69. (*a*) 0, $\pi/2$, 0 (*c*) $-\pi/2$, π, $-\pi/4$ (*e*) $-\pi/6$, $\pi/3$

1. (*a*) A 1-1 function (*b*) Not a 1-1 function (*c*) A 1-1 function
(*d*) Not a 1-1 function

2. (*a*) Not a 1-1 function (*b*) Not a 1-1 function (*c*) Not a 1-1 function
(*d*) Not a 1-1 function (*e*) A 1-1 function (*f*) Not a 1-1 function

4. (*a*) $f^{-1}(x) = \frac{1}{2}(7 - x)$. (*b*) $f^{-1}(x) = -\sqrt{1 - x^2}$, $x \in [0, 1]$.
(*c*) $f^{-1}(x) = 1/x$, $x > 0$. (*d*) $f^{-1}(x) = 1 - 3/x$, $x < 0$.
(*e*) $f^{-1}(x) = x/(1 - x)$, $x > 1$. (*f*) $f^{-1}(x) = \frac{1}{4}(1 - \sqrt{1 + 4x})^2$, $x \geq 0$.
(*g*) $f^{-1}(x) = (b - dx)/(cx - a)$.

8. (*a*) $x = 1$. (*b*) $x = 0$. (*c*) $x = (\frac{1}{7}) \log_3 2$. (*d*) No solutions. (*e*) $\sqrt{2}$, $-\sqrt{2}$.
(*f*) $\sqrt{\frac{3}{2}}$, $-\sqrt{\frac{3}{2}}$. (*g*) No solutions.

10. (*a*) $f^{-1}(x) = \log_4 x$. (*b*) $g^{-1}(x) = \log_2 x + \log_2 14$.
(*c*) $h^{-1}(x) = 1/(\log_a (x - 1))$ (*d*) $f^{-1}(x) = 2^x$. (*e*) $g^{-1}(x) = \frac{1}{2}\sqrt{2}E_3(x/2)$.
(*f*) $h^{-1}(x) = (\frac{1}{7})(E_a(x) - 2)$.

11. (*a*) $\{x \mid x \leq 0\}$. (*b*) The interval (0, 1). (*c*) $\{x \mid x \leq \frac{3}{5}\}$.
(*d*) $\{x \mid x \leq -1\}$ $\cup$ $\{x \mid x \geq 0\}$. (*e*) The interval $[\pi/4, 3\pi/4]$. (*f*) $\{x \mid x > \frac{1}{3}\}$.
(*g*) $\{x \mid x > \sqrt{2}\}$ $\cup$ $\{x \mid -\sqrt{2} < x < 0\}$.

12.

	mantissa	characteristic
(*a*)	log (2.21)	2
(*c*)	log (1.0)	-2
(*e*)	log (5.280)	0

13. (*a*) 0.8531 (*c*) 0.4914 $-$ 3 (*e*) 3.6772 (*g*) 1.3496 (*i*) 3.5591

14. (a) $x = 4.07$ (c) $x = 52.4$ (f) $x = 0.00772$ (h) $x = 5.535$ (j) $x = 4.335$
(l) $x = 22.46$ (n) $x = 1683$ (p) $x = 0.02449$

15. (a) 10.98   (c) 2.065   (e) 85.17   (i) 2.41   (k) 22.39   (m) 9378

1. (a) $7 \cdot 1 + 7 \cdot 2 + 7 \cdot 3 + 7 \cdot 4 + 7 \cdot 5$
(b) $(5 - \frac{1}{2}) + (6 - \frac{1}{2}) + (7 - \frac{1}{2}) + (8 - \frac{1}{2}) + (9 - \frac{1}{2}) + (10 - \frac{1}{2})$
(c) $a^1 + a^2 + a^3 + a^4 + a^5 + a^6 + a^7 + a^8$ (d) $7 + 7 + 7 + 7$

4. (a) 63   (b) 31   (c) 7   (d) 40/27

7. (a) 4   (b) 16   (c) 64   (d) 4^n (e) 18252

8. (a) 676   (b) 17576

9. 9,000,000

10. 100

11. 22,500

12. (a) one   (b) 225,000   (c) 225,000,000   (d) 224,900,000

13. (a) 25   (b) 20   (c) k^2 (d) $k^2 - k$

15. $P(5, 2) = 20$

17. $P(99, 1) = 99$

18. $P(m, 5) = m \cdot (m - 1) \cdot \ldots (m - 4)$

19. $P(35, t) = 35 \cdot 34 \cdot \ldots (36 - t)$

21. $\binom{10}{3} = 120$

23. $\binom{100}{100} = 1$

26. $22! = 22 \cdot 21!$

27. $38! = 38 \cdot 37 \cdot 36 \cdot 35!$

29. $100! = P(100, 50) \cdot 50!$

31. (a) 6   (b) 6720   (c) 10   (d) 5   (e) $m(m - 1) \ldots (m - n + 3)$
(f) $(t - 1)(t - 2)(t - 3) \ldots 3 \cdot 2 \cdot 1$

33. 495

34. 336 (assuming the president has not been chosen)

35. (a) 2,598,960   (b) 249,900   (c) 1287

36. 16

37. 30

38. 28

39. 210

40. 1680

41. 56; 21; 35

42. (*a*) (i) 1000   (ii) 0100   (iii) 0010   (iv) 0001   (v) 1100   (vi) 1010   (vii) 1001
(viii) 0110   (ix) 0101   (x) 0011   (xi) 1110   (xii) 1101   (xiii) 0111   (xiv) 1011
(xv) 1111   (xvi) 0000
(*b*) (i) $\{b, d\}$ (ii) $\{b, c, d\}$ (iii) $\{a, c\}$ (iv) $\{a, b, c\}$ (v) $\{d\}$ (vi) $\{a, d\}$
(vii) $\{a, c, d\}$ (viii) $\{a\}$ (ix) $\{b\}$ (x) $\{c\}$ (xi) $\{a, b\}$ (xii) $\{b, c\}$ (xiii) $\{c, d\}$
(xiv) $\varnothing$ (xv) $\{a, b, c, d\}$ (xvi) $\{a, b, d\}$

43. *abc, acb, bac, bca, cab, cba*

44. (*a*) $a^3 - 3a^2b + 3ab^2 - b^3$ (*b*) $81x^4 - 108x^3 + 54x^2 - 12x + 1$
(*c*) $x^6 - 6x^5 + 15x^4 - 20x^3 + 15x^2 - 6x + 1$
(*d*) $x^2 + 6x^{5/3}y^{1/3} + 15x^{4/3}y^{2/3} + 20xy + 15x^{2/3}y^{4/3} + 6x^{1/3}y^{5/3} + y^2$
(*e*) $32x^{-5} + 40x^{-3} + 20x^{-1} + 5x + 5x^3/8 + x^5/32$
(*f*) $a^6 - 3a^4bx + 3a^2b^2x^2 - b^3x^3 + 3a^4x^2 - 6a^2bx^3 + 3b^2x^4 + 3a^2x^4$
$- 3bx^5 + x^6$

45. $2a^4 + 36a^2 + 18$

46. $140\sqrt{2}$

48. (*a*) $-6435x^8$ (*b*) $840d^4/c^6$ (*c*) 853,125

1. (*a*) $\{(x, y) \mid y = -(\frac{5}{4})x\}$ (*c*) $\{(x, y) \mid x = \frac{1}{2}\}$ (*e*) $\{(x, y) \mid y = x + 10\}$
(*g*) $\{(x, y) \mid y = -x\}$ (*i*) $\{(x, y) \mid y = \pi\}$

3. (*a*) $\{(x, y) \mid x = -1\}$ (*c*) $\{(x, y) \mid x + y = \sqrt{2}\}$ (*e*) $\{(x, y) \mid y = 0\}$

9. (*a*) $\{(x, y) \mid y = 4x - 3\}$ (*b*) $\{(x, y) \mid x - 5y + 4 = 0\}$

10. (*a*) $\{(x, y) \mid x + y + 2 = 0\}$ (*b*) $\{(x, y) \mid 12x + 4y = 29\}$

12. (*a*) $6/\sqrt{2}$ (*b*) $8/\sqrt{5}$

14. Vertex: (0, 0); directrix: $\{(x, y) \mid x = -\frac{1}{2}\}$; axis: the *x*-axis; focus: $(\frac{1}{2}, 0)$

16. Vertex: (0, 1); directrix: $\{(x, y) \mid y = 0\}$; axis: $\{(x, y) \mid x = 0\}$; focus: (0, 2)

18. Vertex: (0, 2); directrix: $\{(x, y) \mid y = \frac{55}{28}\}$; axis: $\{(x, y) \mid x = 0\}$; focus: $(0, \frac{57}{28})$

21. $\{(x, y) \mid y^2 = -8x\}$

Chapter
8

23. $\{(x, y) \mid (x + 1)^2 = 12(y - 5)\}$

30. (a) $(-\frac{1}{2}, \frac{3}{5})$ (c) $(-\frac{1}{2}, \frac{28}{5})$ (e) $(\sqrt{2} - \frac{1}{2}, \frac{11}{10})$

31. (a) $\{(\bar{x}, \bar{y}) \mid 10\bar{x} - 10\bar{y} + 11 = 0\}$ (c) $\{(\bar{x}, \bar{y}) \mid \bar{x} = \bar{y}\}$
(e) $\{(\bar{x}, \bar{y}) \mid (\bar{y} - \frac{6}{5})^2 = 4(\bar{x} + 1)\}$

32. (a) $\{(\bar{x}, \bar{y}) \mid \bar{y} = 0\}$; $i.e.$, the $\bar{x}$-axis
(c) $\{(\bar{x}, \bar{y}) \mid (1 - 2\sqrt{3})\bar{x}^2 + (3 + 2\sqrt{3})\bar{y}^2 + (2\sqrt{3} - 4)\bar{x}\bar{y}^2\sqrt{3}\bar{x} - 2\bar{y} = 28\}$

39. (a) $x^2 + y^2 = 2x$ (b) $3(x^2 + y^2) = 2y$

40. (a) $r \cos \theta + 2r \sin \theta + 1 = 0$ (b) $r^2 = 1$ (d) $r^2 (2 \cos^2 \theta - 1) = 4$

41. (a) The point is in the yz-plane; 1 unit from the xz-plane, is in the xy-plane.
(c) 12 units from the yz-plane, 8 units from the xz-plane, 3 units from the xy-plane.
(e) $4\frac{1}{2}$ units from the yz-plane, 10 units from the xz-plane, 8 units from the xy plane.
(g) $|a|$ units from the yz- and xz-planes and $|b|$ units from the xy-plane.

42. (a) A plane parallel to and 10 units from the xy-plane. (c) The xz-plane
(e) The line $x = y$ in the xy-plane.
(j) A semi-circle of radius 1, center at the origin in the xz-plane.

43. (a) 1 (b) $\sqrt{2}$ (c) $\sqrt{2}$ (d) $\sqrt{3}$ (e) $\sqrt{3}$ (f) 3 (g) $\sqrt{14}$

44. (a) 1, 0, 0 (b) $1/\sqrt{2}, 1/\sqrt{2}, 0$ (c) $1/\sqrt{2}, 0, 1/\sqrt{2}$ (d) $1/\sqrt{3}, 1/\sqrt{3}, 1/\sqrt{3}$
(e) $-1/\sqrt{3}, -1/\sqrt{3}, 1/\sqrt{3}$ (f) $2/3, -2/3, 1/3$ (g) $3/\sqrt{14}, 1/\sqrt{14}, 2/\sqrt{14}$

46. (a) $(\frac{1}{2}, 0, 0)$ (b) $(\frac{1}{2}, \frac{1}{2}, 0)$ (c) $(\frac{1}{2}, 0, \frac{1}{2})$ (d) $(\frac{1}{2}, \frac{1}{2}, \frac{1}{2})$ (e) $(-\frac{1}{2}, -\frac{1}{2}, \frac{1}{2})$
(f) $(1, 0, \frac{1}{2})$

47. (b) 55°, 55°, 55° (c) 45°, 90°, 45° (d) 90°, 90°, 0° (e) 45°, 45°, 90°

48. (a) $x = w, y = w, z = w$ (b) $x = 3 + 2w, y = -5, z = 2 - w$
(c) $x = -1 - 3w, y = 12w, z = 1 + 8w$
(d) $x = \frac{1}{2} - w, y = 3/4 + 4w, z = 11 - w$
(e) $x = 22, y = -7 + 3w, z = 2 - \frac{1}{2}w$
(f) $x = \pi + 2\sqrt{2}w, y = 5 - 4w, z = -1 + 5w$

49. (b) $x = w, y = w, z = w, x = y = z$ (c) $x = w, y = 1, z = w, x = z, y = 1$
(d) $x = 1, y = 1, z = w, x = y = 1$
(e) $x = -1 + 2w, y = -1 + 2w, z = 1, x = y, z = 1$
(f) $x = 2w, y = 1 - 2w, z = 2 - 2w, \dfrac{x}{2} = \dfrac{y - 1}{-2} = \dfrac{z - 2}{-2}$

50. (c) 55° (d) 35°

55. (a) $z = 0$ (b) $y = 0$ (c) $x = 0$ (d) $x = 1$ (e) $y = 1$ (f) $z = 1$
(g) $z = 5$ (h) $y = 6$ (i) $x = -2$

56. (a) $x + y + z = 0$ (b) $2x - y + 4z = 0$ (c) $x + y + z - 6 = 0$
(d) $3x - 5y + z = 0$ (e) $-x + y - 2z + 13 = 0$ (f) $2x + 3y - 5z - 1 = 0$

57. (a) $y + z - 1 = 0$ (b) $3x - z - 6 = 0$ (c) $6x + y - 7 = 0$

58. (a) $y = z$ (b) $x = z$ (c) $x = y$ (d) $x + y + z - 1 = 0$
(e) $6x + 3y + 2z - 6 = 0$ (f) $y = 0$ (g) $3y - z + 3 = 0$
(h) $14x - 21y + 17z - 42 = 0$

59. (a) $3x + 2y - 6z + 1 = 0$ (b) $3x - 3y - z - 6 = 0$
(c) $2x - y + 3z + 2 = 0$ (d) $-x + 2y - 4z - 1 = 0$
(e) $x + y - 3z + 1 = 0$ (f) $2x - 2y + 3z - 2 = 0$

1. (a) $(\frac{3}{2}, 9)$ (c) $(-1, 8)$ (e) $-2 - 4i$ (g) $4 + 23i$ (i) 3

Chapter
* 9

2. (a) $(1, 0)$ (c) $\frac{1}{2} + \frac{1}{2}i$ (e) $-\frac{1}{2}i$

3. (a) $1 + 0i$ (b) $7 + 2i$ (c) $2 + 4i$ (d) $0 + (-1)i$ (e) $1 + (-\frac{1}{2})i$
(f) $-(9/595) - (75/595)i$

4. (a) $-\frac{9}{5} + (\frac{7}{5})i$ (b) $7 - i$ (c) $\sqrt{2}i; -\sqrt{2}i$ (d) $\frac{1}{2}\sqrt{2} - \frac{1}{2}\sqrt{2}i$

8. (a) $\frac{1}{2}\sqrt{2}(\cos \pi/4 + i \sin \pi/4)$ (b) $\frac{1}{2}\sqrt{2}(\cos (-\pi/4) + i \sin (-\pi/4))$
(c) $7\sqrt{2}(\cos (-3\pi/4) + i \sin (-3\pi/4))$ (d) $2(\cos 0 + i \sin 0)$
(e) $2(\cos \pi/2 + i \sin \pi/2)$ (f) $0 \cdot (\cos \pi + i \sin \pi)$
(g) $4\left(\cos \left(-\frac{2\pi}{3}\right) + \sin \left(-\frac{2\pi}{3}\right)\right)$ (h) $2(\cos \pi/3 + i \sin \pi/3)$
(i) $2(\cos (-\pi/6) + i \sin (-\pi/6))$

9. (b) $-\frac{1}{8}$ (c) $-16(1 + \sqrt{3}i)$ (d) $-128 + 128\sqrt{3}i$

10. (a) $i, -i$ (b) $\pm\sqrt[4]{2}i$ (c) $\pm\sqrt{7}i$ (d) $\pm 2i$ (e) $\pm\sqrt{|d|}i$

12. (a) $-\frac{1}{2} \pm \frac{1}{2}\sqrt{3}i$ (c) $-1 \pm i$ (e) $\frac{1}{4} \pm \frac{1}{4}\sqrt{7}i$

13. (a) $1, \cos (2\pi/3) + i \sin (2\pi/3), \cos (4\pi/3) + i \sin (4\pi/3)$
(c) $2^{1/4}(\cos (-\pi/8) + i \sin (-\pi/8)), 2^{1/4}(\cos (7\pi/8) + i \sin (7\pi/8))$

Table I
Trigonometric Functions sin θ, cos θ, tan θ, and cot θ

Angle θ		sin θ	cos θ	tan θ	cot θ		
Radians	Degrees						
0.0000	0.0	0.0000	1.0000	0.0000	—	90.0	1.5708
0.0087	0.5	0.0087	1.0000	0.0087	114.5887	89.5	1.5621
0.0175	1.0	0.0175	0.9998	0.0175	57.2900	89.0	1.5533
0.0262	1.5	0.0262	0.9997	0.0262	38.1885	88.5	1.5446
0.0349	2.0	0.0349	0.9994	0.0349	28.6363	88.0	1.5359
0.0436	2.5	0.0436	0.9990	0.0437	22.9038	87.5	1.5272
0.0524	3.0	0.0523	0.9986	0.0524	19.0811	87.0	1.5184
0.0611	3.5	0.0610	0.9981	0.0612	16.3499	86.5	1.5097
0.0698	4.0	0.0698	0.9976	0.0699	14.3007	86.0	1.5010
0.0785	4.5	0.0785	0.9969	0.0787	12.7062	85.5	1.4923
0.0873	5.0	0.0872	0.9962	0.0875	11.4301	85.0	1.4835
0.0960	5.5	0.0958	0.9954	0.0963	10.3854	84.5	1.4748
0.1047	6.0	0.1045	0.9945	0.1051	9.5144	84.0	1.4661
0.1134	6.5	0.1132	0.9936	0.1139	8.7769	83.5	1.4574
0.1222	7.0	0.1219	0.9925	0.1228	8.1443	83.0	1.4486
0.1309	7.5	0.1305	0.9914	0.1317	7.5958	82.5	1.4399
0.1396	8.0	0.1392	0.9903	0.1405	7.1154	82.0	1.4312
0.1484	8.5	0.1478	0.9890	0.1495	6.6912	81.5	1.4224
0.1571	9.0	0.1564	0.9877	0.1584	6.3138	81.0	1.4137
0.1658	9.5	0.1650	0.9863	0.1673	5.9758	80.5	1.4050
0.1745	10.0	0.1736	0.9848	0.1763	5.6713	80.0	1.3963
0.1833	10.5	0.1822	0.9833	0.1853	5.3955	79.5	1.3875
0.1920	11.0	0.1908	0.9816	0.1944	5.1446	79.0	1.3788
0.2007	11.5	0.1994	0.9799	0.2035	4.9152	78.5	1.3701
0.2094	12.0	0.2079	0.9781	0.2126	4.7046	78.0	1.3614
0.2182	12.5	0.2164	0.9763	0.2217	4.5107	77.5	1.3526
0.2269	13.0	0.2250	0.9744	0.2309	4.3315	77.0	1.3439
0.2356	13.5	0.2334	0.9724	0.2401	4.1653	76.5	1.3352
0.2443	14.0	0.2419	0.9703	0.2493	4.0108	76.0	1.3265
0.2531	14.5	0.2504	0.9681	0.2586	3.8667	75.5	1.3177
0.2618	15.0	0.2588	0.9659	0.2679	3.7321	75.0	1.3090
0.2705	15.5	0.2672	0.9636	0.2773	3.6059	74.5	1.3003
0.2793	16.0	0.2756	0.9613	0.2867	3.4874	74.0	1.2915
0.2880	16.5	0.2840	0.9588	0.2962	3.3759	73.5	1.2828
0.2967	17.0	0.2924	0.9563	0.3057	3.2709	73.0	1.2741
0.3054	17.5	0.3007	0.9537	0.3153	3.1716	72.5	1.2654
0.3142	18.0	0.3090	0.9511	0.3249	3.0777	72.0	1.2566
0.3229	18.5	0.3173	0.9483	0.3346	2.9887	71.5	1.2479
0.3316	19.0	0.3256	0.9455	0.3443	2.9042	71.0	1.2392
0.3403	19.5	0.3338	0.9426	0.3541	2.8239	70.5	1.2305
0.3491	20.0	0.3420	0.9397	0.3640	2.7475	70.0	1.2217
0.3578	20.5	0.3502	0.9367	0.3739	2.6746	69.5	1.2130
0.3665	21.0	0.3584	0.9336	0.3839	2.6051	69.0	1.2403
0.3752	21.5	0.3665	0.9304	0.3939	2.5386	68.5	1.1956
0.3840	22.0	0.3746	0.9272	0.4040	2.4751	68.0	1.1868
0.3927	22.5	0.3827	0.9239	0.4142	2.4142	67.5	1.1781
		cos θ	sin θ	cot θ	tan θ	Degrees	Radians
						Angle θ	

Angle θ		$\sin \theta$	$\cos \theta$	$\tan \theta$	$\cot \theta$		
Radians	Degrees						
0.3927	22.5	0.3827	0.9239	0.4142	2.4142	67.5	1.1781
0.4014	23.0	0.3907	0.9205	0.4245	2.3559	67.0	1.1694
0.4102	23.5	0.3987	0.9171	0.4348	2.2998	66.5	1.1606
0.4189	24.0	0.4067	0.9135	0.4452	2.2460	66.0	1.1519
0.4276	24.5	0.4147	0.9100	0.4557	2.1943	65.5	1.1432
0.4363	25.0	0.4226	0.9063	0.4663	2.1445	65.0	1.1345
0.4451	25.5	0.4305	0.9026	0.4770	2.0965	64.5	1.1257
0.4538	26.0	0.4384	0.8988	0.4877	2.0503	64.0	1.1170
0.4625	26.5	0.4462	0.8949	0.4986	2.0057	63.5	1.1083
0.4712	27.0	0.4540	0.8910	0.5095	1.9626	63.0	1.0996
0.4800	27.5	0.4617	0.8870	0.5206	1.9210	62.5	1.0908
0.4887	28.0	0.4695	0.8829	0.5317	1.8807	62.0	1.0821
0.4974	28.5	0.4772	0.8788	0.5430	1.8418	61.5	1.0734
0.5061	29.0	0.4848	0.8746	0.5543	1.8040	61.0	1.0647
0.5149	29.5	0.4924	0.8704	0.5658	1.7675	60.5	1.0559
0.5236	30.0	0.5000	0.8660	0.5774	1.7321	60.0	1.0472
0.5323	30.5	0.5075	0.8616	0.5890	1.6977	59.5	1.0385
0.5411	31.0	0.5150	0.8572	0.6009	1.6643	59.0	1.0297
0.5498	31.5	0.5225	0.8526	0.6128	1.6319	58.5	1.0210
0.5585	32.0	0.5299	0.8480	0.6249	1.6003	58.0	1.0123
0.5672	32.5	0.5373	0.8434	0.6371	1.5697	57.5	1.0036
0.5760	33.0	0.5446	0.8387	0.6494	1.5399	57.0	0.9948
0.5847	33.5	0.5519	0.8339	0.6619	1.5108	56.5	0.9861
0.5934	34.0	0.5592	0.8290	0.6745	1.4826	56.0	0.9774
0.6021	34.5	0.5664	0.8241	0.6873	1.4550	55.5	0.9687
0.6109	35.0	0.5736	0.8192	0.7002	1.4281	55.0	0.9599
0.6196	35.5	0.5807	0.8141	0.7133	1.4019	54.5	0.9512
0.6283	36.0	0.5878	0.8090	0.7265	1.3764	54.0	0.9425
0.6370	36.5	0.5948	0.8039	0.7400	1.3514	53.5	0.9338
0.6458	37.0	0.6018	0.7986	0.7536	1.3270	53.0	0.9250
0.6545	37.5	0.6088	0.7934	0.7673	1.3032	52.5	0.9163
0.6632	38.0	0.6157	0.7880	0.7813	1.2799	52.0	0.9076
0.6720	38.5	0.6225	0.7826	0.7954	1.2572	51.5	0.8988
0.6807	39.0	0.6239	0.7771	0.8098	1.2349	51.0	0.8901
0.6894	39.5	0.6361	0.7716	0.8243	1.2131	50.5	0.8814
0.6981	40.0	0.6428	0.7660	0.8391	1.1918	50.0	0.8727
0.7069	40.5	0.6494	0.7604	0.8541	1.1708	49.5	0.8639
0.7156	41.0	0.6561	0.7547	0.8693	1.1504	49.0	0.8552
0.7243	41.5	0.6626	0.7490	0.8847	1.1303	48.5	0.8465
0.7330	42.0	0.6691	0.7431	0.9004	1.1106	48.0	0.8378
0.7418	42.5	0.6756	0.7373	0.9163	1.0913	47.5	0.8290
0.7505	43.0	0.6820	0.7314	0.9325	1.0724	47.0	0.8203
0.7592	43.5	0.6884	0.7254	0.9490	1.0538	46.5	0.8116
0.7679	44.0	0.6947	0.7193	0.9657	1.0355	46.0	0.8029
0.7767	44.5	0.7009	0.7133	0.9827	1.0176	45.5	0.7941
0.7854	45.0	0.7071	0.7071	1.0000	1.0000	45.0	0.7854
		$\cos \theta$	$\sin \theta$	$\cot \theta$	$\tan \theta$	Degrees	Radians
						Angle θ	

Table II

Common Logarithms log t

t	0	1	2	3	4	5	6	7	8	9
1.0	0.0000	0.0043	0.0086	0.0128	0.0170	0.0212	0.0253	0.0294	0.0334	0.0374
1.1	0.0414	0.0453	0.0492	0.0531	0.0569	0.0607	0.0645	0.0682	0.0719	0.0755
1.2	0.0792	0.0828	0.0864	0.0899	0.0934	0.0969	0.1004	0.1038	0.1072	0.1106
1.3	0.1139	0.1173	0.1206	0.1239	0.1271	0.1303	0.1335	0.1367	0.1399	0.1430
1.4	0.1461	0.1492	0.1523	0.1553	0.1584	0.1614	0.1644	0.1673	0.1703	0.1732
1.5	0.1761	0.1790	0.1818	0.1847	0.1875	0.1903	0.1931	0.1959	0.1987	0.2014
1.6	0.2041	0.2068	0.2095	0.2122	0.2148	0.2175	0.2201	0.2227	0.2253	0.2279
1.7	0.2304	0.2330	0.2355	0.2380	0.2405	0.2430	0.2455	0.2480	0.2504	0.2529
1.8	0.2553	0.2577	0.2601	0.2625	0.2648	0.2672	0.2695	0.2718	0.2742	0.2765
1.9	0.2788	0.2810	0.2833	0.2856	0.2878	0.2900	0.2923	0.2945	0.2967	0.2989
2.0	0.3010	0.3032	0.3054	0.3075	0.3096	0.3118	0.3139	0.3160	0.3181	0.3201
2.1	0.3222	0.3243	0.3263	0.3284	0.3304	0.3324	0.3345	0.3365	0.3385	0.3404
2.2	0.3424	0.3444	0.3464	0.3483	0.3502	0.3522	0.3541	0.3560	0.3579	0.3598
2.3	0.3617	0.3636	0.3655	0.3674	0.3692	0.3711	0.3729	0.3747	0.3766	0.3784
2.4	0.3802	0.3820	0.3838	0.3856	0.3874	0.3892	0.3909	0.3927	0.3945	0.3962
2.5	0.3979	0.3997	0.4014	0.4031	0.4048	0.4065	0.4082	0.4099	0.4116	0.4133
2.6	0.4150	0.4166	0.4183	0.4200	0.4216	0.4232	0.4249	0.4265	0.4281	0.4298
2.7	0.4314	0.4330	0.4346	0.4362	0.4378	0.4393	0.4409	0.4425	0.4440	0.4456
2.8	0.4472	0.4487	0.4502	0.4518	0.4533	0.4548	0.4564	0.4579	0.4594	0.4609
2.9	0.4624	0.4639	0.4654	0.4669	0.4683	0.4698	0.4713	0.4728	0.4742	0.4757
3.0	0.4771	0.4786	0.4800	0.4814	0.4829	0.4843	0.4857	0.4871	0.4886	0.4900
3.1	0.4914	0.4928	0.4942	0.4955	0.4969	0.4983	0.4997	0.5011	0.5024	0.5038
3.2	0.5051	0.5065	0.5079	0.5092	0.5105	0.5119	0.5132	0.5145	0.5159	0.5172
3.3	0.5185	0.5198	0.5211	0.5224	0.5237	0.5250	0.5263	0.5276	0.5289	0.5302
3.4	0.5315	0.5328	0.5340	0.5353	0.5366	0.5378	0.5391	0.5403	0.5416	0.5428
3.5	0.5441	0.5453	0.5465	0.5478	0.5490	0.5502	0.5514	0.5527	0.5539	0.5551
3.6	0.5563	0.5575	0.5587	0.5599	0.5611	0.5623	0.5635	0.5647	0.5658	0.5670
3.7	0.5682	0.5694	0.5705	0.5717	0.5729	0.5740	0.5752	0.5763	0.5775	0.5786
3.8	0.5798	0.5809	0.5821	0.5832	0.5843	0.5855	0.5866	0.5877	0.5888	0.5899
3.9	0.5911	0.5922	0.5933	0.5944	0.5955	0.5966	0.5977	0.5988	0.5999	0.6010
4.0	0.6021	0.6031	0.6042	0.6053	0.6064	0.6075	0.6085	0.6096	0.6107	0.6117
4.1	0.6128	0.6138	0.6149	0.6160	0.6170	0.6180	0.6191	0.6201	0.6212	0.6222
4.2	0.6232	0.6243	0.6253	0.6263	0.6274	0.6284	0.6294	0.6304	0.6314	0.6325
4.3	0.6335	0.6345	0.6355	0.6365	0.6375	0.6385	0.6395	0.6405	0.6415	0.6425
4.4	0.6435	0.6444	0.6454	0.6464	0.6474	0.6484	0.6493	0.6503	0.6513	0.6522
4.5	0.6532	0.6542	0.6551	0.6561	0.6571	0.6580	0.6590	0.6599	0.6609	0.6618
4.6	0.6628	0.6637	0.6646	0.6656	0.6665	0.6675	0.6684	0.6693	0.6702	0.6712
4.7	0.6721	0.6730	0.6739	0.6749	0.6758	0.6767	0.6776	0.6785	0.6794	0.6803
4.8	0.6812	0.6821	0.6830	0.6839	0.6848	0.6857	0.6866	0.6875	0.6884	0.6893
4.9	0.6902	0.6911	0.6920	0.6928	0.6937	0.6946	0.6955	0.6964	0.6972	0.6981
5.0	0.6990	0.6998	0.7007	0.7016	0.7024	0.7053	0.7042	0.7050	0.7059	0.7067
5.1	0.7076	0.7084	0.7093	0.7101	0.7110	0.7118	0.7126	0.7135	0.7143	0.7152
5.2	0.7160	0.7168	0.7177	0.7185	0.7193	0.7202	0.7210	0.7218	0.7226	0.7235
5.3	0.7243	0.7251	0.7259	0.7267	0.7275	0.7284	0.7292	0.7300	0.7308	0.7316
5.4	0.7324	0.7332	0.7340	0.7348	0.7356	0.7364	0.7372	0.7380	0.7388	0.7396
t	0	1	2	3	4	5	6	7	8	9

t	0	1	2	3	4	5	6	7	8	9
5.5	0.7404	0.7412	0.7419	0.7427	0.7435	0.7443	0.7451	0.7459	0.7466	0.7474
5.6	0.7482	0.7490	0.7497	0.7505	0.7513	0.7520	0.7528	0.7536	0.7543	0.7551
5.7	0.7559	0.7566	0.7574	0.7582	0.7589	0.7597	0.7604	0.7612	0.7619	0.7627
5.8	0.7634	0.7642	0.7649	0.7657	0.7664	0.7672	0.7679	0.7686	0.7694	0.7701
5.9	0.7709	0.7716	0.7723	0.7731	0.7738	0.7745	0.7752	0.7760	0.7767	0.7774
6.0	0.7782	0.7789	0.7796	0.7803	0.7810	0.7818	0.7825	0.7832	0.7839	0.7846
6.1	0.7853	0.7860	0.7868	0.7875	0.7882	0.7889	0.7896	0.7903	0.7910	0.7917
6.2	0.7924	0.7931	0.7938	0.7945	0.7952	0.7959	0.7966	0.7973	0.7980	0.7987
6.3	0.7993	0.8000	0.8007	0.8014	0.8021	0.8028	0.8035	0.8041	0.8048	0.8055
6.4	0.8062	0.8069	0.8075	0.8082	0.8089	0.8096	0.8102	0.8109	0.8116	0.8122
6.5	0.8129	0.8136	0.8142	0.8149	0.8156	0.8162	0.8169	0.8176	0.8182	0.8189
6.6	0.8195	0.8202	0.8209	0.8215	0.8222	0.8228	0.8235	0.8241	0.8248	0.8254
6.7	0.8261	0.8267	0.8274	0.8280	0.8287	0.8293	0.8299	0.8306	0.8312	0.8319
6.8	0.8325	0.8331	0.8338	0.8344	0.8351	0.8357	0.8363	0.8370	0.8376	0.8382
6.9	0.8388	0.8395	0.8401	0.8407	0.8414	0.8420	0.8426	0.8432	0.8439	0.8445
7.0	0.8451	0.8457	0.8463	0.8470	0.8476	0.8482	0.8488	0.8494	0.8500	0.8506
7.1	0.8513	0.8519	0.8525	0.8531	0.8537	0.8543	0.8549	0.8555	0.8561	0.8567
7.2	0.8573	0.8579	0.8585	0.8591	0.8597	0.8603	0.8609	0.8615	0.8621	0.8627
7.3	0.8633	0.8639	0.8645	0.8651	0.8657	0.8663	0.8669	0.8675	0.8681	0.8686
7.4	0.8692	0.8698	0.8704	0.8710	0.8716	0.8722	0.8727	0.8733	0.8739	0.8745
7.5	0.8751	0.8756	0.8762	0.8768	0.8774	0.8779	0.8785	0.8791	0.8797	0.8802
7.6	0.8808	0.8814	0.8820	0.8825	0.8831	0.8837	0.8842	0.8848	0.8854	0.8859
7.7	0.8865	0.8871	0.8876	0.8882	0.8887	0.8893	0.8899	0.8904	0.8910	0.8915
7.8	0.8921	0.8927	0.8932	0.8938	0.8943	0.8949	0.8954	0.8960	0.8965	0.8971
7.9	0.8976	0.8982	0.8987	0.8993	0.8998	0.9004	0.9009	0.9015	0.9020	0.9025
8.0	0.9031	0.9036	0.9042	0.9047	0.9053	0.9058	0.9063	0.9069	0.9074	0.9079
8.1	0.9085	0.9090	0.9096	0.9101	0.9106	0.9112	0.9117	0.9122	0.9128	0.9133
8.2	0.9138	0.9143	0.9149	0.9154	0.9159	0.9165	0.9170	0.9175	0.9180	0.9186
8.3	0.9191	0.9196	0.9201	0.9206	0.9212	0.9217	0.9222	0.9227	0.9232	0.9238
8.4	0.9243	0.9248	0.9253	0.9258	0.9263	0.9269	0.9274	0.9279	0.9284	0.9289
8.5	0.9294	0.9299	0.9304	0.9309	0.9315	0.9320	0.9325	0.9330	0.9335	0.9340
8.6	0.9345	0.9350	0.9355	0.9360	0.9365	0.9370	0.9375	0.9380	0.9385	0.9390
8.7	0.9395	0.9400	0.9405	0.9410	0.9415	0.9420	0.9425	0.9430	0.9435	0.9440
8.8	0.9445	0.9450	0.9455	0.9460	0.9465	0.9469	0.9474	0.9479	0.9484	0.9489
8.9	0.9494	0.9499	0.9504	0.9509	0.9513	0.9518	0.9523	0.9528	0.9533	0.9538
9.0	0.9542	0.9547	0.9552	0.9557	0.9562	0.9566	0.9571	0.9576	0.9581	0.9586
9.1	0.9590	0.9595	0.9600	0.9605	0.9609	0.9614	0.9619	0.9624	0.9628	0.9633
9.2	0.9638	0.9643	0.9647	0.9652	0.9657	0.9661	0.9666	0.9671	0.9675	0.9680
9.3	0.9685	0.9689	0.9694	0.9699	0.9703	0.9708	0.9713	0.9717	0.9722	0.9727
9.4	0.9731	0.9736	0.9741	0.9745	0.9750	0.9754	0.9759	0.9763	0.9768	0.9773
9.5	0.9777	0.9782	0.9786	0.9791	0.9795	0.9800	0.9805	0.9809	0.9814	0.9818
9.6	0.9823	0.9827	0.9832	0.9836	0.9841	0.9845	0.9850	0.9854	0.9859	0.9863
9.7	0.9868	0.9872	0.9877	0.9881	0.9886	0.9890	0.9894	0.9899	0.9903	0.9908
9.8	0.9912	0.9917	0.9921	0.9926	0.9930	0.9934	0.9939	0.9943	0.9948	0.9952
9.9	0.9956	0.9961	0.9965	0.9969	0.9974	0.9978	0.9983	0.9987	0.9991	0.9996
t	0	1	2	3	4	5	6	7	8	9

INDEX